PRAISE FOR
The New Livestock Farmer

"Great practical advice on choosing the species to raise, humane treatment, and marketing. Informative chapters on processing, regulations, and starting a business."
—**Temple Grandin,** author of *Humane Livestock Handling* and
Improving Animal Welfare: A Practical Approach

"Responsible and healthful meat consumption starts on the farm, literally from the ground up, with solid and ethical animal-husbandry practices. *The New Livestock Farmer* provides a clear understanding of how to achieve fulfilling and delicious results. The authors share their proven wisdom to help small-scale, grass-based farmers avoid the pitfalls of an often confusing and intimidating agricultural landscape."
—**Adam Danforth,** author of *Butchering Beef* and
Butchering Poultry, Rabbit, Lamb, Goat, and Pork

"When it comes to raising healthy livestock in harmony with the land and the local community, this book shows why the old ways are new again . . . and why they work better than the industrial methods that are all too popular today. The detailed instructions in these pages are all you need to start raising livestock ethically and sustainably. In fact, I'd say it's a better investment than an agricultural degree from a land-grant university."
—**Mike Callicrate,** owner, Callicrate
Cattle Co. and Ranch Foods Direct

"My husband and I have farmed for a living all of our adult lives, and farmed with our parents before that. So we had the good fortune of being surrounded by people with deep generational knowledge when we started out. I can't imagine how new farmers are making it today without that kind of support. Recently we added Large Black hogs to our small farm in North Central Kentucky. *The New Livestock Farmer* came to us just when we needed it. It is what my father, Wendell Berry, would call the best of books because it is a tool. It fills a cultural need, and will give beginning farmers just the information they need, just the way they need it."
—**Mary Berry,** executive director, The Berry Center

"The livestock farmer of today must master not only the vast skills necessary to be an ethical rancher but also marketing, sales, processing, packaging, and so many others. Rebecca and Jim's book is a humble and deeply informative guide from a couple that has been deep in the metaphorical and literal weeds of this challenging work. Without thriving agricultural-based communities, resources like this book are invaluable substitutes, creating a network of like-minded land stewards. Rebecca and Jim have done a good turn in sharing their knowledge in this straightforward and honest primer."

—**Marissa Guggiana,** cofounder, The Butcher's Guild, and author of *Primal Cuts: Cooking with America's Best Butchers*

"The real question for the reform of livestock agriculture in the US is whether we can move from a media saturated with images of what we can't stomach—digestively or culturally—to a more sophisticated understanding of practical and ethical alternatives that make ecological and economic sense. As a farmer and a teacher, I have yet to find a guiding text that does it so well or so comprehensively. Thistlethwaite and Dunlop take us step by step from the open-ended question of breed selection to the finality of slaughter options, providing clear pathways of decision making for farmers, students, consumers, and advocates."

—**Philip Ackerman-Leist,** professor, Green Mountain College, and author of *Rebuilding the Foodshed*

THE NEW
Livestock Farmer

The Business of Raising and Selling Ethical Meat

REBECCA THISTLETHWAITE
and JIM DUNLOP

Foreword by BILL NIMAN and NICOLETTE HAHN NIMAN

Chelsea Green Publishing
White River Junction, Vermont

Copyright © 2015 by Rebecca Thistlethwaite and Jim Dunlop.
All rights reserved.

Unless otherwise noted, all photographs by Rebecca Thistlethwaite.
Illustrations copyright © 2015 by Kelsey Mosley.
Cover photographs courtesy of (*clockwise from the top left*)
Alicia Jones of Afton Field Farm, John Suscovich of
FarmMarketingSolutions.com, Jennifer Jones,
Jessica Reeder/Wikimedia Commons, and Bonnie Long.

No part of this book may be transmitted or reproduced in any form
by any means without permission in writing from the publisher.

Project Manager: Patricia Stone
Project Editor: Benjamin Watson
Copy Editor: Deborah Heimann
Proofreader: Eric Raetz
Indexer: Lee Lawton
Designer: Melissa Jacobson

Printed in the United States of America.
First printing May, 2015.
10 9 8 7 6 5 4 3 2 1   15 16 17 18 19

Chelsea Green Publishing is committed to preserving ancient forests and natural resources. We elected to print this title on 100-percent postconsumer recycled paper, processed chlorine-free. As a result, for this printing, we have saved:

**98 Trees (40' tall and 6-8" diameter)
44 Million BTUs of Total Energy
8,452 Pounds of Greenhouse Gases
45,839 Gallons of Wastewater
3,069 Pounds of Solid Waste**

Chelsea Green Publishing made this paper choice because we and our printer, Thomson-Shore, Inc., are members of the Green Press Initiative, a nonprofit program dedicated to supporting authors, publishers, and suppliers in their efforts to reduce their use of fiber obtained from endangered forests. For more information, visit: www.greenpressinitiative.org.

Environmental impact estimates were made using the Environmental Defense Paper Calculator. For more information visit: www.papercalculator.org.

Our Commitment to Green Publishing
Chelsea Green sees publishing as a tool for cultural change and ecological stewardship. We strive to align our book manufacturing practices with our editorial mission and to reduce the impact of our business enterprise in the environment. We print our books and catalogs on chlorine-free recycled paper, using vegetable-based inks whenever possible. This book may cost slightly more because it was printed on paper that contains recycled fiber, and we hope you'll agree that it's worth it. Chelsea Green is a member of the Green Press Initiative (www.greenpressinitiative.org), a nonprofit coalition of publishers, manufacturers, and authors working to protect the world's endangered forests and conserve natural resources. *The New Livestock Farmer* was printed on paper supplied by Thomson-Shore that contains 100% postconsumer recycled fiber.

Library of Congress Cataloging-in-Publication Data
Thistlethwaite, Rebecca, author.
  The new livestock farmer : the business of raising and selling ethical meat / Rebecca Thistlethwaite and Jim Dunlop.
      pages cm
  Other title: Business of raising and selling ethical meat
  Includes bibliographical references and index.
  ISBN 978-1-60358-553-8 (pbk.) — ISBN 978-1-60358-554-5 (ebook)
1. Livestock—Moral and ethical aspects. 2. Animal welfare. 3. Animal industry—Moral and ethical aspects. I. Dunlop, Jim, 1969- author. II. Title. III. Title: Business of raising and selling ethical meat.

SF140.M67T45 2015
636--dc23

2015002104

Chelsea Green Publishing
85 North Main Street, Suite 120
White River Junction, VT 05001
(802) 295-6300
www.chelseagreen.com

✸

To our daughter Fiona
for her patience with this process
and for loving venison backstrap

✸

Contents

Foreword	ix
Acknowledgments	xiii
Introduction	1
1: The Meat Landscape	5
2: Poultry Production	19
3: Sheep and Goat Production	51
4: Pig Production	75
5: Cattle Production	103
6: Exotics: Rabbits, Bison, Elk, and Deer	129
7: Regulations	151
8: Sales Outlets and Market Options	165
9: Slaughtering and Butchering Logistics	209
10: On-Farm and Mobile Processing	229
11: Packaging, Labeling, and Cold Storage	241
12: Principled Marketing	255
13: Financial Management, Pricing, and Other Business Essentials	273
Appendix A: Simple Farm Biosecurity Tips	287
Appendix B: Meat and Poultry Processing Rules	289
Appendix C: Sample Cutting Instructions	296
References	310
Index	313

Foreword

Like the authors of this book, and countless other couples in agriculture, we function as a team. Rather than each of us heading out separately every morning to a distant job, on a typical day we live and labor alongside one another. We share household and ranch chores, take our daily meals together, cooperate in child-rearing duties, and consult with each other about our meat business. Occasionally, we share the speaking stage and sometimes even write collaboratively. When we first met in 2001, Bill was the founder and CEO of Niman Ranch, a collective of hundreds of farmers and ranchers raising livestock using traditional methods, and Nicolette was the senior attorney for an environmental group whose top priority was addressing pollution from the livestock and poultry industries. Both of us, in our very different ways, were fully engaged in what we viewed as the essential task of remaking the norms of how farm animals in the United States and elsewhere would be raised in the twenty-first century.

At the time, more than a decade ago now, concentration, confinement, and concrete had become so widely accepted and entrenched that returning animals to grass seemed more than a monumental task; to many, it seemed an impossibility. Other than animal rights activists, who argued completely against animals in farming, few publicly questioned the appropriateness of the ways in which farm animals were being raised.

The industrialized approach had taken hold gradually. At the dawn of the twentieth century, time-honored husbandry methods were starting to be abandoned. Various human inventions were ushering in the era of mass-scale total confinement systems. One such technology emerged around 1880, when incubators came into widespread use for hatching chicken eggs, especially in Petaluma, California, just up the road from where we now live. A history of poultry farming notes (approvingly) that this set poultry farming "on an unerring industrial course." It separated the hen from her eggs and her chicks, usurping the hen's brooding and mothering role, and reducing her to nothing more than an egg producer. By 1930, chicken raising had become quite industrialized: Breeds were specialized into egg-laying and meat types (so-called "broilers") and a single incubating machine could hatch 52,000 chicks at one time; broiler flocks were kept indoors and numbered in the tens of thousands. In 1938, the DeWitt brothers in Zeeland, Michigan, invented automated feeders for poultry, which substantially reduced the human care required.

Within two years of the United States entering World War II, the government set an annual production goal for chicken meat of four billion pounds, an 18 percent increase over the previous year. Farmers were encouraged to contribute to the war effort by raising "defense chicks," which would be raised as food to be shipped abroad to American soldiers and allies. Over these years, the value of domestically raised chickens went from $268 million (in 1940) to over $1 billion (in 1945), nearly a four-fold expansion in just five years.

Novel feed additives soon reinforced the trend toward total confinement. Since the birds were deprived of sunshine, laboratory-manufactured vitamin D had to be included in their feeds. Starting in the early 1950s, antibiotics were added, too. Initially, this was to stave off diseases. Morbidity and mortality had risen sharply as flocks expanded. When poultry farmers first began keeping sizable flocks in confinement, typical rates of death loss quadrupled, jumping from 5 to 20 percent. "With commercialization and greater intensification have come [new] disease problems ... the chances of disease spread are materially enhanced," two professors of poultry husbandry wrote in the 1930s. Adding antibiotics to daily feed and water reduced rates of death and illness. The continual drug dosing was soon discovered to have the added bonus of dramatically increasing the chickens' growth rate as well. By mid-century, some pigs and dairy cows were being kept in large, total-confinement herds.

Antibiotics thus cheapened production and enabled crowded total confinement by keeping animals alive in otherwise unlivable conditions. In the decades that followed, adding antibiotics (and other drugs) to animal feed became standard. The Union of Concerned Scientists has estimated that 97 percent of hog finishing operations continually feed antibiotics.

However, the meat industry's heavy reliance on these pharmaceuticals has come with a high public health cost. In 2010, the Food and Drug Administration reported that over 80 percent of antibiotics used in the United States were going to farm animals, with over 90 percent of them used in the daily feed or water of animals living in confinement operations. Overuse of antibiotics contributes to the rise of pathogens that are resistant to the drugs. Such pathogens travel through the soil, air, or water, leave farming operations, and populate the meat and eggs produced.

Research by Johns Hopkins University School of Public Health and others has connected the livestock and poultry industry's drug overuse with the rise of antibiotic-resistant diseases. A few decades ago, most staph infections were readily curable with antibiotics, but no longer. In 2005, one such infection, methicillin-resistant *Staphylococcus aureus* (known as MRSA) resulted in 18,000 deaths in this country, more than were killed that year by HIV-AIDS. A University of Iowa study demonstrated that 49 percent of pigs and 45 percent of farmers at confinement operations in Illinois and Iowa tested positive for MRSA.

Despite such sobering study results the livestock and poultry industry still strenuously resists Congress's repeated efforts to rein in antibiotic use. The meat industry now declares that these drugs are essential for preventing illness and death for animals living in intensive confinement. For workers and farm animals, exposure to antibiotic-resistant infections is just one of many dangers of confinement operations.

Keeping animals continually confined invariably alters their diets. Where livestock have no access to foraging and grazing, all feed must be planted, watered (in some cases), harvested, transported, and dispensed to them. Because this process is expensive and labor-intensive, there is a strong incentive to feed concentrated feeds (mostly grains and soy) rather than forages high in roughage, and this is how animals in confinement operations are typically fed. Thus, an animal in such an operation loses not only the manifold benefits of sunshine and exercise, but it is also deprived of a diet comparable to that for which its body type evolved.

In the United States, the feeding of concentrates was made especially advantageous after World War II by federal grain subsidies, put in place partly to help feed the hungry masses in Europe. When pigs and poultry were kept in small flocks and herds, those omnivorous animals had been fed from a combination of foraging, excess crops, scrap foods, and farm by-products, supplemented with some grain. Thus, their recycling assisted farms and households in optimally using resources. However, as those animals were increasingly kept in large flocks and herds separated from the land, it became necessary to raise crops dedicated to feeding them. Dairy cows were grazing less and fed more concentrates, and beef cattle, too, were increasingly being kept in feedlots for several months before slaughter, where they were fattened on grains and soy.

In the United States today, about 55 percent of grain produced is fed to livestock and poultry. (Lower, actually, than the average for all industrialized countries, which is around 70 percent.) Omnivorous animals—pigs and poultry—are nearly all confined and fed a diet of grains and soy. An estimated 99 percent of turkeys and chickens (of both the egg-laying and meat types) and 95 percent of pigs live in metal buildings with 24-hour ventilation to manage the fumes, mechanized feeding and watering, and elaborate waste-handling systems for their feces and urine.

For those fortunate farm animals raised free-range, the overwhelming body of scientific evidence confirms what common sense already tells us: Animals are happier and healthier when raised with sunshine, fresh air, and grass, and when given the opportunity to exercise. Not surprisingly, animals raised on pasture also produce healthier, safer, and (many people agree) tastier food.

Compare a dairy cow's situation in confinement to living on pasture. Given the opportunity, bovines spend most of their waking hours in an ambulant state of grazing, walking an average of 2.5 miles a day, all the while taking 50 to 80 bites of forage per minute. But confinement allows virtually no exercise and zero grazing. Standing on cement for hours every day "is murder on cows' legs," as animal welfare expert Marlene Halverson says. For every 100 US dairy cows, for instance, there are 35 to 56 cases of lameness. Conversely, welfare and health markedly improve when cows exercise outdoors. A 2007 study in the *Journal of Dairy Science* found that continual access to pasture cured cow lameness, and a 2006 study in *Preventative Veterinary Medicine* found it also eliminated joint swelling and open sores.

Chickens, too, fare better in both welfare *and* health when they have access to pasture. Caged hens have extremely weak and brittle bones due to their physical inactivity, leading to frequent bone fractures. Animal welfare expert Dr. Sara Shields notes that free ranging has multiple benefits in terms of food safety as well as animal welfare. Disease risks, a report she authored notes, are minimized by factors associated with the outdoor, free-range environment. "Natural sunlight kills many pathogens and virus particles, and the lower stocking densities and access to fresh air typical of free-range flocks lower infection and transmission rates." Compared to battery-cage egg production, studies have found the odds of *Salmonella enteritidis* contamination are 98 percent lower in free-range systems; for *Salmonella typhimurium*—the most common US source of *Salmonella* poisoning —the odds are 93 percent lower in organic and free-range systems; and for the other *Salmonella* serotypes found, the odds are 99 percent lower in free-range birds.

Finally, various studies have documented that food from free-ranging animals is more nutritious. Research, summarized in the Union of Concerned Scientists' 2006 report *Greener Pastures* showed higher levels of vitamin E and omega 3 fatty acids in grassfed beef, and that grazing dairies produce food higher in both the beneficial fatty acids ALA and CLA—an effect that was particularly pronounced in cheese. Likewise, research done by Clemson University and the USDA has found that grassfed meats have more calcium, magnesium, potassium, thiamin, and riboflavin. A 2003 Penn State study found a dramatic difference in the nutritional qualities of eggs from pasture-raised hens: "On average, we saw about twice as much vitamin E and 40 percent more vitamin A in the yolks of pasture-fed birds than in the caged birds," the lead researcher reported. "The longer the animals were on pasture, the more vitamins they produced."

The consensus among credible sources—scientific research, practical farming experience, and common sense—tells us that farm animals are healthiest when provided the opportunity to thrive in environments where they can breathe fresh air, graze, forage, exercise, and soak up the sun. Our parents told us as children to "go outside and run around a while" because they knew physical activity and fresh air would help keep us in good health. And so it is with farm animals. The healthiest, safest, and most nutritious food comes from farms following that wisdom.

Can animal farming be returned to grass? As *The New Livestock Farmer* demonstrates, it's not only possible, it's vitally necessary. And to do it, we need a brave new generation of farmers throwing their hearts, minds, and bodies into the task. Rebecca and Jim have tackled it head on by creating their own Next Generation livestock farms. This book is a distillation of their combined knowledge and experiences, both good and bad. In addition, they have sought out and included the wisdom and examples of many other successful farmers and ranchers, all of whom reflect the best management practices and savvy business sense that define any successful animal farming operation, regardless of its size or location. Now they are sharing what they have learned. Thank goodness.

Bill Niman and Nicolette Hahn Niman

Bill Niman is the founder and former CEO of Niman Ranch. Nicolette Hahn Niman is the author of the books Righteous Porkchop *and* Defending Beef. *Together they own and operate BN Ranch in northern California.*

Acknowledgments

Thank you to Chelsea Green Publishing and our editor Ben Watson for taking the plunge once more to work with us. Although I (Rebecca) had written another book with CGP just two years previously, they trusted us to produce this one together as a married couple. Jim has always been my fabulous personal editor and it turns out he is a pretty great writer as well. Where I am long-winded, he is succinct and to the point. We survived this process of writing a book together and our daughter Fiona doesn't feel totally abandoned—pretty good for our insanely short six-month writing timeline along with working other full-time jobs.

So many people helped us write this book and our knowledge base has expanded considerably as a result. Sausage-slinger Jorgia Jacobs called every single state Department of Agriculture to find out the latest red meat and poultry rules across the country (phew!). Former slaughterhouse manager and meat consultant Chris Fuller gave us the inside knowledge on how to work with meat processors, and he also developed the ever-so-useful custom cut sheets for beef, pork, lamb, and goat featured in appendix C. Butchering educator Adam Danforth gave us the skinny on meat science with his fabulous set of books *Butchering* and workshops he gives around the country on the same topic. Rancher and marketing genius Cory Carman shared her story about developing her meat brand Carman Ranch and reviewed our chapter on sales outlets. Matthew LeRoux, Ag Marketing Specialist at Cornell University Cooperative Extension, also combed through the same chapter and added his experience on finding the most suitable markets for meat producers. Pastured pork and chicken trailblazer Greg Gunthorp reviewed the chapter on regulations with his encyclopedic brain and gave us much inspiration for both the poultry and pig chapters. Heritage Meats owner Tracy Smarciarz answered any question that we threw him and reviewed the practical chapter on packaging, labeling, and cold storage. Graphic-design-loving friend Michelle McGrath did our meat inspection flow chart. An especially large shout-out to Kelsey Mosley for the wonderful pen and ink illustrations featured throughout the book. For a steadfast vegetarian, she drew some amazingly stout meat animals!

People we admire sifted through all of the animal production chapters to give them more boots-on-the-ground expertise. The poultry chapter was thoroughly reviewed by the owner of Feathermen Equipment and a pastured poultry producer himself, David Schafer. Renard Turner of Vanguard Ranch, a meat goat farmer who travels to fairs and festivals selling fantastic goat-centric meals, reviewed our sheep and goat chapter. Longtime sheep producer and Tamarack breed developer Janet McNally of Minnesota also did wonders for the sheep section. Grazers Joe Morris of California and Rod Ofte of

Wisconsin helped "beef" up our cattle chapter with their decades of experience running grass-fed beef operations. Finally, the founders of Northstar Bison, Lee and Mary Graese, reviewed the bison, elk, and deer sections of our exotics chapter. Combining our experience and research with the insight of these experienced and successful meat animal producers makes these chapters even better.

Others we have to thank are those who sat through our interviews while we prodded them with questions. These include Jennifer Core of Hettie Belle Farm, Jennifer Curtis of Firsthand Foods, Robert Doering of Sarver Heritage Farm, Bruce Dunlop of the Island Grown Farmers Cooperative and Lopez Island Farm, Bartlett Durand of Black Earth Meats, Tom Frantzen of Frantzen Farm, Mary Graese of Northstar Bison, Lauren Gwin of Oregon State University and the Niche Meat Processors Assistance Network, Kathleen Harris of the Northeast Livestock Processing Service Company, Will Harris of White Oak Pastures, Doug Knippel of Northwest Red Worms, Rachel Kornstein of Boondockers Farm, Melissa Miller of Miller Livestock Co, Bob Nero of La Montanita Co-op, Rod Ofte of Wisconsin Grassfed Beef Co-op, Bob Perry of the University of Kentucky, and Josh Schaeding of The Maple Grille.

We also picked the brains of countless others on the American Pastured Poultry Producers Association listserve, such as Mike Badger and Jeff Mattocks, as well as folks on the Niche Meat Processors Assistance Network listserve. If you are going to produce pastured poultry or sell meat, you should join these fantastic listserves to learn from your peers. Thank you to all the folks we spoke to at farming conferences over the past year and the Slow Meat convention put on by Slow Foods USA. If you gave us advice or spoke your mind, thank you for that. All of these people mentioned are doing important work at different points along the good meat supply chain. We wanted to talk to hundreds more, but this was a pretty good start.

Many thanks to the following photographers for allowing us to use their photographs to enliven this book: John Deck, Chris Fuller, Jim Gerrish, Evan Gregoire, Jenn Ireland, Alicia Jones, Jennifer Jones, Kendra Kimbirauskas, Northstar Bison, Linda Ozaki, Mike Stricklin, and John Suscovich. The rest of the photos were taken by the author Rebecca Thistlethwaite.

Introduction

Neither one of us is a farm kid, nor are we part of some glorious multi-generation ranching lineage. Jim grew up on a pre-suburban homestead in New York raising small livestock and poultry with his family for their own consumption, before the area became dotted with modest middle-class homes. Rebecca grew up in a homeowner-association-ruled neighborhood in Oregon where a couple of eccentrics grew vegetable gardens but mostly manicured lawns dominated the landscape. Jim went on to do a four-year service with the US Marine Corps and got out wanting to surf and travel. As he learned about conventional meat production and its less than humane practices, he opted to avoid meat in his diet. Rebecca decided in high school that the environmental impacts of conventional meat production were too much for her to swallow, so, right before her grandparents' fiftieth wedding anniversary pig roast, she became a strict vegetarian.

A decade or so later, one winter I (Rebecca) remember myself arguing with a grassfed beef producer, one who has since become my friend. We were at the Ecological Farming Conference, a large gathering of the organic tribe that occurs near Monterey, California, each January. I was making my case for vegetarianism with the rancher, who kindly nodded his head listening to my dogma going around in circles like a broken record. I was probably saying something about the amount of grains used to feed livestock and how the planet could not sustain that increasing amount. Exactly one year later, I was introducing my new rancher husband Jim to that same cattle rancher and gushing about my newfound love for meat. It was like I was making up for lost time, going from twelve years as a vegetarian straight to dedicated meat enthusiast.

Jim too was a vegetarian for almost the same amount of time before he began raising animals, first starting out with poultry and then progressing to raising pigs, sheep, and some cattle on two different farms in California. He prides himself on his understanding of electric fences, building innovative animal shelters out of junk materials, and producing really tasty meats. If there were such a term as "cowboy," except with pigs, that would be Jim. "Pigboy" just doesn't have the same ring.

Newly married and recovering vegetarians, we progressed quickly from raising animals to killing them, cutting them up, and eating them. We started a farm on the central coast of California called TLC Ranch, beginning with broiler chickens, and then including laying hens, pigs, sheep, and some cattle. Our goal was to eventually make a living from farming, and a few years into it we did just that. By our fifth year of production we were grossing nearly half a million dollars—not too shabby on just twenty acres of irrigated pasture.

Slowly, we cultivated our knowledge of cooking meat, relying on favorite books such as *The Grassfed Gourmet* and even *The Joy of Cooking*. We began more complicated meat projects such as curing pig jowls and bacon, brining meats, stuffing and roasting, and making pâtés. We won't claim to be full-fledged nose-to-tail eaters, nor the best home cooks, and there still are a few odd bits of meat that we won't eat, yet. But we have a much greater understanding and appreciation for the whole animal and the logistics that bring meat to our plates. We know now that for us, and probably the planet, finding our way to the "balanced middle" of meat consumption is how we probably should be eating. But this book is not a treatise on meat consumption. You can find that subject extensively researched in Simon Fairlie's outstanding book *Meat: A Benign Extravagance* or Nicolette Hahn Niman's new book *Defending Beef*. We just wanted to give you a little background on our thinking about meat and meat consumption, and how we got to where we are today.

We now live on a patch of ground in Oregon and are trying to figure out the best animal fit for our land in a careful and considered way. We have started with a small pig herd of American Guinea Hogs to take advantage of the copious acorns that drop in our oak woodland. This year we'll add ducks to enjoy our half-acre pond, as well as some chickens and rabbits for our garden and for home meat consumption. Right now, we're rotationally grazing a small flock of Jacob sheep around our property to fireproof our land, instead of using fossil fuels to do that. We are also experimenting with different breeds to find the ones that best complement our geography. We are taking our time to see if commercial production of animals is right for us in our new surroundings, taking into consideration the land base first, then our family's holistic goal, our access to critical inputs such as grains and slaughter facilities, and also market conditions. We also have a huge vegetable garden, are installing an orchard and cane berries in the fall, and own and operate a farmstand in town to sell our produce and other farm goodies from our community.

This book draws upon our mistakes and wisdom gleaned raising animals for close to ten years in California and the knowledge of countless inspiring and innovative meat producers around the country. Just like Rebecca's last book, *Farms with a Future*, we rely heavily upon the knowledge earned by hardworking farmers who have graciously shared their stories with us so that others may learn from their mistakes and perhaps pick up some successful practices to integrate into their current systems. Maybe your current animal management system needs a 180-degree turn: if so, this book is for you, too. We have met producers who are struggling with an entrenched way of raising animals but who honestly crave a way out of the muck, literally. Go climb up to the highest place on your land or get on the roof of your barn and take a peak down. Is this the animal production system you want to continue? Ask yourself the ethical questions described in the next chapter and see where that leads you.

The Purpose of This Book

Our goal for this book is to transform the meat supply chain by making it easier for meat producers to raise healthy animals and get them to market in a way that achieves four equally important objectives:

1. keeps them in business and earns them a fair living;
2. treats the animals humanely;
3. is gentle and even restorative on the earth; and
4. provides nutritious and tasty food to a wider audience of consumers.

Introduction

The first half of the book will focus on raising various livestock and poultry species in the best way we know how (this is a collective "knowing" based on interviewing lots of farmers and some solid scientific references). Each animal chapter will be broken down into the key bits of production information that we think you ought to know, with highlights of best practices. These chapters are not comprehensive production manuals but rather distilled drops of wisdom for pasture-based systems. We figure you already have a bookshelf dedicated to books about single-species production, and we don't plan on duplicating those efforts. This book also emphasizes profitable, commercial raising of animals, not backyard self-sufficiency production: there are many homesteading books out there already. Because the focus is on profitable animal production, we try to steer clear of the hobby-scale advice as much as possible, but hey, we all have to start somewhere.

The second half of the book is about the business of getting those animals to market, including the transportation, slaughter, butchering, packaging, labeling, cold storage, market channels, pricing, financial management, and other pertinent odds and ends of running a farm business.

Book Overview

What will this book cover? We will summarize the best practices of pasture-based animal production from around the country, taking into account the large variations in climate, soils, vegetation, breeds, access to inputs, and other variables. There is no one-size-fits-all approach, and we understand that producers need to build the most sustainable systems that work for them *given where they live and farm*. You may not agree with a particular system or may think there is only one "right" way to raise animals or one perfect breed. You are encouraged to have that passion and conviction, but our goal is to include many different production systems that are balancing sustainability, animal welfare, and economics. We also aim to be broad enough that this book can be as useful for a sheep rancher in New Mexico as it is for a grass farmer in Minnesota as it is for a pastured poultry producer in Georgia. It will not be all-encompassing for one region, one animal species, or one production model. There are other great books, periodicals, blogs, and other online resources that will help you hone in on your specific system. Look to the references section in the back of the book for some of our favorite animal production resources.

This book will delve into how to go about processing and selling your meat in different market channels and ways you can collaborate with others to scale up over time. We will present farmer-based advice to successfully direct-market your meats and some of the significant pitfalls that we hope you can avoid. By addressing those challenges in this book and illustrating that this business is not easy, nor is it for everyone, we hope to interject a serious dose of reality into the conversation. To improve the robustness of this information and get more boots-on-the-ground realism, we sought the advice of hundreds of other meat producers around the country to put this book together. If we left you out, we're sorry, but we are confident you are doing amazing things too.

One thing we have learned in writing this book is how very little we actually know. It has been both a humbling experience and an enlightening one. We wish we could have done this extensive research back when we were producing animals on a commercial level, but at least now we are ready for the future. We hope that you, too, will be better prepared for this "new meat market" by reading our book.

CHAPTER 1
The Meat Landscape

The meatpacking industry in this country is so grossly consolidated that just four firms control 80 percent of the meat supply. Four companies. Yet there are thousands of farmers across the country raising animals as either their sole source of income or as part of a diversified operation. Farmers are powerless to negotiate the terms of meat pricing with those four firms, to demand humane and ethical practices at the slaughterhouses, and to demand food safety and quality control after their animal leaves their ranch. If you raise an animal with care, but then it is treated carelessly in the slaughter and butchering process or ends up wrapped in plastic under the fluorescent light of a meat counter, overseen by an employee who knows very little about animal agriculture, it's likely you want things to change. If you are a consumer or buyer of meat and are sick of the lack of quality, flavor, and transparency in the system, you probably want things to change too. All of us—producers and consumers (and animals too)—need a new meat market. We want to know that animals are treated well while they are alive and are properly slaughtered at the end. That the animals are butchered in a way that makes use of the whole animal, because they give their lives for our nourishment. That the meat is packaged and kept cold in a way that keeps it safe and maintains quality, and that it gets to grateful consumers who know how to cook it to perfection. This is the "new" meat market—one that respects all stakeholders in the system while providing nutritious, safe, quality food.

We want farmers to make the most money on the animals they raise, distributors and retailers to cover their costs and profit margins, and consumers to pay a fair price for a quality product that both tastes great and is nutrient-dense. That is a tall order, and it requires a lot more education, as well as transparency and creativity, throughout the system. This book aims to cultivate a little of all those things, so that meat producers are empowered to raise and market their meat as directly as possible and we all have more access to better meat. We don't expect every farmer to do this alone, so we will cover several collaborative and cooperative models that distribute meat for multiple producers and we will explore some of the markets in the middle, such as grocery stores and institutional markets.

Studies say that somewhere between 54 and 99 percent of the meat we consume in the United States comes from confinement-style operations (it varies somewhat by species and the level of concentration). In cattle, for example, 71 percent of all feeder cattle come from farms selling over 5,000 head annually, considered "large CAFOs" by the US Environmental Protection Agency (MacDonald et al., 2009). CAFO is an abbreviation for "confined animal feeding operation"; we will refer to these large operations as CAFOs throughout the book. Whether this is a feedlot for beef or sheep or a densely packed barn for poultry or pigs, these are operations that cram many animals into a small space for rapid, efficient growth on grain-based diets. Interestingly, the EPA definition of a CAFO does not sugarcoat the practice: "Animal Feeding Operations (AFOs) are agricultural operations where animals are kept and raised in confined situations. AFOs congregate animals, feed, manure and urine, dead animals, and production operations on a small land area. Feed is brought to the animals rather than the animals grazing or otherwise seeking feed in pastures, fields, or on rangeland." Because of the high animal density, manure, exposure to dead animals, and other factors, disease pressures are high among the animals. Thus the CAFO model breeds a dependency on antibiotics. Indeed, estimates are that 70 percent of all antibiotics used in this country go to animals that are being raised for human food. Other chemicals, such as growth promoters and arsenic (in poultry), are added to the feed ration to increase the speed at which the animals get to slaughter weight. Breeding sows and veal calves are kept in crates where they can't turn around. Poultry is packed so tightly in large barns that the phrase "free range" is mostly meaningless; as the birds grow larger they are practically on top of each other.

The most recent 2012 USDA Agriculture Census is showing some positive trends, but some alarming ones too, about the number of livestock and poultry farms in the United States. Even though there are more direct-market producers, there has been an overall decline in the number of farms raising meat animals. On the positive side, the number of farms raising sheep was up 6 percent and the number of poultry farms was up 29 percent over the 2007 census. On the decline (or gobbled up through concentration) were beef cattle producers (down by 5 percent), hog producers (down by 25 percent), and goat farms (down by 11 percent) since 2007.

Despite the census trends, there is a proliferation of the number of direct-market meat producers listed on websites such as LocalHarvest.org, Eatwild.com, and HomeGrownCow.com. On LocalHarvest, we counted 4,625 farms selling beef, 3,373 selling pork, 6,577 selling chicken, 2,548 selling lamb, 3,114 selling goat, 1,571 selling duck, 138 selling buffalo, and even 157 selling venison! On Home Grown Cow, we counted 179 meat products within 500 miles of where we live in Oregon. On Eatwild we found 74 grass-based livestock producers in our region. Livestock and poultry sales account for over half of US farmer income, exceeding $100 billion annually (USDA Economic Research Service website: www.ers.usda.gov/topics/animal-products.aspx). Meat is valuable stuff, and an increasing number of farmers are direct-marketing it.

Why would farmers want to direct-market their own meat? The main reason is that wholesale prices paid to farmers from the meat middlemen (packers, distributors) are typically not attractive and in many cases do not cover the costs of production. In 2013 average broiler chicken prices were just $0.91 per pound, $92.00 per cwt for pork (that's $0.92 per pound), and $202.50 per cwt for Choice grade beef (that's $2.02 per pound) (cwt is an abbreviation meaning hundred weight, or 100 pounds). ERS data shows that the share of retail value that goes to the farmer for pork has averaged around 30 percent for the last eight years and close to 50 percent for Choice grade beef. The share going to the middleman has actually decreased, putting further squeeze on their tight margins, yet the share has increased for the retailers. Retailers often make 50 to 60 percent of the retail dollar on meat. Granted they have costs and losses too, but we think the share that goes to each step in the meat supply chain (producer-processor-retailer) should be somewhat proportional to the work that they do. Maybe more like 50 percent to the farmer, 20 percent to the processor, and 30 percent to the retailer. There are a lot of ways that retailers can be smarter about how they sell meat while minimizing their losses (referred to as "shrinkage" in grocery-speak). For example, we have seen many smaller natural-foods stores keep all their meat in freezers, with no fresh case. This reduces labor costs and shrinkage, and allows the stores to buy from small farmers, who often don't have the option of selling fresh meat because they don't have control over the meat processing and distribution. We will discuss in Chapter 8 ways that meat buyers (stores and restaurants, primarily) can maximize their meat purchases while working with individual farmers. It does not have to break the bank to buy or sell good meat.

The Better System: Pasture-Raised, Ethical Meats

What do we mean by the term "ethical" meats in the subtitle? In our minds, ethical meats come from producers who want their animals to live comfortable lives and to die as quickly and as humanely as possible. These animals get to exhibit their natural behaviors: as Virginia farmer Joel Salatin likes to call it, "the animal-ness of the animal." Poultry get to take dust baths, scratch, and peck; waterfowl get access to water to play in; pigs get to root or make nests; cattle get to eat grass or lie in the shade to ruminate. If animals are kept in cages, a densely packed barn, or an outdoor feed yard knee-deep in muck and manure, they won't be exhibiting their natural behaviors. They will be just barely coping and often going slightly crazy, leading to abnormal behaviors like chewing off each other's tails or excessively pecking the backs of other birds.

Ethical meats come from animals that spend time outdoors and get to eat naturally growing vegetation that their digestive systems are used to handling. Goats get browse; pigs get roots, nuts, or legumes; sheep and cattle get pasture; poultry get some bugs, green growing vegetation, and seeds. We are not saying ethical meats have to be 100 percent grassfed or 100 percent organic, because every animal is different and every production system is different. What works in one region, such as year-round grazing, may not work for a region with a two-month-long active grazing season. However, to the extent possible, the farmer should match the animals to the landscape, not the other way around. Raising broiler chickens in winter, for example, does not work very well in most northern climates unless you have a fully temperature-controlled building. Broiler chickens raised outdoors won't thrive in a cold winter and most likely will all die. But confining them to a barn starts to look more and more like a CAFO as one scales up.

Do ethical meats come from animals fed grain that has been conventionally grown with petroleum-based fertilizers, neonicotinoid pesticides, or glyphosate herbicides? If the system depends on feed grains that are grown using genetic engineering technology or cancerous chemicals, then the system is broken. It is just shifting the negative impacts onto the grain production in order to prop up cheaper meat production. Can ruminant animals be fed some grain and still be "ethical"? We think so. We don't love the feedlot model, but feeding some grain during the last sixty to ninety days of finishing, in addition to providing plenty of forage and healthy living conditions, seems like a pragmatic compromise. Depending on your pasture production, rainfall amounts, and other conditions, some grain feeding might make sense for you.

We are not going to create a complex definition of what constitutes ethical meats, but would rather like you to ask yourself some tough questions. Good agriculture and good business involve continuous improvement. You may start at one point on this continuum and be able to achieve a more holistic system of production after you gain experience and begin to earn the revenues necessary for improvement. Most importantly, if you are going to market ethical meats, educate your customers about your practices and why you have chosen to raise the animals in the way that you do. Don't lie about it and don't hide it. More educated consumers will make the whole food system work better.

Questions to Ask Yourself on Your Path to Producing Ethical Meats

ANIMAL WELFARE

- Are my animals able to express their natural instincts?
- Are my animals comfortable during weather extremes?
- Do my animals have reasonable protection from predators while still getting to live outdoors?
- Do my breeding animals have safe places to give birth and provide milk to their offspring?
- Do I give my animals sufficient time to get to market size or must I resort to other methods such as growth hormones or overfeeding or restricting movement to get them to size?
- Do I have a low stress way to corral, sort, and transport my animals to slaughter?
- Have I visited the slaughterhouse and verified how they pen, handle, and kill the animals?

ANIMAL HEALTH

- If I don't have a closed herd or do all my own breeding, are the animals I'm bringing onto my farm healthy and were they treated in a humane way before they got to my farm (chicks, weaner pigs, stockers, lambs, breeding stock, etc.)?
- Does my animal management system keep animals safe from each other or is there a lot of fighting?
- Do I check on my animals frequently? Can I get close to my animals?
- Do I have a good system in place to prevent illness and disease in my animals?
- Do I have a way to segregate or capture an animal if it appears sick or injured?
- Do I have a way to treat or medicate sick animals?
- Do my animals have access to the outdoors whenever they feel like it?
- Do I keep my animals from lying in or living on top of their own dung?

LAND MANAGEMENT

- Are my pastures and paddocks protected from degradation during weather extremes?
- Does my animal management system lead to soil erosion, water pollution, or air pollution?
- Does my animal management system degrade wildlife habitat or enhance it?
- Do I utilize a variety of preventative practices to deter predation or just resort to lethal control of predators, such as bobcats and coyotes?
- Do I rotate and rest my pastures?
- Do I manage my manure so that it is a valuable resource or is it more like a waste product that can cause pollution?

FEEDS AND FEEDING

- If I buy growing stock, have those animals been fed medicated feeds, given antibiotics, or given hormones prior to arriving on my farm? Do I have any control over that?
- Do my ruminants have continuous access to pasture or forage whenever they feel like it?
- Do my feed grains come from places recently deforested to grow grains?
- Do my feed grains come from genetically engineered seed?
- Does the production of my feed grains require significant fertilizer or pesticide use?
- Were my feeds trucked in from a very long distance (over 500 miles, for example)?
- Do I make an effort to grow my own feeds and forages to the extent possible?
- Do I utilize other locally produced nutritious feedstuffs (not post-consumer garbage) to support my animals?
- Do I prevent my animals from eating animal products of the same species?

COMMUNITY

- Are my employees subjected to toxic fumes, frequent pathogen exposure, unsafe working conditions, and stressful animal handling conditions?
- Is my surrounding community subject to toxic fumes, manure runoff, excessively loud noises, or other unsafe conditions because of my operation?
- Am I producing safe, healthy, high-quality meats for consumers?
- Do I give back to the community in any other ways?

We could ask you a bunch more questions, but this will do for now. This is not a scorecard: If you answer yes or no to some of the questions, that does not mean you can't consider yourself an ethical producer. Just figure out the best way for you to raise animals in a way that is humane, environmentally and socially responsible, makes you money, and produces a high-quality end product *that you can be proud of.* This book will help you to do just that. Don't shift the costs of your animal production onto other places or other people. To the extent possible, *take responsibility for your full operation and strive for continuous improvement.*

We will give you an example from our previous farm to illustrate some of the ethical conundrums that animal producers face: it is neither simple nor easy to come up with a perfectly ethical system. Keep in mind that there is no flawless production system anywhere; there are repercussions to all of our actions. Agriculture is a complex system: if you pull one string, the effects are felt in a lot of different places. The key is to manage those negative impacts and increase the positive ones. So here is a fully honest description of some of our ethical conundrums at our previous farm on the Central Coast of California during the years 2004 to 2010.

Conundrum #1: Irrigation Water. We rented 20 acres of irrigated pasture near the Monterey Bay in California. Land was expensive and 20 acres was all we could find that was not being used for growing high-value berries or vegetables. However, because the land was irrigated we could run animals almost year-round on it, maximizing its total productivity and our ability to be profitable. We had to irrigate to maintain year-round pastures and our irrigation water was groundwater that was being pumped at an unsustainable rate in our particular watershed.

Conundrum #2: Procurement of Feeds. This farm was in an area with very little animal agriculture or forage production. Accordingly, there wasn't anybody growing feed grains locally nor was there much hay production in the region. We did find one gentleman with an ancient set of haying equipment that could hay our land if the forage got away from us. We started out haying once a year for the first few years until our animal numbers were at the point that they could eat all the forage and we no longer wanted to cut hay from the fields. Because we were raising primarily monogastrics (poultry and pigs), we had to feed grains and could not rely solely on pasture or hay for their dietary needs. But because there was little grain production in the region, most of our grains and pulses came from Montana and the Midwest. They were milled in central California and then transported to us in bulk. However, all of our feed was 100 percent certified organic and US grown. We chose organic to avoid GMOs and the pesticide industry—plus, our customers were looking for organic.

As time went on, we tapped into local waste feedstuffs, including clabbered raw milk, tons of organic vegetables, and brewers' grains from local microbreweries. We started picking up all the produce waste from several Whole Foods stores. By about the fourth year of farming, around 50 percent of our pig feed was local food waste, significantly reducing our costs and our dependency on out-of-state grains. We also found a local in-county source of organic barley, which we fed either whole or soaked to our pigs. Our chickens foraged for some of their nutrient requirements in our lush pastures, but they were largely dependent on a formulated feed from the same organic feed mill. Our system relied on summer irrigation to keep our pastures green, although we used less than one-tenth of the amount of irrigation that nearby vegetable farmers were using. It also relied on feed produced elsewhere, because we did not have the land base or the equipment to grow our own grains. We would look longingly at farms in places like Minnesota and Oregon that were producing their own grains for their animals. But our animals had a good life living outside on pasture, with access to shelter at all times, protected from predators with a livestock guard dog and electric fencing.

Conundrum #3: Predators. Our biggest predators, in order of destruction, were feral dogs, coyotes, humans, and birds of prey. We used electrified perimeter fencing, but because we were renting land, we could not install a taller, more sturdy perimeter fence. We also had only one livestock guard dog at a time, which was probably not enough. We had troubles with our dogs getting out and bothering neighbors since we did not have a perimeter fence to contain them. We also didn't have the luxury of living where we farmed, so we could not be as vigilant as we wanted against predators. We only had to resort to lethal control of predators in the case of a couple of sick, mangy coyotes and some feral dogs that animal control would not deal with. We focused on management over lethal control and, for the most part, it worked out.

Conundrum #4: Processing. The pigs had to be transported a long distance to slaughter, but we would do that in the middle of the night in a well-bedded trailer where they would sleep. The slaughterhouse would receive the animals early in the morning, and they would all be killed that morning. It was quick and efficient. Fully counted, our pigs traveled around 800 miles to be completely processed, turning our "local" pigs into long-distance road trippers. We slaughtered our poultry on the farm and would also load up the chickens the night before, not into crates but into the back of the truck or trailer, with bedding and water, so at least our chickens stayed "local." But our meat processing in general was a long distance affair.

Conundrum #5: Picking Animal Species. Although our customers wanted us to produce all their favorite meats, for two main reasons we did not get heavily into ruminant animals: (1) California already has a lot of sheep and cattle producers, and (2) we only had 20 acres of rented pasture, not nearly enough to raise a profitable amount of ruminants. So we picked animals that fit with our land base, our mild climate, our market niche, and what we enjoyed (we love pigs). We periodically raised feeder sheep or a handful of Jersey calves depending on how our pastures looked and whether we had the forage to support them to finish. We loved the idea of raising ruminants so we would not be dependent on outside feeds, but it was not possible to raise very many with our land base. We longed for a more diversified system, including other animal species, access to woods, and rotating in crops, another impossibility on our leased land.

Those were some of the ethical and practical decisions that guided our system for six years. While we negotiated those issues and others, we built a very strong business that nourished thousands of people in our region with healthy meats and eggs. We wrestled with the ethical decisions constantly and chose a system that worked best given the parameters of what resources we had. You will have to do the same and will surely change systems over time as you learn from your mistakes and successes.

Truth and Verification

There are a wide range of certifications that you can obtain for your animal production system that will help you market your meat, just as there are for field crops or vegetables. Of course, you can raise animals in any legitimate manner you choose, but certain supply chains and consumers will be looking for verification of those practices you say you are using. Your word may not work in those situations. We met an egg farmer a couple years back who told us, "I'm basically organic; I don't need to get certified." When we visited his on-farm store to buy some eggs, we noticed his Purina chicken feed bags lying nearby. We asked if that was what he was feeding his hens and he said yes. We didn't comment any further, but we then knew that the "basically organic" claim was untrue due to the conventional feeds he was using. Unfortunately, this happens throughout the food system, with meat in particular. False claims abound, and most consumers are not equipped with the knowledge and understanding to decipher otherwise. Sometimes those false claims are not egregious lies by the producer—often the producer just doesn't have the full information.

If food producers want to demonstrate their commitment to certain practices and have markets that are requesting verification, then it may make sense for them to get third-party certification. By "third party" we mean that the certifying organization is supposed to be neutral—that is, not representing the producer and not representing the buyer. However, these certifying organizations receive income from their certification services, so they may be swayed by who pays the bills. Some organic certifiers have lost their accreditation because they were willing to overlook violations of the organic law on certain operations that were paying them hefty fees. However, there are a couple of certification programs that are free or inexpensive to the farmer: Animal Welfare Approved (free) and Certified Naturally Grown (under $200) are two that we've encountered. Getting certified through these could eliminate the bias that is inherent in a fee-for-service certification system and create more trust with your consumers.

Nonetheless, some farmers are vehemently opposed to getting any sort of certification: They say it is too cumbersome a process, costs too much money, or that people should "just trust their word." We agree with many of the arguments against certification, but still find a lot of value in going through with it if you plan to build your business to a commercial, income-producing level. The recordkeeping involved is just the same good recordkeeping you should be doing if you want to be a profitable farmer. The certification cost is a tiny fraction of your gross sales and may help you command higher prices from buyers. And as for "trusting the good word" of the farmer, there just are too many lies and obfuscations going on in the food system for us to inherently trust anybody. From people calling their animals pasture-raised when in fact they are raised in dirt lots to others calling their meat "natural" when it was finished on conventionally produced GMO feeds in a feedlot environment, honesty is not always the *modus operandi*. The myth of the "honest farmer" is like the myth of the "noble savage"—it makes us feel good, but it is not reflected in either the present or the past reality.

If you are honest and believe in what you do, how will you convey all of that to your consumers? One way is to use a lot of words and pictures describing all of your systems; another way is to get certified and use one word or phrase to encompass all of those practices. Here are some of your options:

Certified Organic requires that animals are fed 100 percent certified organic feeds or forages, the pastures they graze on are managed organically, no antibiotics or hormones are administered, and some outdoor access (the amount varies widely between ruminants and non-ruminants) is provided. Most animals must be raised organically in their last trimester of life gestating; poultry from their second day of life on. If the animal products are to be sold as "organic," they have to be processed in a certified organic facility as well. An increasing number of slaughterhouses are getting organic certification, so they can process organically one day a week or in the morning; this certification means that they have to use different cleaning agents on the equipment and different sanitizers on the animal carcass. You could help your meat processor get certified if they are hesitant about the process. We know of a slaughterhouse whose organic certification fees were paid by the producer who wanted them to be certified.

Certified Naturally Grown follows most of the organic standards explained above with a few modifications. It requires pasture access for all animal species, including poultry and pigs. It requires organic feed but that feed does not have to be *certified* organic. And it requires that animals brought into an operation be raised organically or CNG from the day they arrive, not the day they were born.

100 Percent Grassfed by the American Grassfed Association applies to ruminant animals only. Animals are fed only grass and forage from weaning until harvest; animals are raised on pasture without confinement to feedlots; animals are never treated with antibiotics or growth hormones; and all animals are born and raised on American family farms. It should be pointed out that the USDA Agricultural Marketing Service allows a label claim of "grassfed," which is different from the AGA 100 Percent Grassfed standards. Though this USDA-approved "grassfed" claim requires that the animal has to be eating grasses and forages exclusively since weaning age, it does not prohibit feedlot conditions, does not require the ruminants actually be on pasture, nor does it prohibit the use of antibiotics or growth hormones. Therefore it is a very inadequate label claim because it does not meet consumers' expectations of what grassfed means. 100 Percent Grassfed certification by the AGA is better for that.

Animal Welfare Approved. Every AWA-certified farm provides their animals with continual access to pasture or range, as well as the opportunity to perform natural and instinctive behaviors essential to their health and well-being. AWA is one of only two labels in the United States that require audited, high-welfare slaughter practices too, and is the only humane label that requires pasture access for all animals. AWA prohibits most body modifications except ear tagging and early castration and does not require organic feed but encourages the use of non-GMO feeds.

Certified Humane provides more freedoms to the animals than conventional production, longer time before weaning, and some access to the outdoors at all times. This certification meets the Humane Farm Animal Care program standards, which include providing a nutritious diet without antibiotics or hormones, and seeing that animals are raised with shelter, resting areas, sufficient space, and the ability to engage in natural behaviors.

American Humane Certified is primarily for confinement-style producers to ensure what are known as the "Five Freedoms" and other veterinary-approved confinement practices to make the lives of the animals a little more "comfortable." An example of this is that traditional caging of laying hens is not allowed under American Humane Certified standards, where six to eight hens are placed in a small cage, each getting around the size of a sheet of paper to live their poor, short lives in. American Humane is now endorsing "enriched colony housing" in which 10 to 100 laying hens have access to perches, nesting boxes, and some scratching material but are still in a cage. The roomier spacing is about a sheet of paper plus a postcard, so not much more than the traditional battery-style cages. This system is being criticized by many as being no more than a slightly larger cage, and yet it is allowed under this certification system. This illustrates how the American Humane certification may work for large, CAFO-style operations but may not enforce high enough welfare standards for where your values, or those of your customers, may lie.

Global Animal Partnership (GAP) is a humane animal care program piloted by Whole Foods Market, but producers can now use this label to sell to other retailers. It has a tiered set of animal welfare standards, with level 1 being the lowest level, meeting certain threshold standards, and 5 being the highest level. The program was designed to recognize and reward producers for their welfare practices, promote and facilitate continuous improvement, and better inform consumers about the production systems they choose to support. If you plan to sell meat to Whole Foods, they require that you meet at least level 1 standards.

Wildlife Friendly or Predator Friendly are small and relatively unknown certifications; only eight animal producers in the country are certified Predator Friendly as of this writing. These farms agree to coexist with wildlife and use management practices to prevent predator problems with their animals. Tens of thousands of predators die in this country each year in the name of food animal production, and these certified farms are trying to change that. Certified producers use things like electric fencing and livestock guard animals, as well as herd management and protections during susceptible life phases such as giving birth. Animal Welfare Approved is now auditing farms for the Wildlife Friendly certification, so more farms are likely to pursue this certification in the future.

There are other programs that are not necessarily certifications but they are label claims that can be authenticated in some way. Read chapter 11, on labeling, to get a better understanding of what you can say on a meat label.

Using the term **"Heritage Breed"** refers to purebred animals from old-world breeds that were not typically bred for superfast growth, leanness, double muscling, or big frames like the modern breeds of today. If you are claiming purebred registration, be sure that your animals are actually registered. Hybrids of heritage breeds are not true heritage breeds, but you could claim to use "heritage breed genetics" or something like that. Likewise, you could claim to be raising a critically rare breed according to the Livestock Conservancy (formerly known as the American Livestock Breeds Conservancy, or ALBC). Again, if you make that claim, make sure it is true. We can't tell you how many times we have seen certain breeds of animals referred to as heritage when they are not heritage breeds. Duroc hogs or Broad-Breasted Bronze turkeys, for instance, are not considered heritage breeds. And just because you might raise an animal on pasture or using "old-world" techniques, it does not make the animal a heritage animal. Also, there is no such thing as an "heirloom" animal. The word *heirloom* is used for plants and the word *heritage* is used for animals. Don't confuse the two or you'll confuse your customers. We saw Whole Foods labeling some Broad-Breasted Bronze turkeys as "heirloom" turkeys when they are neither heirloom nor heritage. That, my friends, is called purposeful marketing obfuscation—clouding the truth to confuse consumers into making a purchase. Although certain marketers might encourage you to do that, we never will. If we can't be honest with our consumers, we are not creating an ethical meat system. We are just the same as the big guys who hide behind an opaque meat supply system to maintain their market position. We will cry "Pink Slime!" on any false claims.

Another marketing claim that is slowly growing in awareness are Slow Food **"Ark of Taste"** animals. They are rare breeds with usually a small geographic range that also have excellent taste

The Meat Landscape

and eating qualities. Navajo-Churro sheep are one such breed, as are Corriente cattle or Ossabaw Island hogs.

Keep in mind that there are some very practical reasons why some heritage and rare animal breeds fell out of favor, and these breeds may not work for the economic viability of your farm. A chicken that takes 16 weeks to get to a 4-pound size might not be practical; neither is a sheep that has a very low meat-to-bone ratio. Just because an animal is a heritage breed, it does not mean it will be a good fit for your system or your marketplace or make you money. These issues will be described in chapters 2 through 6, which focus on various animal species and selecting the best breeds for your production system and marketplace. Likewise, many successful producers have chosen to cross heritage breeds for their feeder animal stock. The breeding stock is often kept purebred, but each parent may be a different breed, making the offspring a mix. There is nothing wrong with this practice; in fact it may produce a superior animal for your system. Just make sure you label it correctly. Don't call it a purebred Berkshire pig when the dam was a Large Black sow. We have seen some farmers come up with their own names for cross-breeds, which can be kind of fun but requires buyer education. A Berkshire/Large Black pig cross could become a "Blackshire." Or an Ossabaw/Guinea Hog cross could become a "Guineabaw" or a "Southern Cross." A Red Devon/Angus beef cross could become a "British Red." It's up to you to get creative with the names—just be sure to be accurate with your labeling.

What all will this book cover? Hopefully the keys to your success as a direct-market meat producer. These include:

KEYS TO SUCCESSFUL DIRECT MARKETING OF MEAT

1. Produce a great product (chapters 2 through 6)
2. Know your product (chapters 2 through 6 and chapter 9 on cutting instructions)
3. Know your market and clientele (chapters 8 and 12)
4. Know your local, state, and federal regulations (chapter 7)
5. Develop necessary infrastructure

   - Processing relationship and logistics (chapters 9 and 10)
   - Transport, cold storage, packaging, labeling (chapter 11)
   - Distribution avenues (chapter 8)
   - Market segments (chapters 8 and 12)
   - Cooperative arrangements (chapter 8)
   - Cash flow requirements (chapter 13)

6. Manage your business well (chapter 13)
7. Price your products appropriately (chapter 13)

Marketing your own meat can be extremely challenging; indeed, many who try to don't last long. We are sort of an example of that ourselves. We just could not get all the pieces of the puzzle to fit together seamlessly enough to continue farming in the high-cost, drought-stricken area of California we landed in. Any successful farming business must seek to obtain the best fit of puzzle pieces and put them together in the most orderly fashion. We could also use a pie analogy to describe a successful farming business: You have to have several components to make a complete pie. They include: a good crust (the land, the soil, the location), a delectable filling (great animals, the people, the product, the finances), and creamy topping (the market, a reasonable level of competition). If you don't get all of the components right, the pie does not taste very good. It may still serve as a pie, but it's not going to win any taste contests at the county fair. We had a less than stellar pie in our previous farming situation and figured it was time to get out of the average pie business. Now we are building a new pie with new ingredients that will manifest in the years to come. With that candid personal reflection, there are several pitfalls that direct marketers of meat can fall into that we would like to warn you about.

Direct marketing your meat is not a panacea; it comes with a long list of hurdles. These include:

CHALLENGES OF
DIRECT-MARKETING MEATS

1. Possible large cash outlays or investments
2. Higher financial risk (and risk of product loss)
3. Cost/profit ratio of products (i.e., margins can be slim)
4. Lack of marketing knowledge and skills
5. Lack of processing infrastructure or good processor relationship
6. Liability issues
7. Regulatory requirements
8. Logistical challenges: transport, cold storage, packaging, labeling, and so on
9. Volume/scale dilemma (having the right volume to penetrate the markets you are seeking at a price that is fair to you and viable for the buyer)
10. You have to slaughter and process animals to make money—you can't hoard animals and you shouldn't get too attached. You also have to be ruthless with culling out bad genetics and poor performers.

These challenges will present themselves in many different ways at numerous times during business. Discussing them is the first step. Preparing to overcome them is the next. A good way to protect yourself from the sting or bite (or worse) of these problems is to create a solid concept and execute it. At a recent presentation given by organic hog and beef producer Tom Franzten of Iowa, he talked about the hedgehog concept, outlined in the book *Good to Great* by Jim Collins (2001). The idea of a hedgehog is that it rolls itself up around a central core, defending itself from predation or attack. This concept can be used to build a strong business model for an integrated pasture-based farm and meat business. To build this strong core, ask yourself these three questions and find out where they best intersect:

1. What am I most passionate about?
2. What can I be the best in the world at?
3. What drives my economic engine (i.e., what are my most profitable enterprises)?

It is at the intersection of these three circles that you find that sweet spot, a spot that will theoretically propel you to sustained growth and success; indeed that is the hypothesis of the book *Good to Great*. The idea is that instead of focusing on just one of those questions and putting all your energy into that sphere of influence, you focus equally on all three in the places they intersect. I (Rebecca) did this exercise recently for my own life, not necessarily for a specific business. Here is what I found:

1. I am passionate about a clean environment, biodiversity, sustainable agriculture, economic justice, and equality for all humans.
2. I can be best at writing about sustainable animal production and farm business manage-

ment for the masses. I can also be best at growing nutrient-dense vegetables, meat, and eggs in the town where I live. (This question is the hardest one for me because most of the time I feel pretty mediocre about my skill sets).
3. My economic engine is driven primarily by my salaried nonprofit job, my private consulting with farmers, and my conference speaking and book sales. My least lucrative activity is growing vegetables for market (go figure!).

So where is the intersection of these three spheres for me? Probably in doing nonprofit work around sustainable agriculture, particularly business development for niche livestock and poultry producers. Ha! Hence the genesis of this book. Maybe I am onto something here . . .

Try doing this exercise yourself and see where your intersection lies. It's also a helpful process to see what things you should *stop* doing, especially if they don't hit all three spheres for you. A lot of folks try to do too many things when farming, many of which they aren't particularly good at. Are you raising some animal species that you don't enjoy or that don't fit your landscape well? Do you lack passion for some things you are doing (don't we all)? Are you running enterprises that don't generate adequate profit margins? Can you ditch those things without suffering too much? This will help you build your impenetrable, spiny hedgehog armor to protect you from competition, economic fluxes, and even burnout. In addition to building your hedgehog concept, we will describe ways to overcome the specific challenges listed above throughout the book.

With that, let's delve into raising animals for profit, keeping the market in mind the whole time. Starting with the smallest animals (poultry), moving to the largest animals (beef), and then even in a chapter on exotics (rabbits, bison, elk, and deer), we will cover breeds, feeds, fencing, production systems, handling, health, and more. We have included lesser-known meats, such as rabbit and venison, because they present certain niche market opportunities, profit potential, and an expanding demand as consumers seek out healthier and safer forms of protein. In a book on niche meat marketing we felt it was important to include these niche meats too.

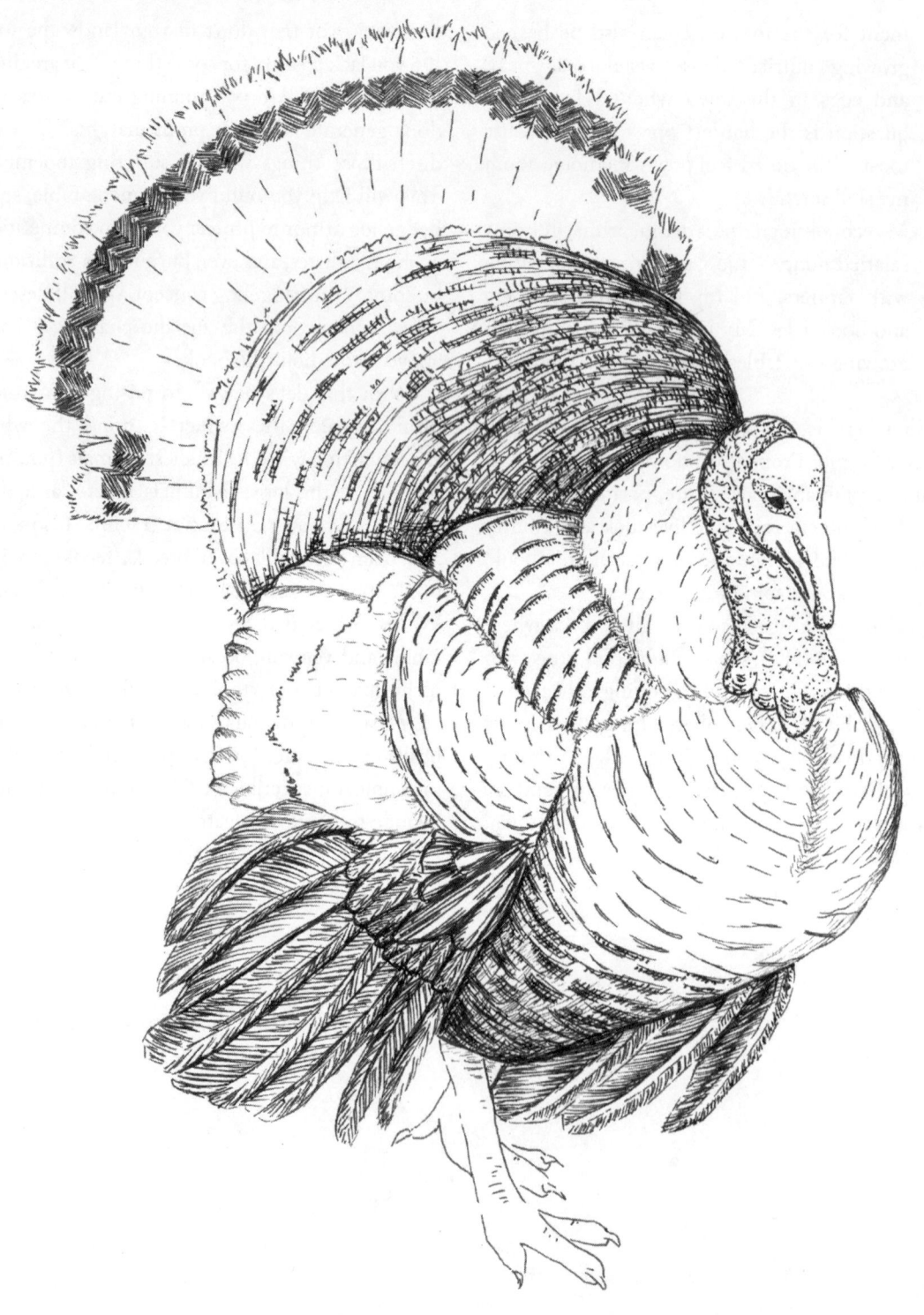

CHAPTER 2
Poultry Production

Pasture-based broiler production is often an entry point for farmers looking to produce meat for sale. Broilers, and other poultry species, don't require big investments in equipment and infrastructure, nor do they need a large land base. What is needed, however, is a lot of labor and sometimes the willingness to process the birds yourself. As will be described in the chapter on regulations, most states allow producers to process annually between 1,000 and 20,000 birds on their own farm without USDA inspection, as long as they are selling direct. Since many states have no or only a few USDA poultry processors, this exemption creates an entry point for small to mid-scale producers. But on-farm slaughter can also get old really quickly. In fact, for us, even though we had a great market at the time, the main reason we stopped producing broilers is that we didn't want to spend so much of our time killing chickens.

Compared to the other types of meat animal production, we think that outdoor poultry production can be the most challenging, but the market is the best and the overall investment the smallest. Potential challenges include multiple broodings, management of birds at different life stages, predation, high feed costs, and the logistics and labor necessary to deliver feed and water on pasture, in addition to safe and quality processing. That said, Americans eat a lot of chicken (currently more than 60 pounds per year per person, and this is increasing) and many are looking for an alternative to the standard fare. In 2013 chicken consumption surpassed beef consumption for the first time in US history. Pasture-raised turkeys, ducks, and geese are also increasingly popular. An

DEFINITIONS

APPPA American Pastured Poultry Producers Association
Broiler a chicken that grows to between 3 and 6 pounds in weight in 6 to 12 weeks; used for meat
Chick a newly hatched or young chicken
Cockerel a male chicken that is less than one year old
Drake a male duck
Fodder system a hydroponic system of sprouting grains to grow dense mats of young forage for livestock
Poult a newly hatched or young turkey

especially great niche for restaurant sales these days is ducks and geese. In fact, *Food & Wine* magazine labeled duck and all its parts as the hottest new food trend in 2014.

In this chapter we will discuss three general models for outdoor poultry production, including what we will call a "stationary housing with pasture access" model. With the term "pasture access," we don't mean the typical "conventional organic" model where the poultry don't even go outside, but a system that includes a well-managed pasture area from which the birds derive a portion of their nutrition. We list the pros and cons of each of the three models in comparison to each other. We then go into more detail about some of the types of shelters that can be utilized in each model. Losses and poor performance in the brooder are the bugaboo for poultry production, so we give good brooder management a lot of attention in the next section. From brooder to grow-out, we discuss some of the management issues involved in raising birds on pasture, as well as some general types of broiler breeds that you might want to consider for your operation. Poultry need to be fed concentrated feed for the majority of their diet, so we discuss dietary needs and the feeds that can fulfill them. At the end of the chapter we discuss the production of turkeys, ducks, and geese, mainly in terms of the differences in raising them versus broiler chickens. Although we don't cover game birds and other rare poultry species due to space constraints, there is a growing market for those species too (quail, chukar, partridge, pheasant, and a few others). We will also focus attention on how to perform efficient, quality, and clean on-farm processing of poultry.

We don't go into incubating eggs and hatching chicks, nor will we include information about keeping a breeding flock. More power to the folks who want to take on this aspect of poultry production, but it is outside of the scope of this book, and is not generally practiced on most family farms at this time due to the intense management and expertise that it requires. It is also really difficult to scale up farm-raised breeding flocks to accommodate viable, commercial-scale meat production on the same land and with the same management. There are a few folks out there trying it, but not many. (Our producer profile of Boondockers Farm examines one such operation.) Instead our emphasis will be on starting out with day-old chicks from a hatchery and how to manage the birds from that point on.

Management Systems

There are three common models for outdoor poultry production that are widely used on pastured poultry farms. While these models fall into distinct categories, there is often a lot of variation and overlap within them. There is no production rulebook, and we encourage you to adapt a model to make it work better on your farm or for your lifestyle. In no particular order these systems are: portable bottomless shelter with no free range access (aka chicken tractor); semi-portable day range shelter; and a stationary shelter with free range access. All three production models have their pros and cons, which we will discuss to enable you to better understand which method will work best on your farm.

Portable Bottomless Shelter

The portable shelter that is used in this model is often referred to as a chicken tractor or a field pen. A chicken tractor is a small, lightweight, and bottomless poultry shelter that will usually house between fifty and one hundred broilers that don't leave the shelter. Chicken tractors are moved once or twice a day using human, ATV, or truck power, in order to provide fresh forage and clean ground for the birds. Chicken tractors can be operated with or without electric fencing. Surrounding the shelters with electric netting adds another layer of protection in areas with high predation. Some farmers use electric fencing to keep a guard dog close to the pens, but the fencing doesn't have to be in the form of netting and can be whatever configuration will confine the dog.

Chicks are most often brooded for the first 2 to 4 weeks in a more secure shelter that is easy to service. The birds are then crated and delivered to the field pens where they spend the rest of their short lives. The young birds can be stocked at higher densities at first and then dispersed to more shelters as they grow larger in order to reduce the density.

PROS:

- **The risk of predation is relatively low** if the shelters are secure, although clever nocturnal predators will kill the birds through the wire if they can, or get under them if there are gaps.
- **The shelters are lightweight and can generally be moved by hand.** No tractor or other type of expensive farm equipment is necessary if they are built to be lightweight. Some farmers use a modified dolly to facilitate easier movement of lighter shelters or an ATV or truck if heavier. Shelters can easily be moved to other parts of the farm or even other farms, since they are so portable.
- **Chickens are confined, which can make management easier.** Chasing chickens can be kept to a minimum. You don't have to worry about chickens sleeping outside or finding their way into your vegetable garden.

- **The manure load is spread very evenly on the pasture** where the shelters are moved. The pastures respond well to the short-duration, intense impact from the confined birds.
- **The pastures are grazed more evenly by the birds,** so you can prevent both overgrazing and undergrazing. You can use the birds to revive tired pastures.
- **Groups of birds stay together; they don't mingle with other age groups,** which would require sorting at harvest time, and could allow some birds to avoid successive harvests and get way too big.
- **This model allows farmers an easy entry into broiler production with very low-cost infrastructure.** The shelters themselves can be made from scrounged materials or can be built from purchased materials for less than $200 each. They can also easily be repurposed into other farm uses should you change direction in the future. We turned ours into laying hen shelter materials after using them for broilers for a couple years.
- **Portable field shelters don't require bedding.** Supplying bedding to the birds requires labor and expense. The chicken tractor relies on regular moves to fresh ground, not bedding, to keep the birds off of their manure.

CONS:

- **This model requires many individual shelters,** each with its own utilities (drinkers and feeders). Shelter costs on a per bird basis can be high.
- **The shelters—especially the low "Salatin-style" ones—are single-purpose**, and will often sit empty during the off-season. Shelters made of PVC often become junk after one or two seasons and should be avoided.

It might require two moves per day to keep birds off heavily manured ground. While the extra move will increase the amount of forage consumed by the birds, it will also add quite a bit of labor. Failure to move the birds as often as necessary can lead to foot sores, breast blisters, and coccidiosis.

The birds are not as protected from severe weather events and heat. In very hot regions the birds can overheat under a low tin roof. The shelters are floorless, so flood events can be devastating on poorly drained land. Tarps can protect from rain, but they fall apart in the sun. Strong winds can destroy lightweight shelters.

Chicken tractors require a very well-maintained pasture. Chicken tractors can be difficult to move over rough ground or hilly terrain and through tall grass.

Feed and water logistics can be difficult. Chores are once or twice a day and include moving the pen, feeding, and watering. Feed must be carried to each pen by hand and watering logistics can be difficult to streamline. Large range feeders that can be filled once a week can't be utilized in this system.

Chickens will only eat grass for a short period of time when the shelter is first moved; they will quickly trample and soil the remaining grass, making it unpalatable to them. They will then lose interest in the forage until the next move.

It takes a few pasture moves for the young birds to get trained, and they can get crushed by the trailing edge of the shelter while learning the ropes. This is especially true when the grass is too long.

Young chickens can escape, either while the pen is moved or from gaps created by uneven ground along the bottom of the shelter. Rounding up the escapees can be time-consuming and frustrating. Birds that escape while you aren't around may fall prey to aerial predators.

Shelters can be difficult to move if they are too heavy. Difficult-to-move shelters often equate to shelters that aren't moved frequently enough. The shelters need to be lightweight, with a lot of their strength in the form of bracing to withstand the abuse of being dragged over the pasture. If you can't build a lightweight and strong structure yourself, find someone to help you and set yourself up for success. A shelter that is easy to move, yet strong, is the cornerstone of this system.

Semi-Portable Day Range Shelter

Day ranging allows broilers access to pasture, usually with an electric netting perimeter fence. The shelter or mini-barn is more substantial and larger than a chicken tractor and is moveable, but not by hand. A tractor, truck, or ATV (or even a team of draft horses) is used to pull the shelter to a new location between batches or a couple of times during the grow-out of a single batch, as pasture conditions dictate. The chickens use the shelter for sleeping and to get out of bad weather, but are free to come and go throughout the day. Farmers that have high predator pressure will choose to close the birds in at night and then reopen the houses in the morning.

The day range shelter can be floored or floorless. Farmers that use the chicken manure for their crop production will prefer a floored model with lots of bedding that makes it easier to harvest the manure. The mix of bedding and manure is ready to compost as is. A floored shelter will have an advantage for brooding and for use during wet weather to keep the birds off wet ground. The mini-barn is moved much less often than a chicken tractor, but the electric netting can be moved around it in order to provide fresh green pasture. Pasture right around the shelter will usually become denuded and heavily used while the outer reaches of the fenced-in areas are less used. They are not as good for controlled pasture renovation as the bottomless shelters.

PROS:

The shelters can have less square feet per bird than a chicken tractor, since the birds are allowed outside access during the day.

Chickens can forage for a larger percentage of their feed depending on the amount and quality of the pasture.

Birds are fairly well protected from severe weather events. Day range shelters are usually built a bit stronger and can withstand higher winds, rains, and other weather events. Birds can also choose to go in or out to moderate their body temperatures during heat.

Day ranging requires a less maintained pasture because you don't have to move pens by hand on a daily basis. Grass can be longer without causing as much grief.

Feed and water logistics are easier than with a chicken tractor. You can use larger feeders that only need to be filled once or twice a week, or you can use daily fill trough feeders that are easy to service out on the pasture. Pasture impact can be spread by moving the feeders daily or between fillings. You can get away with using fewer drinkers and they can be located centrally, reducing the need for feeder lines to each individual field pen.

It might be possible to brood and grow the broilers in the same shelter. In addition, the shelters can be used for other purposes outside of the broiler growing season, since they are essentially just small barns.

CONS:

Chickens will apply manure to the pasture less evenly. There will be an uneven impact in and around the shelter. With or without a floored shelter, you will most likely need to supply some bedding to keep the birds from spending too much time on their manure.

It can be difficult to get some birds inside at night, but generally they want to go inside as the sun goes down. Birds sleeping unprotected at night on the pasture can be killed by great horned and other large owls, and everything else that hunts nocturnally. Confining the chickens to the shelter for the first few days helps bond the birds to the shelter and makes it more likely that they will go inside to sleep at night.

Birds are more susceptible to daytime predation from raptors and dogs. The freedom that the birds have in this system comes with a higher likelihood that they will get attacked by hawks. Dogs can crash through their netting and have a field day.

Shelters have to be opened in the morning and closed at night if there is a high likelihood of predation. This might seem trivial, but it locks you into a routine that has to be performed at the same time twice each day, unless you install an automatic door that opens and closes on a schedule.

It's harder to integrate this model with other livestock in the same field because it is easy for other livestock to get inside and wreak havoc or eat the chicken feed. For example, you could not raise sheep, goats, or pigs in the same pasture at the same time as a day range model: They will gorge themselves on the chicken feed. Georgia farmer Will Harris has had success running cattle in the same pasture as his broilers by using a single strand of poly wire to exclude the cattle from the broilers and their feed.

Free Range Stationary Shelter

This system is characterized by the use of a stationary building to shelter the birds and pasture access directly adjacent to the building. The chicken-rearing building could be a retrofitted barn or hoophouse, or any outbuilding that can serve as a brooding area and grow-out shelter. It requires several subdivided pastures in order to keep the birds on fresh ground with growing vegetation, and in most cases the building will have pop-out doors in several locations to provide access to the multiple pastures.

Pasture fences can be permanent chicken wire fence or temporary electric netting, or most likely a combination of both. The easiest method for raising large numbers of birds year-round might be to permanently fence the perimeter and then use electric netting to subdivide the paddocks. The most challenging parts are managing the manure, nutrient, and pathogen buildup, and maintaining good vegetative cover.

PROS:

It is usually easy to provide utilities and perform most daily chores. The stationary shelter might already have water and electricity or these can be installed. Automating the water system is easy.

Housing is very safe from predation, with the exception of rats that might thrive in this system and cause losses in the brooder or steal a lot of feed. Rats could be the biggest bugaboo in this system.

The shelter can be a retrofitted barn or hoophouse, or any existing outbuilding. It need not be fancy. Unless you are using it for year-round production, it does not have to be insulated or completely weatherproof.

The birds can be brooded in the same building that they are grown in, which removes the tedious chore of catching and crating the chicks to move them to their grow-out shelter.

Birds are well protected from severe weather events.

Feed logistics are easy and can be fully automated if desired, with a feed tank right outside the barn. Feed can also be provided in large capacity feeders. However, it's critical to keep the feed area clean and to trap for rats.

CONS:

Stationary shelters require large amounts of bedding and bedding management. Because of the quantity of bedding and the fact that most of the chicken manure is deposited in the barn, these shelters might require machinery for bedding removal.

The outdoor pasture provided for the chickens might get destroyed after one batch of broilers, which would require a long rest to regrow. It is difficult to maintain the pasture in a green and growing state in this system, and it will often require irrigation. The impact within individual paddocks will be uneven. Nutrient accumulation is highest next to the building, leading to potential runoff, salt accumulation, and pathogen buildup. It might be hard to call this system "pasture-raised" as the pasture can quickly become degraded. Rotation is key.

Diseases and parasites are more likely to build up, which will require sanitation and rest between batches of birds. Chicken mites can also become severe in stationary buildings.

Broilers might fall prey to raptors, rats, or small mammalian predators like raccoons, but usually the barn is located in an area with more human traffic than the middle of a pasture. The increased human presence helps to deter predators.

Breeds and Breeding

Cornish Cross Chickens

Even for pastured poultry producers the white Cornish Cross is still the bird of choice. They can

TABLE 2.1. Characteristics of Different Poultry Species

Breed	Length of avg. grow-out on pasture	Production model most suited for	Strengths	Challenges
Cornish Cross Chickens	7–8 weeks	confined intensive meat production, either summer only, spring and fall only, or year-round with climate-controlled barn	• fast grow-out • efficient feed converters • meaty carcasses	• poop factories • require a high level of management • less hardy
Rainbow Broiler or Freedom Ranger	10–12 weeks	day range model, seasonal	• better foragers • decent grow-out • decent carcasses • can roost	• smaller breasts • meat can be tougher
Heritage Breed Chickens	16 weeks	day range model, year-round production with shelter, dual-purpose	• better foragers • dual purpose if you want • can roost	• limited breast meat • meat can be tougher • can fly
Heritage Turkeys	28 weeks	day range or barn range model, year-round with breeding if desired	• great dark meat • can breed naturally	• can fly • slow-growing • smaller carcasses
Broad-Breasted Turkeys	18 weeks	day range or barn range model, seasonal	• large carcass • large breasts	• non-breeding
Ducks	8 or 18 weeks	day range or barn range model, year-round or seasonal dual-purpose	• good flocking instinct • friendly • better foragers than chickens	• pin feathers hard to pluck • messy, mud creators
Geese	10–12 weeks or 24 weeks	day range or barn range model, year-round or seasonal, dual-purpose	• great foragers • weed control • hardy • large carcass • value-added products include livers and down	• pin feathers hard to pluck • can be loud or aggressive • like to follow you everywhere • messy, mud creators

be extremely challenging birds to raise unless you are on top of your game. They will teach you to create the most robust management system possible, because of their high maintenance nature. Cornish Cross have the type of large-breasted carcasses (with no dark pin feathers) that are preferred and expected by the majority of consumers. We have raised thousands of CCs and found that the high level of management did not work for us, especially because we lived 15 minutes away from where we farmed. However, they are the most economical to raise and consumers love the meat. Below we will break down the pros and cons of raising them and let you decide. Maybe we will save you a few headaches or lessen your potential death losses.

PROS:

- **They grow very fast** when given plentiful feed, and they have a great feed-to-gain ratio. Their shorter grow-out allows you to have fewer pens and to cycle the birds through them faster.
- **They are easier to process** than most other types of chickens because they have white feathers and there are fewer of them.
- **They are extremely slow and docile, which makes them easier to catch** than most types of chickens. Their docile demeanor also makes it easier to keep them in their pens and behind fences, since they don't have a very strong drive to escape and roam around the farm.
- **These chickens have the type of carcass that many consumers have come to expect.** You don't have to do a lot of customer education when selling these broilers, as their carcasses are just like the birds sold in the grocery store.

CONS:

- **They may develop leg problems.** They grow so fast that the rest of their body can't keep up with the rapid weight gain if they are overfed during the later part of their grow-out.
- **They require a high level of brooder bedding management.** They grow fast because they eat a lot of feed, and yeah, they poop a lot, too. This large amount of manure has to be mitigated with lots of bedding. As they get larger and closer to leaving the brooder they will need to have their bedding refreshed more often.
- **They are prone to having heart attacks and heat stress** as they reach harvest weight. Their feed needs to be restricted as they get older, even though they will act as if they are starving to death. Also they need to have plenty of shade and airflow on hot days.
- **They are generally poor foragers** when compared to just about any other type of chicken. As they get older and heavier they become more lethargic and will often lie in front of the feeders for most of the day.

Privett Hatchery sells a "Slow Cornish" which is a little slower-growing than the standard CC, yet feathers out better and doesn't have the same health problems, due to slower growth. Other hatcheries have their own slow Cornish or colored Cornish type varieties too.

Alternative or "Pasture-Type" Hybrid Broilers

Many pasture-based broiler producers want an alternative to the Cornish Cross and now there are some good types to choose from. While these hybrid broilers won't give you the same rapid growth as the Cornish Cross, they do have an acceptable growth rate without a lot of the headaches. Broilers like the Rainbow Broiler and the Freedom Ranger retain a lot of the characteristics of a heritage-type bird, but will grow a meaty carcass in a reasonable length of time. Will Harris of White Oak Pastures, profiled later in this book, raises Rangers on pasture and processes several thousand weekly. He prefers them to the Cornish Cross despite their longer grow-out. However, if you are using expensive feed, such as certified organic feed, the extra 3 to 4 weeks of grow-out may make them economically unfeasible unless you can pass those extra costs on to your customers.

PROS:

- **They are more active and better foragers** than the Cornish Cross. They don't spend all their time sitting by the feeders.
- **They produce a meaty carcass**, a bit more slowly, but with less intense management than the Cornish Cross. Expect around a 10- to 12-week grow-out, as opposed to a 7- to 8-week grow-out for the Cornish Cross.
- **They are good for ethnic markets** that want a colored bird.

CONS:

- **They have a higher desire to escape from pasture pens**, which can be frustrating and time-consuming.
- **The colored feathers can leave dark spots on the carcass.** While this isn't a big deal, some consumers will be put off by it. They also have more feathers, which means more plucking.
- **They are more active and harder to catch for harvest.** They will struggle and try to escape quite a bit more than the Cornish Cross. During processing they will have a lot more energy and may try to get out of the cones.
- **These birds often can fly**, which means they may be able to fly over your fences or up on top of the shelters and bulk feeders.
- **The breasts are a little less meaty than CC breasts and the meat may be darker** and require more chewing because they take just a bit longer to grow out and exercise more. Some customers won't appreciate that.

Heritage Breed Chickens and Dual-Purpose Chickens

This is the traditional type of farm chicken and it is the most active of all. These chickens are often referred to as dual-purpose, meaning that they are good egg layers and also have decent meat qualities. They aren't as popular as the previously mentioned birds, but might be useful for a diversified farm or to fill a niche. The cockerels can be purchased very cheaply from most hatcheries, since the majority of the males of these breeds are killed at hatch. If you purchase your chicks as straight run (non-sexed birds) you can harvest the males and keep the females for layers. Some of the so-called dual-purpose breeds include Barred Rock, Buckeye, Delaware, Dominique, Jersey Giant, New Hampshire, Rhode Island Red, and Wyandottes. The cockerels may take 16–18 weeks to grow to acceptable size, again something you have to factor into your enterprise budget. It may be next to impossible to raise these slower-growing birds on more expensive organic feed: The higher cost may not attract enough customers.

PROS:

- **They are great foragers** and can fulfill more of their dietary requirements from the pasture.
- **They can perform on a high-energy diet.** They don't have the same protein requirements as

faster-growing broilers and can do well on cheaper, high-energy feeds.

There is a niche market for these breeds. Some foodies and people interested in preserving heritage breeds may be into supporting you, along with recent immigrants looking for this type of chicken to use in the recipes from their homelands, which often involve long, slow cooking. As a bonus, the chickens can often be sold live, since processing the chickens is part of the cultural experience for these consumers.

CONS:

They are the slowest-growing and small-breasted. Some folks won't know how to prepare this type of chicken in the kitchen or may consider it tough.

They are hardest to catch for harvest. They are very active and strong, and will resist being caught and handled more than other types of broilers.

Heritage breed chickens are able to fly, which can pose challenges, especially in a day range model. In a day range model they may try to roost on top of your lightweight housing, and if enough of them do it, your shelter could be damaged or destroyed.

Life-Stage Considerations

Brooding Phase

Many poultry producers agree that the brooding phase is the most important aspect of broiler production. Any setbacks that occur in the brooder will add up in the form of poor feed efficiency, slow growth, and diminished vigor in the birds. A case of coccidiosis, for instance, might be treated and cured in short order, but damage has been done to the birds, and often this damage is difficult to quantify. A situation that kills a lot of chicks will surely negatively impact the survivors. Just keeping the chicks alive through the brooding phase is not enough to ensure success. The chicks must thrive and be managed in a way that prevents disease and promotes health.

Losses during brooding can most often be attributed to piling, dehydration, coccidiosis, and predation from rats. Some losses might be attributable to bad conditions during shipping or chicks that are not received on time. Poultry expert and author Joel Salatin recommends doing some troubleshooting of your brooder situation if you have more than 8 percent death loss in the brooder.

DAY-OLD CHICKS

Make sure that the brooding area is warm prior to the arrival of the chicks. Some producers mix and refresh the bedding between chicks, instead of removing it and sanitizing the brooder, in order to encourage the composting action. This provides a nice heat source for the chicks in addition to the brooder lamps or propane heaters. Be ready to pick up the chicks as soon as you get a call from the post office. Make sure your hatchery does not ship chicks on Thursdays or Fridays as they may end up stuck in a closed post office over the weekend. High death losses will result. Buy chicks only from reputable hatcheries that are breeding for *production* characteristics, not show characteristics. If you have high death losses during shipping, the hatchery may give you a credit or send you a new batch for free.

We like to buy straight run chicks because they are not only cheaper, but that way you will have some faster-growing males and slower-growing females that give you a couple of weeks of harvest from the same batch. When we raised Cornish Cross broilers, for example, we would harvest the males at around 7 weeks of age and then harvest the females a week or so later.

WATER

APPPA board member, long-time chicken producer, and Featherman poultry equipment manufacturer David Schafer recommends using nipple waterers from day one in the brooder. Properly adjusted, this type of system will reduce spillage and prevent chicks from drowning or getting chilled in their waterers. Always check your waterers for leaks and keep the bedding around them as dry as possible. This is one of the most important disease prevention steps you can take.

HEAT

Chicks should be evenly spread out under the heat source with a small, unoccupied circle in the middle. If there is no circle in the middle, you need to lower the brooder or increase the heat. If there is a large circle in the middle, the chicks are too warm and the brooder should be raised or the heat should be reduced. Farmers who brood more than a couple hundred birds at a time will want to consider investing in propane or natural gas–type pancake brooders. Equipped with a thermostat and a remote temperature sensor, one pancake brooder can brood up to 1,000 chicks. If you have enough heating capacity, usually lowering or raising the brooder heat lamps or hover will allow you to find the sweet spot for the chicks. We hung our brooders from the ceiling so we could easily lower and raise them as the birds grew. Daily observation of the chicks' behavior in the brooder is key.

If you use a heat lamp for brooding you might want to use more than one bulb, so that you have a backup if one bulb burns out. You can use two bulbs of half the wattage to get the total wattage you are looking for. Heat lamps use a considerable amount of energy—our energy bill always went way up during brooding season. A real brooder lamp fixture will have a guard wire over the bulb to prevent it from touching the bedding if it falls. Fire is a serious concern in the brooder and flammable bedding material should always be kept a safe distance from any heat source. Make sure that your brooder is safely suspended and can't accidentally fall onto the bedding. Barn fires started by brooders are not uncommon in farm country.

During the summer, David Schafer uses a propane brooder to heat 500 chicks at a time. He uses one propane tank that runs out after three days. After this time the chicks are on their own. During early spring and fall, you would need to provide heat for a longer time period.

Drafts can be particularly hard on young chicks. Make certain that the brooding area is draft-free. A brooding ring creates a draft-free space within the larger brooding house. The circular shape of the brooding ring gives it the added benefit of eliminating corners, which are often the sites of chicks piling on top of one another. It also keeps the chicks confined to a smaller space for the first week or so, where they can get acquainted with their surroundings while not traveling too far from their food, water, and heat. Brooding rings can be purchased or homemade and do not necessarily have to be round. You could probably get away with an octagon made of plywood strips or curved tin roofing, for example.

While you want to avoid drafts, it is still necessary to have some ventilation to exhaust the ammonia that will develop from the bedding pack. Ventilation should be high in the building with no drafts at ground level where the chicks are brooding.

BEDDING

Use bedding that the chicks can easily walk on, like wood shavings, chopped straw, or rice hulls. As the birds get older you can add a little more coarse material as a component of their bedding. You can use straw (or other coarse material) as a thick base layer for your bedding pack, but it does have a tendency to get matted with manure when used alone, and it doesn't provide good footing or ease of travel for young chicks.

Bedding material can be a significant expense for your broiler operation, especially if you are buying retail wood shavings. While you really can't skimp on bedding, you can source it more cheaply by buying in bulk or dealing directly with local sawmills. A thick underlayer of cheaper coarse material will give you some absorbency and add bulk to the composting bedding pack, at a lower price. When buying wood shavings, try to get actual shavings rather than sawdust, which is too dusty and is hard on the poultry respiratory system. Chicks might even eat the sawdust under certain conditions. David Schafer recalls a batch that suffered heavy mortality while bedded with sawdust. The brooder was large and the feeders were spread out too far, which encouraged the chicks to consume the bedding. David found he was able to use the free sawdust only by introducing it after four days of newspaper or shavings bedding. Hardwood shavings are fine too, as long as you can source them cheaply. Some farmers even like to use sand or shredded newsprint in the brooder. Experiment to see what works best, is the most affordable, and can be best used as compost on your farm.

During cold times of the year, when chicks must be kept longer in the brooder, bedding costs can increase drastically as the bigger birds produce more waste. Daytime outdoor access for birds at this stage of growth can reduce crowding while also lowering bedding costs. This can provide a buffer until the weather cooperates and the birds can be moved to their field pens. Day rangers solve a lot of these problems by brooding and growing the birds in the same shelter. Most pasture-based broiler farmers are not feeding medicated feed, which is a crutch for filthy conditions, and they must use brooder bedding management as one of their most important tools in preventing health problems. Faster-growing breeds, especially Cornish Cross or Freedom Ranger chicks, will need a lot of bedding; depending on the chick density you may have to refresh the bedding a couple of times per day. Never let your chicks' bedding become completely caked with manure. The bedding around the waterers should be kept as dry as possible. This might require spreading a layer of dry bedding around the drinkers on a daily basis or placing the waterers on top of a piece of wood or other material off the bedding. Any moist area in the brooder will increase the risk of coccidiosis, explained later in this chapter.

FEED

In the brooder you can hand-add fishmeal to the chicks' feed to increase protein and methionine and then decrease it as the birds mature. This is an alternative to buying a separate starter ration with higher protein ratios in addition to the normal grower ration. Sometimes it can be difficult to purchase and store a variety of different feeds.

You can also feed crumbled-up hard-boiled eggs, shell and all. Hard-boiled eggs are excellent food for young poultry and can be fed at a rate of four eggs per one hundred birds per day for the first week. After 2 weeks increase the amount of hard-boiled eggs to six eggs per one hundred birds. We recommend continuing this regime for the first 4 weeks.

Spilled feed can cause trouble in the brooder. The chicks' feet are dirty, and they stand right where the feed will spill when they are eating. This creates a contaminated space alongside the feeder, which can harbor and concentrate disease-causing organisms. Adjust feeders to the correct height for the size of the chicks and only fill the feeders three-quarters full in order to reduce spillage. Don't let the chicks run out of feed in the brooder or they will start to pick through the bedding to find spilled feed. Feeders should be of a design that doesn't allow chicks to walk on the feed. Spinners mounted on top of the feeders work well.

Be on the lookout for what poultry breeder Timothy Shell refers to as "exotic droppings." By this he means diarrhea and odd-colored poop that might be your first warning that there is a health concern about to occur. Once some of the birds have diarrhea, it easily spreads to the other chicks because it sticks to their feet and is tracked all over the brooding area. At the first sign of exotic droppings you should troubleshoot your brooder management. It might also be wise to give the chicks some raw milk to soothe their digestive tracts and introduce good bacteria. Shell recommends mixing the milk with their feed in an amount that will be consumed within 24 hours and to only feed the milk for a total of 48 hours. If problems persist Shell advises that the birds only receive milk using a 2 days on, 5 days off schedule.

Some farmers also feed raw milk, clabbered milk, or whey during the brooding phase as a way to increase the beneficial bacteria in the guts of the birds. Scientific studies are not conclusive about this, but we found that feeding clabbered milk seemed to prevent and treat coccidiosis in the brooder—plus it was free from a local dairy, so we were also able to save on feed costs.

PREDATION

Rats are often the culprits when predation is suspected in the brooder house. Rats can kill many chicks: We have heard of producers losing up to fifty in a night. Chicks that have been predated on by rats are easily spooked and this might be your first notion that something is amiss. Poultry farmer Dan Bennett says "for my money this is the worst predator issue I've experienced." Bennett recommends a tight brooder house and a good population of cats around the farm. If you have a rat problem, it must be addressed immediately, by trapping rats, exposing tunnels, and checking the brooder at night. Some folks use poison, and although we wouldn't want to, we certainly understand why a farmer with high rat pressure would choose to use it. With a rat problem immediate action must be taken: No matter which method you choose, time is not on your side.

HARDENING OFF

Chicks don't need to be weaned from their mothers, but they do need to be weaned from the cozy conditions in the brooder, and should be "hardened off" gradually over the course of a week or so. This allows the chicks to better tolerate the colder temperatures outside of the brooder, when you move them to their field shelters. Over the course of the last week in the brooder, slowly decrease the amount of heat available to the chicks from the brooder lamp. During this period the lamp should be off during most of the day and only on at night. You should be able to turn off the lamp entirely for the last two days and nights in the brooder.

A body-heat brooder might be a good option if you are concerned about the chicks being too cold in their field shelters during the transition period.

Plans for this type of brooder are all over the Internet, but basically this is a lightweight wood or PVC frame (about 4 feet by 4 feet—whatever works for you) covered with bubble foil insulation. The insulation goes over the top of the frame and hangs down, up to a foot on the sides. Slits cut in the sides allow passage of the young birds in and out of the brooder. The reflective insulation reflects the birds' body heat back at them, creating a cozy space for the chickens. The body heat brooder must be raised during the day or the chicks will scratch the top and poop all over it. It should be removed from the shelter as soon as the broilers can stay warm without it for this same reason.

Some farmers brood their chicks in field pens that have been modified for this purpose. The pens are equipped with lights and a source of heat; usually a propane or electric brooder heater. When the young chickens are ready to live without supplementary heat they are moved to new pens or the heaters are removed. Broilers are divided into smaller groups as they grow and need more space.

Grow-Out Phase

Once the young chickens survive the brooder phase, you need to move them into their grow-out shelters. Depending on the ranging model that you decide upon, there are a variety of ways you can build their shelters. We will describe the styles of housing below.

HOUSING

Joel Salatin–style shelter. This is an example of a floorless chicken tractor that doesn't allow for free ranging—the chickens spend their lives inside the shelters but with access to the pasture growing below them. The shelter is about 2 feet tall and 10 to 12 feet square. It's worth taking the time to understand all of the cross bracing that is required to keep this type of shelter from wracking and twisting over time. Cabling is also a good technique that will strengthen the structure without a lot of weight. Many people make these structures out of PVC, but this is neither cheap nor long-lasting. Many PVC shelters are garbage after one year, and then become a liability and more fodder for the landfill. Also, really lightweight shelters can blow apart in windy areas.

We've talked with farmers who have lost birds in this type of shelter when the temperature gets near 100 degrees F (38 degrees C). The sun beats down on the low tin roof and overheats the chickens. Farmers in regions with very hot summers should consider overheating when deciding on pen design. In areas with very hot summers, farmers will sometimes stop growing broilers for the one or two hottest months, regardless of the shelter they use.

This shelter is very portable when used with a dolly. It may require filling gaps under the shelter that will become escape routes when on uneven terrain. Tall grass and clumpy pastures can make this type of shelter difficult to move. Chicks can get crushed by the trailing edge of the pen.

A 5-gallon bucket on top of the shelter provides water that gravity-feeds an automatic bell waterer. Ideally, a garden hose will be used to fill the buckets. A tank on a trailer is often the easiest way to supply water to remote pastures, because it has volume and gravity-supplied pressure and it is easy to move. It can also be enclosed with siding of some sort in order to keep the water temperatures down. Broilers should not be forced to consume hot water.

Feed is generally provided in a trough feeder that is filled manually with a 5-gallon bucket. This makes feeding labor-intensive and necessitates having a feed supply that is close to the shelters, so feed doesn't have to be carried as far. (Hauling 5-gallon buckets of feed and water can be hard on a person's shoulders.) You may want to devise an easier system with larger feed or water storage nearer to the pens.

One-quarter of the pen top is removable to allow access to the feeders and to load and harvest chickens. Broilers can be sectioned off to this quarter of the pen to facilitate harvesting. Bending over the top of this structure to grab chickens can also be challenging and hard on the back.

Cattle panel hoop structure. This structure is usually 8 or 10 feet wide by 10 feet long. It is a rectangular wooden skid base with cattle panels bent in a half circle to make the roof frame and walls. The roof is generally a tarp, which will only last a couple of years. Some farmers use recycled billboard material, which is stronger than most tarps, for this purpose. The ends are framed with lumber and a door is framed into one end. It is usually a removable hatch-type door and can also be used as a sorting barrier. The farmer can walk into the shelter to feed, water, and harvest the birds. The higher roof and headspace of this shelter makes it generally safer for the birds during extreme heat events. The cattle panels make the shelter a little heavy, but they also do well in windy environments.

A similar house that is a bit lighter is an EMT conduit house (EMT conduit is metal tubing used to house electrical lines). It also has a wood base, but uses homemade, conduit bows instead of heavier cattle panels. Bows can be bent into the required shape using an inexpensive electrician's conduit bender. It is tall enough to enter and is covered with a tarp for the roof and chicken wire for the sides and endwalls. Some farmers build these shelters with a couple of wheels or use a dolly for easier movement. Feeders and waterers can be hung from the roof, but it might be necessary to reinforce the hanging points with a top purlin of conduit.

Moveable mini-barn or day range shelter. This is a long-lasting shelter that is heavier and a bit more substantial than the other structures. The designs vary, but most mini-barns are built on skids and have either corrugated metal roofing, or plywood walls and roofs. The roof can be gable-shaped, angled shed-style, or flat. You can also build a skid base for a poly-type hoop structure and use it as a moveable mini-barn. The mini-barn is tall enough to walk into and is moved with a small tractor or ATV. You may be tempted to build one of these on a trailer, like a hay rack or cotton trailer, but most meat bird types are not flyers nor can they even get up a ramp in their last few weeks of life. More than likely they would try to sleep under the trailer, which would not work well to keep them safe from predation. Save the trailer-mounted house for your laying hen flock (unless you are raising dual-purpose meat and egg birds, in which case a wheeled trailer with ramps may work great).

This type of housing gives the birds a lot of protection from the elements and can easily be modified for inclement weather. It can be made with or without a floor and can be multiuse since it is really just a lightweight shed on skids. Multiple doors can make this structure even more versatile by allowing the shelter to remain in one position while the electric netting is moved to accommodate a different access point. In this manner, at least four different paddocks can be made from one pasture move, from each side of the shelter.

Pastured poultry farmer Will Harris of Georgia uses many large plywood skid houses for his broiler production. He will fill a house with 500 broilers out of the brooder and then lock them in it for the first day and night to get them to learn that it is their new home. He then allows them access to pasture and uses a single strand of hot wire around the chicken area to keep the cattle in the same pasture out of the chicken feed. The chickens are in a large pasture and separated from the other chicken shelters by enough distance so that they don't venture away from their own area. The shelters are moved when the birds have used up the pasture. Harris's shelters have a roll-up door on one end of the house and a human door on the other end to facilitate catching the birds for harvest. Livestock guard dogs keep most predators away and the guinea hens sharing the pasture also let the dogs know if there is trouble. Will loses some birds to aerial predators, but he just calls this "tithing to nature."

Hoophouse. A hoophouse is a multipurpose structure that can be used year-round or seasonally. In the winter this makes a cozy shelter for any type of livestock or poultry. These are mostly fixed structures, but can also be mounted on skids or sliding tracks to make them more portable. Hoophouses can be covered with greenhouse plastic or longer-lasting woven poly tarps. The woven poly comes in many colors and thicknesses, and some are clear or dark. Woven poly covers often carry a 10- or 15-year warranty. You might need to cover the hoophouse with shade cloth during the hottest months of the year. For ventilation, hoophouses can be equipped with end doors, roll-up sides, and ridge vents. Some folks will install an inside wall and door, so the house can be entered without the chickens getting out and then they can be accessed by an interior man-sized door. This also creates a separate area in which feed can be stored without the birds getting into it.

Grit

Poultry nutritionist and APPPA board member Jeff Mattocks says that grit is so important in a poultry diet that "if you are not going to provide it you should not raise poultry." You can start with sand in the brooder for the first week or two and then transition to the smallest size of grit. As the birds grow you increase the size of the grit. If the particles are too small they will not stay in the gizzard and will pass through the digestive tract. Grit should be made available to poultry as free choice and should be located in a high-traffic area. Grit is available in several sizes; the type you use depends on the size of the bird. You start with a finer grit and move to a coarser grit as the birds get larger. If you are not exactly sure what size to provide, try mixing all three sizes and the birds will choose what they need. All poultry species, including waterfowl, need grit to efficiently process feeds. The more you want them to forage for their diet from plants, the more grit they will need.

Watering

Chickens drink about 8 ounces of water per day, which is about twice the amount of feed by weight per day. Chickens drink 4 to 10 gallons of water per day per one hundred birds. Their water should be clean enough that you would drink it. As a starting point you will want to at least have water supplied to the point of use in a pressurized garden hose. Carrying water to field pens in 5-gallon buckets should be avoided and employed only as a temporary measure.

Automatic poultry drinkers are one of the industrial farming innovations that pasture farmers have adopted with great success. Bell, drip, pan, and cup-type automatic drinkers all work well and can be hooked up directly to a low-pressure water line or connected to a 5-gallon bucket mounted to the field pen. When connected to a water line, drinkers just need to be checked and cleaned, which is much easier than filling a manual poultry waterer.

Automatic waterers can be as simple as a water pan with a float valve attached to a garden hose. If pressurized water isn't readily available, you can use a trailer with a water tank and poultry drinkers suspended below. We used this type of setup on our farm. We mounted tote tanks inside an enclosed trailer and connected them with a simple PVC manifold and then reduced the water lines to ¼ inch in order to gravity-feed Plasson bell waterers that were hung below the trailer. This design works best in a day range scenario. The dark, enclosed water trailer also helped keep the water cool and prevented algal buildup in the water tanks.

Manual waterers have to be opened and filled, which can't be done from outside the pen. They are often dirtier because they rest on the ground instead of being suspended. Filling them usually means that you are going to get chicken poop on your hands. Most folks find this type of chore to be tedious and time-consuming. It can also be expensive if you are paying an employee to do it. The money spent on upgrading your infrastructure to enable you to use an automatic water system will often pay for itself in the first season. We'll give you a quick example to show you the numbers.

We volunteered on a poultry farm where the farmer used garden hoses to fill 5-gallon waterers for thousands of chickens. She estimated that in the summer it took 4 hours a day to perform the watering chores at a cost of $15 per hour for an employee. This equates to $400 per week or $1,600 dollars per month. That's expensive water! Installing a water line and automatic drinkers would have paid for itself rather quickly, and it would have freed up time to work on other improvements. If you find yourself in this type of situation, do the numbers and take action.

Feeding Poultry

All poultry need to be fed a complete balanced ration that is specific for their life stage in order to produce well and in most cases in order to live at all. The general rule of thumb for all poultry, chickens included, is that protein needs are reduced as they mature and feed rations should be adjusted accordingly.

Poultry are not ruminants and cannot digest a pure forage diet. They are omnivores, and do best with a diverse diet of grains, wild seeds, green plants, insects, worms, and whatever other critters they can find. Raw legume seeds should never be used, except for peas. The rest of the legumes have too high levels of phytic acids, which are only removed through extruding, roasting, or steaming.

Broiler chickens are often fed free choice throughout their entire life in order to gain weight quickly. The exception to this would be the Cornish Cross chickens, which need to be limit-fed as they mature in order to prevent ascites, heart attacks, and leg problems. If limit-feeding, make sure there is plenty of feeder space so that all birds have equal opportunity to get the feed they need. Additionally, constant access to fresh, clean water is a must.

Jeff Mattocks recommends feeding grains within 30 days of being ground. He says that poultry can tell the difference and when presented with very fresh feed they will eat quite a bit more than they would of the older stale feed. Whole grains can be stored for years and maintain their quality, but once the inside of the grain is exposed to air the quality begins to decrease. Ground up feeds can actually go rancid and their nutrients and fatty acids begin to degrade immediately. Some producers choose to grind their own feed to maintain maximum nutrient density and quality of their feed grains. This can also be a way to potentially save money by buying individual grains in bulk, but it obviously requires feed bins and a grain mill.

It is possible to mix your own feed starting from scratch. However, it can be tricky to get all the nutrient levels correct. Most grains will need to be cracked or rolled: They are not digested well if left whole and can sometimes contribute to impaction in the bird's crop. If you are mixing your own feed, make sure you include some sort of vitamin/mineral premix, such as Fertrell Nutribalancer or others, that is appropriate to the species you are raising. For example, turkeys and waterfowl have a higher need for niacin than broiler chickens and thus must be supplemented or you will see weak or bowed legs.

Jeff encourages farmers to get the test weights of any whole grains that they purchase in order to ensure that they are high-quality. Test weights are an indicator of the producing farm's fertility program, growing conditions, and the plants' stress levels during the growing season. He advises against buying low test-weight grain: It is not worth the money you may save.

Another way to harvest grains is to have the poultry or livestock do it themselves: even broilers can do this. Jeff recommends planting the crop more sparsely than normal so the birds can navigate through it. If the crop is too dense and tall you can run pigs or ruminants through it first and then have the birds come through and clean it up afterwards. It is important to wait until the grain has reached the dough stage in order to maximize the nutritional value.

Jeff is not a big advocate for fodder systems for monogastric species because it reduces the protein levels (why pay for high protein and then lose it through sprouting the grains?). He does agree that egg quality can be improved by feeding fresh fodder to laying flocks and that it is something to keep the poultry entertained so that they won't kill each other when in seasonal confinement. For poultry and swine Jeff recommends a 3- or 4-day sprouting period, which doesn't reduce the dry matter protein percentages as much as longer grow-outs of the fodder. He does see the value of fodder systems for ruminants, though.

If you are raising both poultry and swine, it is possible to have one base mix for poultry and hogs and then supplement the rest of the ingredients based on the species being fed. For example, if you wanted to up the protein levels over the base ration, you could add 5 to 10 percent fish or crab meal for the poultry (remove it during the last two weeks of life to eliminate any "fishy" flavor). Medium coarseness of the grind is okay for hogs, but for best results feed along with hay in order to slow the digestion and increase the absorption of nutrients. Also, if feeding spent brewers' grains to poultry and pigs, Jeff recommends that you only allow access to them for a maximum of 36 hours so that the animals are not eating moldy feed. After that, move the animals to a fresh pile of grains.

Consumers are increasingly becoming selective about what kinds of feeds are being given to poultry (and other livestock). They have a growing list of things they don't want you to feed your animals, sometimes not based on scientifically valid information. The ideal feed (for these ardent consumers) would be 100 percent certified organic, certified non-GMO, no soy, no corn (for some reason), and now even no wheat or gluten-containing grains. Pretty soon there won't be much left to feed your animals. The important thing is to feed your poultry a high-quality, fresh, balanced feed ration and give them ample access to pasture, bugs, worms, and whatever else they can forage for. Explain why you use the feeds you choose and how your raise your animals and then let your customers decide. Do not let a few of them drive you to use feeds that make your poultry enterprise unprofitable or less sustainable or that compromise your birds' health.

Broiler Chickens

Jeff recommends feeding a complete ration to poultry and then count the rest, such as cracked corn or pasture grasses, as supplement. We recommend that you use three types of feeds depending on the state of life of the broilers. Zero to 2 weeks is usually a 22 to 24 percent protein starter ration, 2 to 4 weeks is a 20 to 22 percent protein grower feed, and 4 weeks to finish weight is an 18 to 20 percent protein finisher ration. You could probably get away with using just a starter ration and a finisher ration to make things easier. You can always supplement the protein with some added fish or crab meal or other source of protein (canola meal, pea meal, soybean meal, meat and bone meal, insect meal, etc.) to bump it up a little without resorting to buying a whole different ration. It can be tricky to store several feed formulations but you will save money by dropping the protein percentages as the birds mature.

Supply kelp to chickens for micronutrients at 0.5 percent of their ration. Feeding linseed meal will increase omega 3 fatty acids up to a maximum of 7.5 percent linseed meal in the ration. Flax meal will do the same but can cause diarrhea in chickens. Of course, chickens will absorb omega 3 fatty acids from vibrant, growing grasses and forbs without any need for special supplementation. Provide those to the extent possible.

Heritage chickens need less energy than Cornish Cross broilers. They also have a longer feed throughput than the Cornish Cross and are better able to utilize peas and small grains because of their slower digestion. If you are going to raise heritage chickens affordably, you should use a different feed ration than that of the Cornish Cross.

During the colder months free-choice alfalfa hay gives chickens something to do besides peck at each other and it will improve egg and meat quality.

Turkeys

Broad-Breasted turkey poults need to start with 24 to 28 percent protein and have methionine demands that are 20 percent higher than chicks. Indiana pastured poultry producer Greg Gunthorp says that the first 8 weeks are very important when raising turkeys and that they need their highest protein and methionine during that time. After the first 8 weeks they are very hardy and can be fed a regular grower ration with increased energy and decreased protein (around 16 to 20 percent). Turkeys also have a much higher niacin need than chickens.

Ducks

Ducks can be fed much like chickens, starting with a higher protein ratio (18 to 20 percent) and reducing it as the birds get larger (to 16 percent). However, they do have a much higher need for niacin. All duck feed should be pelleted or given as a wet mash. A dry powdery mash can cause choking, and much of it will be wasted by the ducks. If you want to prevent your ducks from being overly fatty (and hence greasy), grow them slower on a lower protein feed with more access to forage. You can also limit-feed them in the last few weeks of life with full access to pasture. Just for a little perspective and budgeting, a typical Pekin duck started on 20 percent protein feed then finished on 16 percent feed will eat around 40 pounds total feed by 12 weeks of age when raised inside (Holderread, 2011). That produces around an 8.5-pound live weight bird.

When ducks are in the brooder or in seasonal confinement, add green chop to their waterers instead of throwing it on the ground. This keeps the greenery fresher and gives the ducks something to do—fish for blades of grass in their water. The same is true for geese.

Oyster shells are not needed for ducks unless they are laying hens.

Geese

In the brooder phase (first 3 weeks), give the goslings a starter ration of around 20 percent protein along with some green chop. Once they move out of the brooder, provide them a 15 percent protein ration. If you want them to forage more, limit-feed them to whatever they can eat in 15 minutes twice a day. If you don't mind longer grow-out times, geese can forage for 100 percent of their diet once they are past the brooder stage. Make sure they always have plenty of clean water to drink in a waterer that allows for them to fully dip their heads into it. They also need plenty of coarse grit to keep their crops healthy.

Keeping Healthy

There are all sorts of poultry diseases, especially in confinement-style operations. However, just like the rest of the animal chapters in this book, we will focus mostly on *prevention* rather than discussing treatment. To avoid some of the nasty diseases that poultry can obtain, such as Marek's disease, Newcastles, or Fowl Pox, you might consider buying vaccinated chicks, poults, or goslings from a hatchery. This may be less important for birds you are going to slaughter at a young age before most diseases develop, but if you are going to raise dual-purpose flocks, breeding flocks, or longer-lived heritage breeds, consider buying vaccinated birds. Also, follow the simple farm biosecurity steps outlined in appendix A to limit disease.

Ascites (broiler chickens). This is fluid buildup in the chest cavity. It occurs more often at high elevations (over 3,000 feet). It can be reduced through good ventilation and slowing growth by limit-feeding, especially in last few weeks of life.

Aspergillosis (all poultry). Several fungus species cause this disease, which can be seen in birds as young as 7 to 40 days old. Symptoms include nervous system problems, labored breathing, emaciation, and increased thirst. Mold-covered litter or moldy hay as bedding is often a cause.

Blackhead (turkeys). Although chickens can be carriers of the protozoa that cause this disease, they don't usually get blackhead themselves. Turkeys are much more susceptible. This is one reason why some producers don't raise chickens and turkeys together. These protozoa will live in your soil for up to three years, so if your turkeys come down with blackhead, you need to rest those pastures for three years before you put turkeys back on them.

Coccidiosis (all poultry). This is a protozoan parasite that warrants special attention because it is so prevalent in poultry (nine species affect chickens and seven affect turkeys, but they are not common in waterfowl). Signs of coccidiosis include white loose droppings and stringy blood in the droppings. When chicks are in really sad shape they will have foamy brown droppings. Avoid coccidiosis by refreshing bedding often and as needed. Wet areas around the waterers are often a source of trouble. Chickens can survive this, but need to be treated immediately and the root causes should be addressed. Even if they get over it, their growth will be negatively impacted for the rest of their short lives.

Mites or Lice (all poultry). These bugs suck on blood or feed on skin and feathers. Symptoms include feather loss, baldness, redness or scabs on skin, drop in egg production, and weight loss. You can help prevent external parasites by thoroughly cleaning your poultry houses between batches, even the chicken tractors. If you provide bedding to their housing or coops, remove and refresh this between batches. Provide dust baths with diatomaceous earth, wood ash, or wormwood (*Artemisia*) leaves in them. If your poultry species roost or nest, consider painting all the roosting bars and nest boxes with linseed oil, which will suffocate any mites that might be living on them.

Parasites (all poultry). Included here are roundworms, tapeworms, and protozoa. Symptoms often include lack of appetite, listlessness, not gaining weight, and abnormal feces. Prevention includes rotating pastures; cleaning out bedding; and feeding strong herbal plants such as oregano oil, garlic, wormwood, mint, and others.

Wet Feather (ducks). The ducks begin to look waterlogged and dirty all the time, and they can't replace the oils on their feathers. Continuously wet conditions exacerbate the problem. Birds should be able to get out of the rain and not have to live in saturated, muddy conditions all the time.

Fencing and Predator Protection

Portable electric net fencing is the go-to product for fencing poultry. Folks that used some of the earlier versions of electrified fencing with mixed results might want to give some of the newer nets a look. Recent innovations include: stronger ground stakes, thicker and stronger posts, stiffer vertical strands, and reduced electrical resistance. These improvements have made it easier to keep the fence upright with less sagging, which keeps the fence from grounding out. The nets are easy to move and effective at keeping predators out and poultry in. The downside is that the lower strands can be difficult to keep free of weeds and grass and they require a strong energizer. If your fencing is remote enough that you can't use a plug-in energizer, you may have to purchase an even more expensive solar-charged battery energizer. Although the initial investment in electrified net fences is high, if well cared for they should last 6 to 10 years. Roll them up properly and stash them away in your barn when not in use. This keeps them out of the elements that will degrade the plastic and keeps weeds from getting entangled in them. They also can be used for a variety of animal species, so they are easily adaptable to other livestock production systems. A more extensive discussion of electric fencing components can be found in chapter 3.

Some farmers who have an extensive land base will raise their broilers in a big pasture with no

Predator Prevention

According to the most recent USDA data on the subject of predation losses, in 2010, cattle and calf losses due to predation were 5.5 percent of their total death loss. Of this number, 53.1 percent were attributed to coyotes (also the number one predator in goats and sheep). Dogs were responsible for 9.9 percent, mountain lions and bobcats for 8.6 percent, vultures for 5.4 percent, wolves for 3.7 percent, and bears for 1.3 percent. Predation losses were far fewer than losses due to health problems, which represent 94.5 percent of all death loss in cattle and calves. The most common cause of death in cattle is respiratory problems—responsible for 28 percent of death losses.

Coyotes, dogs, and feral dogs account for a high percentage of poultry losses too, although there isn't much data available, since pasturing poultry is not the norm in the industry. When we were raising large numbers of pastured chickens, roving feral dogs were our number one predator, followed closely by coyotes; humans third; and birds of prey (including owls) were a distant fourth. Coyotes will usually kill one bird at a time (sometimes once or twice a day), but dogs will often kill and wound dozens of chickens in a single predation event. In our experience, losses in the brooder and other occasional health problems were a far greater challenge than predation. Instead of eliminating predators through lethal methods we focused heavily on predator *prevention*, and that is where we recommend you focus your efforts as well.

There are five main ways to prevent predation, in no particular order: (1) good fencing; (2) guard animals; (3) night penning; (4) other deterrents such as range riding, fladry (see below), lights, and noises; and (5) cultural practices such as culling weak animals, removing afterbirth and carcasses of dead stock, and rotational grazing.

Fencing is covered in other parts of this book, and electrified fencing is especially effective in preventing ground predators, but will not stop aerial predators. A single strand of barbed wire attached at ground level to the outside of your perimeter fence will stop most predators from digging under it.

Guard animals can include guard dogs, llamas, donkeys, and even guinea hens. Guinea hens can be especially useful in alerting guard dogs of a predator's presence. You can also raise larger animals such as cattle with smaller animals like sheep and goats, even poultry. The presence of the cattle herd will help protect the flock, now affectionately called a "flerd."

The use of guard dogs is the most common form of livestock protection, and for good reason. Livestock guard dogs have a strong instinct to guard the animals that they have bonded with. They are often used to protect sheep, goats, and poultry. It's best to choose a dog that fits with your climate, vegetation, topography, and predator type. Dogs with shorter coats do better in the heat and have fewer problems with matting and weed seeds. Dogs with long coats might need shearing, especially if they are in contact with blackberries and spiny weeds. If your dogs will regularly be around people, consider a dog with a more mellow temperament, like a Great Pyrenees. Other breeds to consider are Kangal, Maremma, Akbash, and Anatolian Shepherd.

Starting out with a proven adult dog is the easiest way for most farmers to begin. A mature dog that is already bonded to the type of animal that you raise is ready to go on the day you bring it home to your farm. A puppy, on the other hand, is too young to effectively guard your animals and will need training. An adult dog is a great help in teaching a young dog the ropes. The adult dog can also be a companion and protector for a vulnerable puppy. Make sure the dog you purchase is bonded to the type of stock that you want it to guard: You will pay more for a trained dog, but the investment is well worth it. Neutered males are less likely to wander than intact males. If your predators are other canines that like to hunt in packs, make sure you have more than one guard dog. Otherwise, part of the pack will lure your dog off to one side of the ranch while other canines in the pack come in from the opposite direction to hunt your stock.

Llamas and donkeys can also be used as guard animals, but it is important to understand their strengths and weaknesses. They are less likely than a dog to kill your stock and they are easy to keep, since they eat the same forage as the stock they are protecting. They aren't nocturnal, however, which makes them less effective at guarding the herd during the most active hours for predators. Only one donkey should be used per herd, as they are likely to bond to each other if there is more than one. Llamas and donkeys will not stand up to bears and mountain lions and will leave your stock vulnerable to these predators when encountered. Males should be castrated in order to be effective. Both of these animals should be carefully introduced to the herd.

Night penning involves bringing your stock into a secure shelter in the evening and letting them out after the early morning predation period is over. You can do this for all animals, or just for mothers with their offspring. Young animals are the most vulnerable in your herd or flock, so it makes sense to bring them in at night if predation is a concern in your area. If you have poultry, your night pen has to be pretty sealed off to prevent small predators such as raccoons, foxes, and weasels from killing birds during the middle of the night. We have even heard of a farmer penning his sheep at night only to lose one a night to a stealthy mountain lion that managed to squeeze through a small hole in the sheep barn. Depending on your predators, your night pen may have to be adequately sealed off.

Other deterrents include range riding, in which people roam around on foot, ATV, or horseback, especially at dusk and dawn when ground predators are most actively hunting. Many predators, such as wolves and mountain lions, are wary of humans and will stay out of areas that have had recent human presence. The key for range riding to be effective is for it to be unpredictable. If you ride the same direction at the same time every night, the predators may get used to you. Mix it up: come by foot one evening, come by horse another night at a different time, arrive by ATV another morning from a different direction.

Another deterrent to wolves is the use of **fladry**, which is a rope or wire with flags attached to it. Originally used in wolf hunting, fladry has been shown to create a barrier that wolves are unwilling to cross. Some ranchers deploy fladry on a large scale (more than a mile), using ATVs or other vehicles equipped with reels. The effectiveness of fladry can wear off over time if it is maintained in the same area. Fladry can also be used with electric fence and the electric shock acts to reset the fear of the fladry in the wolf. In addition, some farmers employ fright tactics such as strobe lights or acoustical alarms that can be motion-triggered. Apparently, coyotes can become used to timed devices or other predictable fright tactics, so again you need to be unpredictable.

Cultural practices that help prevent predation include culling weak or diseased animals and prompt disposal of carcasses of dead stock. Bury, compost, or incinerate dead animals. If you are composting dead animals, make sure your compost pile is not an attractant for predators that may come to scavenge through the pile and then look for some nearby live animals to kill at the same time. We once found some neighborhood dogs rummaging through our compost pile looking for dead chickens. They could have easily sauntered over to our laying hens and made a few kills at the same time. Either fence in your composting area or locate it far away from where your live animals are. Also, during calving and kidding season it is important to remove afterbirth from the pasture as much as possible. Spend time around your young and newborn stock, which is not only important for monitoring their health but to also make your presence known to predators. If you are doing any on-farm slaughter, make sure you promptly clean up the area of blood, offal, and other parts that could attract predators. Moving your animals frequently through rotational grazing will also throw predators off a bit, so they don't get used to where your stock are. Again, it's about creating an unpredictable system so that predators are less likely to habituate.

As a last resort, lethal methods to control predators include poisoning, trapping, and shooting. Although we won't fully cover lethal methods in this book, there may be special cases, such as feral dogs caught in the act, in which killing an animal is your only option. Killing wild predators is often ineffective over time and has negative ecological consequences. For example, simply killing coyotes has not proven to be an effective method over time. A 13-year study of predation by coyotes showed that 89 percent of lamb kills were done by alpha males (Conner et al., 1998). Killing the alpha male in the area opens up territory to new coyotes, and they will take over killing livestock. In addition, killing wild predators may increase the nuisance of deer, elk, rabbits, and other herbivores on your land because their movement and population is no longer kept in check by predators. It's certainly something to consider seriously.

chicken fence. Often there is a fence for the cattle that range in the same pastures, but the chickens are free to range as far away from their shelter as they want. It is important to place neighboring shelters just far away enough so that the birds don't leave their group. These shelters can be surrounded with one strand of electric fencing wire to keep the cattle out of the chicken feed. Having no fence is much easier than having to move poultry netting all the time. When the area gets impacted the shelter can be moved to a new spot in order to spread the impact and fertility.

However, a good livestock guard dog or two is pretty essential in these fenceless situations.

Handling and Slaughtering

The easiest way to transport chickens is in poultry crates. The low height of the crate doesn't allow the birds to pile up and smother each other. Catching them at night seems to reduce stress on the birds and is easier than doing it in the daytime when they are more mobile. Crates can easily be moved close to the kill cones and placed in a shady area while the birds wait for processing. We have made crates out of wood in the past, but it's hard to compete with lighter weight plastic crates. Manufactured plastic crates are easy to disinfect, stackable, and free of any sharp edges that might injure the birds and damage the carcass.

Be sure to take the birds off feed for at least 8 hours before slaughter. Poultry with full digestive systems are more difficult to process cleanly. They should, however, still have access to water if possible. Water is not accessible while they are crated, though, and this should be fine as long as the birds are kept relatively cool with good ventilation.

The poultry slaughter can be broken down into two parts: the dirty side and the clean side.

The Dirty Side

This consists of killing, bleeding, scalding, and plucking.

KILLING AND BLEEDING

The bird should be removed from the crate and kept calm. It should be handled in a way that doesn't allow it to flap its wings. David Schafer recommends spraying the kill cones with vegetable oil to keep the blood and feathers from sticking. Place the bird in the kill cone and hold its feet with one hand while grabbing the head with the other and gently pulling it out of the bottom of the cone. Let go of the feet with the other hand and grab your knife. Pull the skin tight across the jawbone and cut the neck right below the jaw just on one side and not so deep as to cut the windpipe. With practice and a sharp knife you will be able to sever the carotid artery, which is deeper than the jugular vein, and cause an instant loss of blood pressure and consciousness. With the brain still communicating with the heart and lungs the bleed out is complete. The cut should ideally be performed in one stroke with a super-sharp knife and not a sawing motion. Avoid cutting bone and feather, which will prematurely dull your knife and possibly result in a poor bleed out. The bird will be completely limp when it is dead.

SCALDING

Scald the birds in 145°F (63°C) water with soap in it. Keeping the scald water at proper temperature is one of the most important tasks in order to keep the process going smoothly. Automated and temperature-controlled scalders are a good investment for anyone processing large numbers of birds. A little soap added to the water makes it more effective in loosening feathers and cleaning up the bird. When tail feathers can be easily removed the bird is sufficiently scalded. Over time you will develop a feel for when a bird is properly scalded. Do not over-scald the bird. Over-scalding is actually cooking the skin of the chicken, and when you pluck it the skin will tear and the result is an unsellable bird. Under-scalding (too cool or too short a scald) results in stuck feathers but is easily remedied by rescalding and picking.

PLUCKING

Pluck the birds in a rotary chicken plucker or picker. Most pluckers allow for several chickens to be processed at one time, and have integral water sprayers that will rinse away the bulk of the feathers. The majority of the feathers should be gone after about 20 seconds with the exception of some of the more stubborn tail feathers. Most of the skin from the feet should also be removed except for some of the toenails. Remove any remaining feathers and feet skin by hand and move the chickens to the clean side of the system.

The Clean Side

This consists of the feet, head, neck, and oil gland removal; evisceration; giblet harvest; and chilling. Neck removal is optional, and time-consuming, but may be dictated by your market.

FEET, HEAD, NECK, AND OIL GLAND REMOVAL

Remove all of the feet by cutting at the joint, unless you have a market for feet-on chickens, which are popular among some ethnic groups. If you do have a market for feet-on chickens, at least clean the outer membrane off the feet. If the feet have nitrogen burn from poor management, they should be removed. After each chicken has its feet removed, turn it so that its neck and head is hanging off the table (this readies it for the next step). After you cut the feet off the last chicken in the batch, go down the line and make a cut in the neck skin of each bird. Put down your knife. Pull the head off each bird by popping it backwards and then pulling it off. With the head off the bird, expose the neck by pulling off the neck skin and peeling back the trachaea and esophagus. Loosen the crop from the neck. Do this to each bird in the batch. Pick up your knife. Cut the neck off each bird (save and chill the neck) and trim back the neck skin (skip this step if you are leaving the neck on). We like to leave enough skin on the neck to cover up the neck hole. Cutting too much neck skin will expose some of the breast meat, potentially drying it when it is cooked. After you cut each neck/neck skin, place the bird on its breast with its tail facing away from you. This sets each bird up for oil gland removal. Oil gland removal is optional, and many producers leave it intact. At this point in the process we like to do a little quality control and remove any leftover feathers and skin and give each bird a good spray with water.

EVISCERATION, GIBLET HARVESTING, AND CHILLING

With one hand resting on the keel bone, pinch the belly skin just above the vent and make a long shallow cut across the skin on the belly. Pull open the cut with your hands in order to make it large enough to get your hand into the cavity. Put down your knife. Squeeze the gizzard between thumb and forefinger to separate it from the belly fat. This leaves the belly fat attached to the carcass and increases the weight of the chicken as opposed to removing the belly fat (plus more belly fat makes the bird roast up juicier). Insert your hand and gently break all the adhesions that are holding the internal organs in place. Tug on the gizzard to loosen the esophagus and the crop from the neck hole and pull it out of the carcass. Sometimes the gizzard will break off while you are tugging it and will leave behind the crop and the esophagus. This is no big deal; just remember to pull the crop out of the neck hole prior to putting the bird in the chill tank. Scoop and slowly and gently remove the rest of the internal organs, being careful to leave all of the belly fat in place. While holding the intestines with one hand make a U-shaped cut around the vent to complete the evisceration. Harvest the liver by pulling it apart from the bile sac. Place the bird in a chill tank or a walk-in cooler for air chilling.

David stresses that efficient processing is one of the most important aspects to making a poultry enterprise sustainable. He has come up with what he calls his "Six Tips for Efficient Poultry Processing" and was willing to share them with us. These include:

1. **Proper station layout.** Do not add any extra steps anywhere; in other words, get your crates close to the kill station, which is close to scald, with chill tanks nearby, and so on.
2. **Rotary kill station (with round cones).** This saves many steps over a cone line (and is less traumatic to the birds). Birds will not back out of round cones.
3. **Automatic scald.** Both in temperature control and agitation. The scald is *the* critical step so accuracy here pays big dividends. The automatic agitation feature, whether dunker, rolling shelves, or baskets, basically adds a person to the evisceration line where manpower is most needed.
4. **Matched equipment/manpower.** Twice as many kill cones as scalder space since bleed out takes twice as long as scald. Enough eviscerators so that the front end team never has to wash their hands and pitch in. Bottlenecks kill line speed.
5. **30-second evisceration.** At least one person in the crew has to be this fast at gutting.
6. **Use eviscerating shackles.** Just as fast or faster than table evisceration plus cleaner, less chance of cross-contamination, more low-back friendly, and gravity is assisting.

Cutting Up the Carcass

If your state allows you to cut up the poultry carcass under exemption, you may want to consider doing so, at least with some of your birds. Or if you pay to have an inspected plant process your birds, find out if they will do cut up as well. Cutting up the carcass has the potential to double the value of the carcass in a couple of minutes of labor. It also allows you to have more variety for your markets, including whole birds, breasts, thighs, legs, and wings. You can sell the backs and feet for stock making or pet food.

Post-Harvest Handling

After the carcasses are chilled to around 40ºF (4ºC), they should be checked one last time to make sure they are clean before they are packaged. The lung bits, fascia, and pieces of crop are often easier to remove when the birds are chilled and less slimy. Birds can be drained at this point and weighed, bagged, and labeled. You can make a broiler drain rack out of short pieces of PVC mounted to a drain board or even a clean metal shoe rack. A rack of this sort will greatly speed up your post-harvest handling. After the birds are bagged and tagged they should be put on ice or refrigerated. If the birds are to be parted out, now is the time to do that as well. If the birds will be frozen, wait at least 24 hours before freezing so all the rigor mortis is released. Also, don't sell the birds until they have been dead for at least 24 hours or at least let your customers know how long to wait before eating them. If you sell a bird while it is still undergoing rigor mortis and somebody eats it, they will have a tough eating experience. That may scare off future sales.

BOONDOCKERS FARM

Boondockers Farm is situated on 75 acres in Beavercreek, Oregon, just 30 minutes south of Portland. Around 35 of their acres are in mixed species woodland with a creek and ponds, and the rest are lush, green pastures. Owners Rachel Kornstein and Evan Gregoire were asked by duck expert David Holderread to help preserve the Ancona breed of duck. They started with a small original flock and now have the largest flock of Ancona ducks in the world. A British breed, the Ancona is a medium-sized duck that is a prolific egg layer and the best forager of ducks of their size. They are extremely hardy and adaptable to varied conditions. They are listed by the Livestock Conservancy as "critical," the classification for the most endangered breeds.

Boondockers sells hatching duck eggs, table duck eggs, fresh and frozen whole ducks (Saxony and Ancona), and heritage breed Delaware chickens. They sell most of their duck meat to Portland-area restaurants utilizing a weekly delivery route, and they also attend a couple of farmers markets that round out their sales. Rachel says that chefs prefer their ducks to the typical Pekin duck, because theirs are less fatty and the quality and mouthfeel of the fat that they do have is superior to the mass-produced Pekin duck. The couple also raise Saxony ducks that are a large breed of duck valued by chefs who want to have a larger carcass bird for presentation purposes and for other dishes that require a bigger duck that is still an alternative to the Pekin.

The Anconas are known to forage better than the Pekins and they are also one of the best egg-laying breeds of ducks, even rivaling chickens in the number of eggs they produce (210 to 280 per year) and outperforming them in terms of the overall volume of eggs produced because the size of the eggs is quite a bit larger than that of a chicken. Boondockers uses the females for egg production and the males for roasting birds. They do keep a few drakes around for breeding purposes, too. Utilizing the excess males for meat and selling less desirable females as utility layers allows for strict selection of breeding stock.

During the brooding phase the ducklings are quite a bit hardier than chicks. They can make a mess out of the brooder if their water isn't set apart from the heated area. When they are young, ducks can get chilled if they get wet and aren't able to dry off and warm up. Ducklings that get chilled in this manner will lose conditioning and can be set back for their entire grow-out. Rachel provides water to young ducklings in a shallow pan filled with marbles to prohibit the ducklings from getting in the water and getting wet. Later in the brooding phase the ducks can get wet as long as they have a spot to dry off, and this will actually stimulate them to secrete feather oils, which will make them more resilient to getting chilled. Ducks don't pile up on each other like chickens do in the brooder. Rachel has had good results using straw in the brooder as long as it is chopped straw. Bedding must be constantly refreshed or replaced in the brooder.

Ducks should have some shade from the sun when the temperature gets above 70 degrees F (21 degrees C). A suspended shade cloth or some type of moveable pasture shelter will work. Boondockers is hoping to obtain a Natural Resources Conservation Service grant to build a hoophouse for winter shelter, but up to now has relied on makeshift pasture shelters for protection from the weather during the coldest months of winter. This is one area of improvement that they are actively pursuing.

The ducks can make quite a mess of the pasture, especially around the water source. For waterers, Rachel and Evan use blue kiddy pools that they empty and refill twice a day. The ducks use the pools as their source of drinking water, a place to dip their food before they eat it, and as a pool to cool off and splash in. Moving the pools each time they are emptied and refilled is important to spread the impact and reduce the amount of heavily impacted spots on the pasture. We were actually surprised that there weren't more wet and muddy areas on the pasture, but this is the beneficial effect of moving the water source twice daily. The pools only last for a year or so before they get leaks, but then they are repurposed as feeders for the ducks.

One of the main problems that folks run into when raising ducks for meat is the growth of pin feathers when the bird is ready to be processed. If the birds are not processed before this new growth of feathers emerges, the value of the carcass will be diminished because of the tiny feathers that are almost impossible to remove. Rachel and Evan avoid this issue somewhat by waiting until the birds are past the pin feather stage and slaughtering the birds after the pins have developed into full feathers and can be more easily removed. Rachel will feel the breast of each bird prior to processing to ensure that there are no pin feathers. It is also important that the birds don't have big feather pores that can absorb chill water and also make the carcass less suitable for proper presentation when served at restaurants.

The birds are processed for restaurants with the head and feet on. The crop is removed carefully to ensure the carcass looks good. Giblets are saved and the livers are in demand from chefs. The birds are processed on-farm using the USDA exemption for on-farm processing for direct-to-consumer sales and processed USDA inspected (at a processor) for restaurant sales. Rachel asks the chefs to advertise their farm on the menu and the fact that the breed is an Ancona, which is a selling point for some. Most importantly, this recognition gives critically endangered breeds a tasty reputation, which may in the end help save them. "Eat them to save them" sounds like a positive way to build support for heritage breeds conservation.

Differences between Ducks and Chickens

Although you may have gotten some sense in the Boondockers narrative above that ducks are a bit different than chickens, we will explain more of the differences here. Ducks require less substantial housing and protection from the elements. Ducks can be confined with a lower fence than chickens; a 2- to 3-foot-high fence will work fine for ducks. Ducks, like broiler chickens, roost on the ground (thus they are susceptible to predation and should probably be locked up at night), but they are more resistant to disease and parasites. Likewise, ducks are more resistant to cold and wet weather and are better foragers than chickens. Ducks can also be started on a lower protein feed (15 percent) than chickens. Ducks can be processed at a younger age than heritage chickens (7 to 12 weeks). Duck feathers have more usefulness and may be valuable to clothing or bedding companies. More feed is required to produce a pound of duck meat than is required for a pound of chicken meat, but this may be offset somewhat by their better foraging ability.

Young ducklings are hardy and easier to raise than other poultry. It is not uncommon for all members of a brood to survive to adulthood. They can chill rapidly if they get wet, so it is important that their waterer be of a type that they cannot enter during the brooding phase. Access to fresh drinking water is very important for ducklings, especially right after they have eaten. For large-breed ducks the capacity of the brooder might only be 50 percent of the recommended capacity for chicks. Ducklings require less heat than chicks do, but exactly how much is based on ambient temperature, humidity, drafts, and the breed of duck being brooded. Ducklings that appear quiet and are comfortably resting under the brooder have the proper amount of heat. If the ducklings are restless and bunched up under the heat source, they probably need more heat. If they are scattered away from the heat source they need less heat. Overheating is almost as damaging to ducks as chilling, so it is necessary for them to have a way to get away from the heat source.

Ducklings consume three to four times as much water as chicks do. Ducks like to wash excretions from their eyes several times a day, so it is good practice to have a watering system that allows them to exhibit this behavior. Ducklings have a higher requirement for niacin than most breeds of chickens. Adding brewers' yeast is an effective way to provide for niacin in the ducks' diet. Ducklings can overdose on calcium. Grit should be provided to ducks at all life stages, just as it should be provided for other poultry. Ducklings thrive when supplemented with green feed in the brooder. An effective way of supplying green feed to them in the brooder is to place it in the water where it won't be trampled and soiled, but instead will remain clean and succulent for a longer period of time.

A big difference between waterfowl and chickens and turkeys is their amount of feathers. According to Metzer Farms, a large waterfowl and game bird hatchery in California, ducks and geese are normally more difficult to process than chickens or turkeys because they have more down and feathers and nature made them waterproof!

Since waterfowl have so many feathers, you want to pluck them when it is the easiest to get them all out. This is when the feathers are all mature and there are no pin (or immature) feathers, meaning they are not in the midst of a molting period when new feathers are coming in. For ducks this is at about 7, 12.5, or 18 weeks of age. For geese it is normally at 9, 15, or 20 weeks of age. If you try to process between these maturation dates you will encounter large numbers of pin feathers that may double or triple your processing time and effort. That is why in the breeds listing at the beginning of this chapter it says ducks: 8 or 18 weeks of age, geese: 10–12 or 24 weeks of age. These dates differ a little from the Metzer website because Metzer

Farms raises their birds in barns with full feed, so they grow faster than the pasture-raised harvest dates in our table. According to the owners of Boondockers Farm, you just need to pick up a few birds in your flock and feel for pin feathers. If there are none, it's time to process. If the birds are not as big as you want them to get, wait until after the molt for the next window of opportunity.

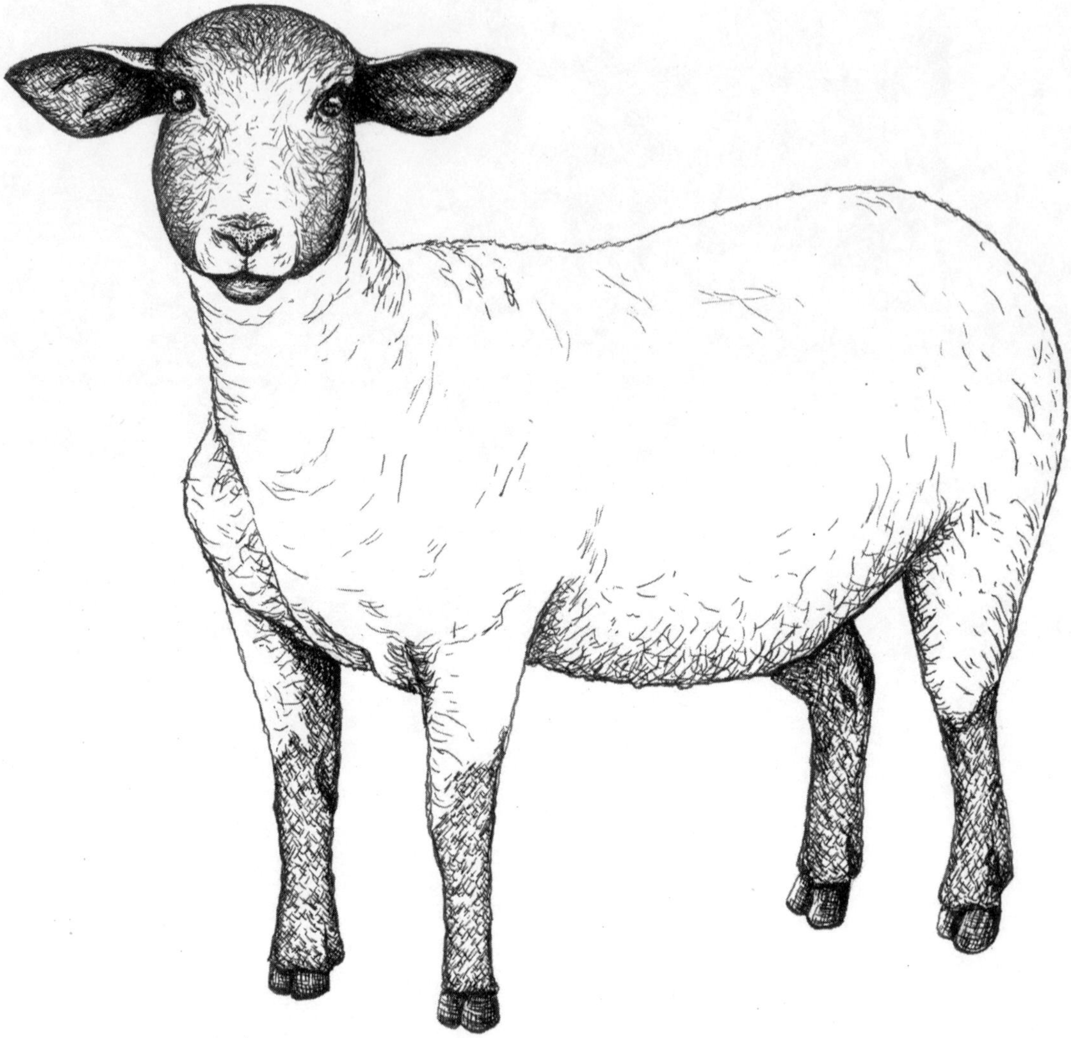

CHAPTER 3
Sheep and Goat Production

Sheep and goats can benefit the farm by harvesting weeds and clearing brush. They can add diversity to your existing markets, and they are easy to sell direct because of their small carcass size, although not necessarily for a good price. They can also be really fun and entertaining or a total pain in the rear depending on your personality and patience and how well you set up your

DEFINITIONS

AI artificial insemination of semen into the oviduct of an animal; used in lieu of natural breeding for a variety of reasons

Anthelmintics drugs or plants that stun or kill worms, commonly referred to as a "dewormer"

Buck a male goat over 1 year of age

Buckling a male goat less than a year old

Caprids mammals in the *Caprinae* subfamily, which includes goats and sheep

Doe a female goat over 1 year of age

Doeling a female goat less than a year old

Ewe a female sheep at least a year old

Ewe lamb a female sheep less than one year old, sometimes called a "eweling"

Flock a group of goats or sheep

Jug small pens to temporarily confine a ewe and her lambs to facilitate bonding

Kid a baby goat of either sex

Lamb a young sheep (usually less than 14 months old)

Ram an intact male sheep at least a year old

Ram lamb an intact male sheep less than a year old, sometimes called a "ramling"

Sheep a mature ovine of at least 1 year of age, either sex

Wether a castrated male (sheep or goat)

Yearling a sheep between 1 and 2 years of age, still used for meat

TABLE 3.1. Characteristics of Different Sheep and Goat Breeds

Breed	Size, optimal finish weight	Production model most suited for	Strengths	Challenges
Kiko goat	large, 100 lbs	meat	• low maintenance • fast growth • good mothers • performance-based breed	• hard to find genetics • expensive
Boer goat	large, 100–120 lbs	meat	• fast-growing • good for cross-breeding	• low parasite resistance • more seasonal breeders
Spanish goat	medium, 60 lbs	meat	• hardy • can be raised on poor-quality rangelands	• can be scrawny • large variability in size
Myotonic or Fainting goat	medium, 60 lbs	meat, but also pet trade	• easier on fences • extended breeding season	• can fall over ("faint") when stressed
Savanna goat	large, 100–120 lbs	meat	• hardy • more parasite resistant than Boers • extended breeding season • prolific	• harder to find genetics than Boer goats • can be expensive
Dorper (hair sheep)	medium, 100 lbs	meat	• polled • mild flavor • don't need to dock tails, no shearing needed	• slower growing than wool breeds
Katahdin (hair sheep)	medium, 100 lbs	meat	• normally polled • high parasite resistance • less seasonal for breeding • good mothering abilities • don't need to dock tails • no shearing needed	• slower growing than wool breeds • low flocking instinct, so not well suited to extensive production

farm to deal with their behavior. From a consumer standpoint, buying an entire lamb or goat is not as cumbersome, nor as expensive, as an entire pig or steer. They are also a great stand-alone enterprise for smaller farms because they require less land, infrastructure, and investment while producing a variety of products such as meat, wool, hides, milk, and grazing services. Goats and sheep are easy to haul and don't require a lot of equipment to get started. Because they usually have multiple births they offer a fast return on investment and a ewe can raise lambs whose weaning weight will equal her own within a year. Goats and sheep grow, mature, and reproduce quickly. For cattle ranchers interested in multi-species grazing, one ewe and her offspring can be added for every cow without lowering the stocking rate.

We will discuss sheep and goats together since they are very similar and belong to the same subfamily, the *Caprinae*. Where there are differences, though, we will point them out.

Breed	Size, optimal finish weight	Production model most suited for	Strengths	Challenges
Columbia sheep	large, 100–130 lbs	dual purpose (wool and meat)	• good gains on grass • suited for Western rangelands • polled	• heavy wool breed • will need to shear • slower growing
Targhee sheep	large, 100–130 lbs	dual purpose (wool and meat)	• hardy • suited for Western rangelands • good mothers • twinning normal • polled	• medium wool breed • will need to shear
Rambouillet sheep	large, 100–125 lbs	dual purpose (wool and meat)	• good gains on grass • ewes are prolific • strong flocking instinct (good in extensive rangelands or high predation areas)	• rams may have horns • heavy wool breed • will need to shear
Suffolk sheep	large, 120–150 lbs	dual purpose (wool and meat)	• most common breed in United States • easy to find genetics • large carcasses • fast growing	• more susceptible to scrapie • large heads can result in lambing problems • will need to shear

Breeds

Kiko and Myotonic goats are more hardy and parasite-resistant than other meat goats, but the breeding stock can be more expensive. Boer and Boer crosses are easy to find and grow to a large, meaty size, but they are the least parasite-resistant. Spanish are well adapted to the scrubby rangeland conditions of the West and are ideal in extensive ranging situations. Savannas seem to have the benefit of being large like Boers but are much hardier and are prolific breeders.

According to Minnesota sheep producer Janet McNally, British and European continental breeds work best in Midwestern and East Coast grazing conditions and climates. Tall Americanized versions of these breeds are not what she is suggesting, but the older phenotypes, which are more compact. Primitive sheep breeds like Jacob and Navajo-Churro may work well in drier, scrubbier rangelands of the West; they grow slower but are less input-intensive. If you are raising sheep on extensive rangeland conditions, a strong herding instinct is a must. For smaller farms with strong fences, this is less of an issue.

For sheep there is also the major decision to go with a wool breed or a hair breed. If you don't have a market for wool and don't want the added chore of annual shearing, then choosing hair breeds is an easy decision. Hair breeds do better in hot climates, have fewer problems with flystrike and parasites, and because most of them originate from regions closer to the equator, they have longer breeding seasons. Another advantage is that some say their meat is milder-tasting: Research has shown a correlation between wool fiber diameter (fineness) and lamb flavor, with fine wool lambs having a more intense ("muttony") flavor than coarse-wooled lambs like hair breeds.

Differences between Sheep and Goats

Goats require a more nutritious diet than sheep, their digestive system does not retain food for as long and therefore doesn't digest nutrients fully. Goats prefer to browse and eat twigs, leaves, vines, and shrubs. Sheep are grazers and prefer to eat forbs and many plants that might be considered weeds. For this reason they work really well together to clean up and regenerate pastures, especially for cattle but even for horses, by controlling invasive and undesirable plant species.

Goats can't digest the cellulose in plant cell walls as well as sheep, so sheep can convert feed more efficiently as a result. Lambs generally grow faster than kids. Grain feeding is less likely to be profitable with goats because they grow more slowly and are less efficient converters of feed.

Sheep have a very strong flocking instinct and being separated from their flock can be stressful for them (although not all sheep breeds have as strong of an instinct). Goats are more independent and curious. Sheep are easier to fence than goats, because they don't climb on fences like goats do. However, if either of them have horns, they are both susceptible to getting their heads stuck in woven fences.

Most goats are naturally horned and many sheep breeds are polled. However, many old-world and heritage sheep breeds still have beautiful curved horns much like wild sheep.

Goats like to have access to shelter more than sheep do. They love shelters that are slightly elevated. One farm in Minnesota, Paradox Farm, actually built a mobile shelter affectionately called the "Milk Star Gallactica" that is on a wheeled hayrack. The goats climb up inside it when they seek shelter, whereas the sheep prefer to hang out in the shade underneath.

Goats will usually dominate sheep when grouped together, especially if the goats are horned and the sheep are polled. Bucks will raise up and come down forcefully to butt heads, but a ram will back up and charge to butt heads. Rams will butt bucks in the belly while the buck is rearing up. The fighting behavior of the ram is often more effective than that of the buck when the two species engage each other. It's probably best to keep the breeding males separate for this reason.

A ewe's estrus cycle is about 17 days while that of a doe averages 21 days. Goats are much easier to artificially inseminate than sheep. Ewes have a more complicated cervix, which makes it difficult to pass the insemination rod. It is more difficult to tell if the ewe is in heat than it is with the doe. A teaser ram is needed to detect estrus in a ewe or timed inseminations are performed after hormonal manipulation.

Goats tend to be more prolific and less seasonal breeders than sheep. Bucks have a really bad odor during breeding season and rams don't.

Sheep deposit external fat before depositing internal fat. Hair sheep deposit their fat like goats. Goats deposit internal fat first and then deposit external fat.

CLOVER CREEK FARMS

Chris Wilson, of Clover Creek Farms in northeastern Tennessee, has 25 years of sheep production experience. We had the privilege of hearing her present at the Southern Sustainable Agriculture Working Group (SSAWG) conference in 2014, where we learned a great deal about her production system as well as using livestock guard dogs, an essential tool for any sheep farm. Chris raises between 125 and 200 ewes per year in a grass-based, rotationally grazed system. She says that sheep need to be watched and managed more than cattle and should be checked at least once a day. Chris has found that her pastures can handle a stocking rate of up to five ewes and their lambs per acre in a good year and only one ewe and her lamb per acre in a drought year. Chris reduces or raises her ewe numbers based on pasture quality and cuts pastures for hay when the grass gets away from her. She uses this hay to feed the animals during drought and over the winter.

Chris raises Katahdin hair sheep. While the hair sheep produce smaller cuts of meat than the wool breeds, Chris appreciates the milder flavor of the hair sheep and the ability to process animals of older ages that don't have a strong mutton flavor.

Chris recommends the employment of a livestock guard dog if you have more than five sheep. She has seen neighbors lose many sheep in overnight predator attacks and says that a good team of livestock guard dogs will more than pay for themselves by keeping your animals alive. Chris also uses a border collie to herd the sheep and to keep them out of her space when she is feeding.

Chris uses a perimeter fence made of woven wire and subdivides her pastures with electrified netting. The sheep don't climb on the non-electrified fence like goats do.

Chris allows her sheep to find shelter under the trees in wooded areas of her pastures or uses three-sided shelters to provide shade and protection from the wind and rain.

For sheep-handling equipment, Chris recommends that you at least have a squeeze and a stop gate in order to separate and sort sheep. She sets up her chutes in a horseshoe configuration that allow the sheep to enter and then exit in the direction they came from. This helps to reduce stress while handling, because sheep are more comfortable returning the way they came. She gives a feed reward to the sheep the first time they are run through the squeeze in order to make the experience more positive and encourage the sheep to go through the system the next time.

Chris allows her ewes to lamb on pasture and then brings them into jugs in her barn. She uses lamb carriers to bring in the newborn lambs. By carrying lambs low to the ground, the ewe can still smell them and will follow them to the jug. Once in the jug, Chris ear tags and castrates the lambs, makes sure the lambs get colostrum and are eating well, and deworms the ewes. She keeps them in the jugs for two days, and then they are released into a holding pasture for another two days where they can be monitored for any problems. Chris uses a bander for castration, and spray paints the ewe's number on the lamb(s) so she can easily see if they are with the right ewe.

> Chris separates ram lambs from the ewes before they can begin to breed at about four months of age. Chris has found that a yearling ram can breed up to twenty ewes and a good mature ram can breed thirty to sixty ewes. The rams will fight if a breeding ram is put back with other rams that did not breed; the breeding ram has to lose the scent of the ewes before they can be penned together. Chris has found that penning them tightly with the pen floor covered with car tires keeps the rams from butting each other.
>
> Chris recommends feeding hay up and off the ground in order to reduce waste. She uses cattle panels to meter out hay from round bales and supplies minerals in farm-made hanging feeders constructed from plastic drums. She provides free-choice kelp along with the minerals.
>
> Chris doesn't trim hooves, but relies on rocks to wear them down. She has surrounded her water troughs with gravel in order to wear down the sheep hooves, and at the same time reduce the amount of mud around the trough. Chris likes black hooves, which are harder, and she selects for them by breeding a Dorper ram with dark feet to her Katahdin ewes.

Sheep have a lower tolerance for excess copper. It is important for this reason to feed a mineral supplement that is formulated specifically for sheep. Goats have a tolerance to copper that is higher than sheep and just about equal to that of cattle. When grazing goats and sheep together, sheep minerals should be used and the goats should be supplemented with copper if there is a deficiency.

Sheep and goats are generally susceptible to the same types of diseases. Both goats and sheep should have a USDA identification tag for scrapie if they are to be moved off of the farm. The coccidia of sheep and goats are different. Goats are more susceptible to worms than sheep, because they didn't gain parasite resistance from their native lands. Goats generally need higher doses of dewormers than sheep do, but they can also be managed so that they consume mainly browse and stay away from nematode eggs located lower down on grass blades.

Life-Stage Considerations

Breeding

Some people get their start raising meat goats or sheep by buying in just-weaned kids or lambs of 3 to 5 months of age, but the most profitable route is raising your own breeding stock and managing them well. Buying weanlings might be a good option for those just starting out, establishing their market, learning how to manage their vegetation, and creating the necessary infrastructure (like fencing) on their land. Once you get the hang of it and decide that raising caprids is for you, adding breeding stock will be the next logical step.

The most important aspect of meat goat and sheep production is the reproduction efficiency of the animals. The buck or ram is the most important individual in the herd. In 1 year the buck is responsible for 50 percent of the genetics in the entire herd (assuming you have one sire), whereas the does are only responsible for the genetics of the one to three kids they raise. As a result, expect to pay much more for a good-quality buck or ram.

The main difference between breeding goats and breeding sheep is that sheep are more seasonal breeders than goats. This means the sheep ewes come into estrus for a more defined period of time, often in the fall, whereas many goat breeds can come into estrus for a longer season (every 21 days on average) or sometimes year-round. However, some hair sheep and other sheep breeds that come from closer to the equator are less seasonal breeders (they aren't affected by day length as much). For a typical British or European breed, the estrus cycle is just 17 days and may happen three or four times in the fall. Having a short estrus season is beneficial if you want lambing to occur at a defined season of the year, usually coinciding with spring. Having a longer kidding season with goats means that you can move toward a year-round meat production model more easily, but it may take more hay and other feeds to get them through colder parts of the year. You could conceivably kid twice a year with your goat does; their gestation length is around 5 months.

BREEDING METHODS

There are five main approaches to breeding goats and sheep.

1. **Continuous Service.** Rams or bucks are turned out with females at all times during the breeding season (typically the fall or whenever the females are in estrus).
2. **Intermittent Service.** Rams or bucks are given access to the females temporarily, usually at night. This is a good way to keep younger males from overworking themselves and to supplement the feed of the males. A bred female or a wether can be kept with the buck or ram for company.
3. **Rotational Service.** These methods involve using a portion of the bucks or rams at different times of the breeding season. This plan can keep young males from overworking themselves.

    - Half and half. Turn out half of the breeding males for about 3 weeks and then remove them and introduce the other half of the males for the rest of the breeding season.
    - Another method is to use one-third of the males for the first 2 weeks and then remove them and introduce the other two-thirds of the males for 2 weeks. After 2 weeks, reintroduce the first third of the males so that all of the males are with the breeding females during the last 2 weeks of the season.

4. **Hand-Mating Service.** This is the practice of bringing the females to the buck individually as they come into heat. Often used for purebred breeding, this practice requires employing a method of heat detection such as a teaser ram or buck. This method is labor-intensive.
5. **Artificial Insemination Service.** AI requires a method for heat detection or synchronizing all the ewes or does at once with progesterone. A trained inseminator is necessary as AI is a minor surgical procedure in sheep and goats. As such, it is not popular for caprid breeding because of the amount of labor needed and expense. However, some small to mid-sized goat farmers are using this method to improve the genetics of their herds. Both fresh and frozen semen can be used.

DOES AND EWES

Ewes and does should weigh at least two-thirds of their full size when you decide to breed them for the first time. They can be bred back immediately after weaning as long as their body condition has recovered to around an average condition score of 3 out of 5. Different breeds have different seasonality for coming into heat, so that should be a consideration when selecting your females. Ewes that wean their lambs in August can regain their body condition before December breeding just by cleaning up pastures after the growing lambs.

Check the body condition score (BCS) of the does at weaning to ensure they will have a high BCS when rebreeding. Thin ewes that are fighting to maintain their own body weight are not able to ovulate as many eggs as ewes in a more desirable condition. This is the reason producers flush ewes (increase their nutritional intake for two to three weeks) prior to breeding. Flushing is necessary to increase the weight and BCS of your does and ewes, which usually allows them to ovulate more eggs. The number of eggs ovulated by the ewe at the time of breeding creates the maximum number of lambs that can be conceived, born, and marketed. On the opposite spectrum, you don't want your does or ewes to be too fat (BCS of 4+) or they will have a higher risk of ketosis.

Other data to track for your does and ewes or those you are considering purchasing are: 6- and 12-month post-weaning weights and FAMACHA score (see the Keeping Healthy section on page 64) over time, reproductive rate (twins or triplets), and longevity in the herd. Compare contemporaries on farm to be accurate, since animals born in different years will face dissimilar conditions. You can add value to sales of breeding animals with this data. For instance, if you have a young doeling for sale, you can sell her for more if you can show records of good FAMACHA, multiple births, fast growth rate, and long productive lives in her lineage.

How long a doe or ewe lasts in terms of breeding is another important factor to consider. Kiko goats have the longest retention rates, followed by Spanish, and lastly Boer. Boer goats generally have the least resistance to parasites, which partly explains their lack of longevity. Kikos have the strongest maternal traits. Boer on Kiko produces the largest offspring.

However, teeth condition, not age, is the most important indicator of doe or ewe longevity. Goats and sheep have teeth on the bottom and a hard palate on top. Goats start with baby teeth, or milk teeth. At 1 year of age their first adult teeth emerge. Two teeth emerge every following year until they have a full mouth of adult teeth at 4 years old. As the goats age further these teeth become flatter and less sharp. The roots become thinner, and gradually the teeth loosen. As they get closer to the end of their productive years they develop a gap in their front teeth or a broken mouth. As their teeth become worn out and loose they are less able to obtain the nutrition they need to reproduce, and they lose their usefulness to the farm. Broken-mouth does are good candidates for culling.

Non-lactating (dry) does and ewes can be maintained on lower-quality forage or hay. Once they are gestating, they need much higher-quality feed. At birthing, they should ideally be on the best forage of the season.

BUCKS AND RAMS

Polled bucks should not be pastured with horned bucks, as the horned bucks will tend to dominate the polled bucks. Bucks and rams should be in good body condition at the time of breeding. Poor nutrition can decrease sperm reserves. You can increase sperm reserves in bucks and rams by beginning supplementary feeding 8 weeks prior to the start of the breeding season. Hot weather can increase body temperature and lead to temporary infertility. Rams should be sheared about 2 months prior to breeding and care should be taken to ensure that all wool is removed from the scrotal area. Rams are not able to breed as many ewes out of season, so for out-of-season breeding you will want to use more rams than you would for in-season breeding. (For more information on this, see Alabama Cooperative Extension, 2007.)

SIRE SELECTION

To choose a sire, first you have to decide if you want a terminal sire (breeding for meat animals, such as Suffolk rams or Boer bucks) or a maternal sire (breeding for replacement ewes and does and breeding stock). Once you have made that decision, look at the physical conformation of the animal and potentially do some testing. Look at phenotype, genotype, and actual performance. Was the sire a single or a twin? What was his birth weight and weaning weight? Was his mother a good milker? Also look for adequate frame size given his age and diet. He should be well muscled and have no genetic defects. Around 10 to 15 percent of rams are homosexual and will not breed at all.

You might want to pay for a physical exam and semen test (or find a ram or buck that has had that done) before you invest in a sire. Be willing to pay more for a breeding ram or buck than an ewe. A good ram is worth around five times the value of a market lamb. A ram should be at least 6 to 7 months old before his first servicing. If you buy a ram or buck that is older than a couple of years, you may want to pay a vet to check his semen to make sure he is still useful as a stud animal. The benefit of buying an older ram or buck is that he should have a track record of successful breeding, whereas a ram lamb or buckling is "new to the job."

The price of a sire may seem high, but it pays off with heavier weaning weights. If a sire breeds fifty does and they produce seventy-five kids with weaning weights that are 10 pounds each heavier than kids from a lower-priced sire, that equals 750 extra pounds of goat on the scale. If goats are going for a conservative price of $5 a pound at markets, this equates to an extra $3,750 for the flock. And that is just the extra revenue for 1 year. Use an Excel spreadsheet to keep records like this.

RATIO OF FEMALES PER RAM OR BUCK (VIA INFOVETS WEBSITE)

Ram lambs or bucklings (8 to 10 months of age): fifteen to thirty ewes/does per ram/buck

Yearlings (12 to 16 months of age): twenty-five to fifty ewes/does per ram/buck

Mature rams or bucks: one hundred ewes/does per three rams/bucks

KIDDING AND LAMBING

Gestation length is around 150 days for both goats and sheep. Mark the time on your calendar for when you think birthing will occur. Take everything else off your calendar, too, while you are at it. This is the primary source of income on a goat or sheep farm and should be a time of high management.

You will need to determine whether your ewes will lamb in a barn or out on pasture. Pasture lambing requires less investment in infrastructure but can present some challenges to the producer. Inclement weather can chill the newborn lamb or kid and predation could be a cause for concern at this time. Lambing on the pasture can also make it more difficult to process the newborns. Tasks such as weighing, vaccinating, marking, or intervening during a difficult birth are all more challenging out on the pasture compared to a barn set up with jugs. A kidding pen or jug allows the producer to easily process the kids and to lock up the doe if she is unwilling to allow the kid to suckle. Grafting kids onto another doe is easier in a pen for this reason as well. Weather is not a factor with a pen, which ensures that goats and sheep aren't giving birth in a rainstorm or on frosty ground.

PASTURE LAMBING

Many producers steer away from prolific breeds of sheep for pasture lambing because they don't want to have to walk all over their pastures to find ewes that need intervention. Janet McNally of northern Minnesota has found that by selecting for the best mothering abilities and utilizing specific management tools, she can successfully lamb triplets on the pastures of her farm. Janet points out that where winter feed costs are high, it is more important to work with prolific breeders to cover winter feed costs. High lambing rates per ewe keep the number of ewes overwintered as low as possible while sending the most lambs to market. Lambing on pasture requires less infrastructure, and therefore lowers the cost of sheep production.

Janet cautions that it might take some time to transition your flock from barn lambing to lambing out on pasture. The ewes themselves will have to change their tendency to lamb in a crowd, but Janet contends that the ewes have the ability, they just need the right situation to express their natural mothering instincts, and the producer will have to learn a few new tricks as well. Pasture lambing usually is done with drift or set stock methods.

Drift lambing is when the ewes are moved every one to three days, and the ewes that have already lambed are left behind in a group. This style of lambing requires more management and skill. Drift lambing concentrates all of the newborns into one paddock, where they are easier to find and to process. This is especially helpful during inclement weather.

Set stock lambing is when ewes are separated into groups on pasture and lamb in that group. This system requires the farmer to check every paddock, every day, for newborns, so it will probably take more time and more walking around the farm. This method is usually used with large numbers of ewes and reduces the chance that there will be mix-ups between litters of triplets since the newborns aren't all concentrated in one area.

After lambing, rotational grazing can begin at 30 days. Groups of ewes should be combined to make three groups: those with triplets, twins, and singles. The ewes with the triplets should be allowed to graze the pastures first, followed by the twins and singles. At some point the twins can be grouped together with the triplets to ease in management.

Janet's three main reasons that a producer might want to try to convert their prolific flock from barn lambing to pasture lambing are economics, nutrition, and labor.

> **Economics.** Pasture-based systems can lower feed costs by reducing the amount of harvested feeds. Farmers can invest their capital in land and livestock, which appreciate in value, and not in buildings and equipment, which depreciate in value.

Nutrition. Ewes raising triplets in a barn system require free-choice high-quality hay and 2 pounds or more of corn per day. Ewes raising triplets on high-quality pasture only need to be supplemented with 1 pound of corn per day to improve weaning weights and maintain good body condition. Feeding hay indoors is labor-intensive and costs between $0.53 and $0.60 per day per ewe. Compare that to about $0.10 per day (or around 1/5-1/6th the going price of hay) on pasture without the labor of feeding and handling hay. Also, good pasture is more palatable than hay, so more milk production could be expected on well-managed pasture.

Labor. Janet thinks that labor savings might be the best reason to move your prolific flock lambing to pasture. In a pasture environment there is no time spent feeding hay to sheep and very little time spent moving them. Observation in this system is much more important, and daily rounds include processing newborns. Janet admits that a shepherd will have to do quite a bit more walking in this system, but contends that this is a more pleasurable way of working, when compared to handling hay and hauling buckets.

In northern states, the high overwintering feed costs make it economically necessary to keep the bare minimum number of animals through the winter. With prolific ewes that lamb at the start of the late spring grass season, the producer has a population explosion at the time of the most abundant feed. Prolific sheep help the producer match the spring forage curve.

It is important that the kid receive colostrum within 6 to 12 hours of being born. Frozen colostrum can be gently warmed and fed to the kids if they don't get colostrum from their mother.

Kids can be breach-born successfully if it happens quickly. If it goes slowly and the umbilical cord is broken there is a chance of suffocation. The normal birth position is upright with the front legs first and the head between the legs. It can take a few minutes to a few hours, but it is important to see progress or else you may have to intervene.

WEANING

Kids and lambs can be weaned at between 90 and 120 days. They should be weaned as a group and monitored closely to ensure their health at this time. Coccidiosis is more common at weaning and can cause dehydration, poor performance, and even death if not treated.

Janet expects her triplet-rearing ewes to eventually drop to a body condition score of 1.5 (out of 5 for sheep). Early weaning of the triplets to enable the ewes to regain body condition for winter weather and to rebreed is the most economical method. As mentioned earlier, your ewes and does should get back to a BCS of between 3 and 3.5 for the best rebreeding success. Early weaning on pasture is about 70 days as opposed to 110 days for normal weaning. Janet recommends vaccinating for enterotoxemia and worming at 6 weeks old and just prior to weaning. Move the flock into the weaning paddock the night before weaning. Keep stress to a minimum and move the ewes to a mature pasture or feed hay for a week. Weaning will go more easily if the lambs and ewes are separated enough so that they cannot hear each other.

Lambs can then be run on the best pasture with the ewes cleaning up behind them. Lambs should only graze the top one-third of pasture: This ensures they are steering clear of parasite eggs located lower on the grass blades. The ewes and lambs should be separated for at least 3 weeks before they can be grazed within sight of each other. In order to get the lambs to slaughter weight in the fall, supplementation may be required. This could include a half-pound of corn or corn distiller's grains per animal daily, which will go a long way without reducing the amount of forage consumed. Janet has moved away from grain finishing as she has improved her forage quality over time and now will feed turnips if her animals have an energy gap in the fall.

Housing

Some shelter is necessary in northern regions for times of rain, wind, and cold. Access to trees and forest is good, but is often not adequate shelter for your animals in the wintertime. Without shelter, most animals will survive, but they will not gain weight and can lose body condition. Pregnant ewes and does should be given priority for winter shelter. A simple three-sided structure that sheds water works just fine. A mobile structure is even better, so you can periodically move it to limit animal impact and parasite buildup in one area.

If predation is a serious issue where you farm, you may want to create a system in which at night the animals come into a barn that is secure from predators. However, you will have to watch the manure and parasite buildup in and around the building.

Handling Equipment

Sheep and goat handling in "make-do" corrals is not only hard, difficult work, it is outright unpleasant, and often results in important jobs like vaccinating and deworming being delayed or not getting done at all.

To ensure that the handling facility will accommodate all the required processing and sorting, make a complete list of the operations that will be carried out, and plan how these jobs will be done. A useful checklist includes: tagging, castrating, shearing, milking, sorting, deworming, vaccinating, body condition scoring, pregnancy scanning, hoof trimming, hoof bathing, weighing, loading, and selling.

Compared to sheep, goats are more gregarious and inquisitive because of their superior intelligence, but they also tend to stress more easily. Use of dogs should be held to a minimum. Working goats is definitely not a "hurry up" task. In fact, the faster you go, the longer it takes. Well-trained herding dogs can be quite useful, however, when handling sheep.

For separating goats or sheep, a cutting gate can be mounted at the head of the working chute, or a cutting chute can be erected for this purpose. More elaborate designs for handling large herds are available but probably should not be attempted until you have enough experience to know what will work for you. This will allow a particular fit between the farm, the goatherd, and the owner.

A good handling system for sheep and goats will include a larger gathering or holding pen, then a forcing pen where they begin to narrow down the flock into a smaller group, then a treatment chute where they are forced to walk single-file and smaller gates can drop down to treat individual animals, followed by various sorts of pens where animals go after their various treatments. These can also be used to pick which animals are going to slaughter or sale, and a scale can be built into the chute for weighing.

Feeding and Watering

Since sheep and goats have different food preferences, they can be raised together successfully. Sheep prefer forbs (broadleafs) and goats prefer browse (shrubs, trees). Graze them with cattle that prefer grasses and you can raise more animal units per acre than you can with any one animal alone. Sheep and goats consume 1 to 2 gallons of water per day per 100 pounds of body weight.

A longer discussion of managed grazing can be found in chapter 5, on cattle production. However, there are a couple of key differences when managing forage for sheep and goats. First, the animals don't live as long as cattle. Many farmers slaughter their feeder stock at between 6 and 12 months of age. That may mean you can get away without using any stored forages at all, or with using just a little.

Sheep prefer and gain weight better on pastures dominated by grasses, forbs, and legumes, although they will eat some weeds and brush. Goats are the opposite: Throw them into a field of weeds, brush, and sprouting trees and they will be in their element. If your pastures are a little bit of both vegetation types, you can easily graze both animals together and they will work different strata of the vegetation. You can encourage the growth of browsing plants by letting your pastures rest longer than you would with cattle. You could also coppice older trees (that readily resprout) to encourage the growth of new stems. Goats especially love this, and sheep and cattle will also eat many sprouts of trees and shrubs. Don't worry about what your neighbors think: A manicured lawn is *not* what a sheep or goat pasture should look like.

In many areas where sheep are produced it is very difficult to finish lambs on late-season pastures, which have lower nutrition and palatability. Grazing annual plants is a great way to provide high-quality late-season forage that will allow for the gains needed to finish the lambs before winter. Tilling the ground for planting will also provide a future forage crop that is much more free of parasites. For farmers who are stuck in a situation of carrying an entire group of lambs through the winter, finishing some or all of the lambs on planted annual forage will likely be a good option. Of course, to do any planting, you need equipment and either ample summer rainfall or an irrigation system. Crops such as alfalfa, ryegrass, turnips, and small grains can be utilized for this purpose, as well as residues from crops, including corn. Stockpiling forage can be done for sheep and goats to help them get through the winter. They don't however dig through as deep of snow as cattle. Finishing lambs in irrigated orchards and vineyards with grass and clover alleyways can work well during certain times of the year. Many sheep farmers in Oregon finish lambs on grass seed farms in the Willamette Valley during the off-season for seed production. In both of these cases the situation produces a win-win for the farmers involved.

Fencing

Fencing is the number one priority for goat and sheep production. The goal is not only to keep your animals in where they should be, but also to keep out the numerous predators that love to attack and kill caprids (dogs, coyotes, and cougars especially). Goats and sheep both can jump over fences that are not tall enough, and goats will climb over just about any type of fencing. Goats are especially destructive on fencing and can use their horns to pull it apart. Electric fencing for both of these species needs to be very hot. Woolly-faced sheep can often push under electric strands without experiencing a shock, and so can heavily haired goats. We once watched as a woolly sheep pressed up against a five-strand electric fence while all the rest of the flock jumped through it. Talk about a "sacrificial lamb."

Without offset energized wires, goats will climb on woven-wire field fencing, and bend it and eventually destroy it. That is a big investment to have fall apart. However, you can put an electric offset on field fencing to keep goats from climbing it. Premier 1 Supplies has a good design for a goat and sheep fence, which consists of a field fence with an offset hot wire down low and a hot top wire. This prevents the animals from climbing on it and keeps predators out more effectively. A physical barrier like this, with an offset wire, works great for perimeter fencing. Electrified netting is often the "go to" solution for temporary fences for both of these species.

Welded panel fencing can withstand the climbing, but it is very expensive and needs to have smaller spacing than cattle panels so the goats and sheep don't get their heads stuck in the squares. We have seen welded wire fences and even welded panels completely torn apart by large Boer goats. This could have been prevented with the addition of one or two strands of offset electric wire.

Electric fencing works well, but you should have at least 6,000 volts to give the goats and sheep a good shock and use thicker wire to conduct the charge better. Some producers, such as Paradox Farm in Minnesota, use five strands of electrified poly wire for both goats and sheep. Their system works because their animals are trained to it when they are very young, so they have a healthy respect for the fence. In areas with potentially high predation potential, such as riparian corridors, the Paradox Farm owners use an electrified net fence with horizontal and vertical strand instead.

Keep in mind that for any electric fence to be hot, you need to keep the weeds off of the lowest electrified strand and you need to be vigilant in removing any tree branches that may fall on it. A couple of ways to keep the weeds down are periodic weed whacking, using a brush hog to mow a line before you erect the fence, or periodically turning off the fence so the animals can graze the vegetation growing along the bottom strand. We used to do the last method with our Dorper sheep and it worked pretty well. Snow and wind can also push down an electric fence, so you may need to reinforce it during the wintertime if you are still running animals outside.

Keeping Healthy

In most regions of North America, and especially in areas with long periods of warm and humid weather, internal parasites are the leading cause of disease in sheep and goats. Goats are less immune to parasites than sheep. Sheep gain some immunity after their first 10 to 12 months of age, but they lose their immunity to worms for about 4 weeks after lambing, so post-lambing might be a good time to worm infected ewes. There is no benefit from worming the mothers while they are gestating: The worms won't be passed onto their offspring.

Parasites

The following are the main types of parasites affecting goats and sheep.

> **Barber pole worm** (*Haemonchus contortus*). This parasitic nematode causes anemia, bottle jaw, and sometimes death in sheep and goats. It has a relatively short life cycle of

Electric Fencing

Electric fencing can be used for all the livestock and poultry species covered in this book. It is our preferred method of interior and temporary fencing for several reasons, although permanent perimeter fencing is better if you can afford it. Electric fencing's advantages are that it is portable, cost-effective, flexible, and good for predator prevention. It is **portable**, which means you can easily move it to accommodate rotational grazing, multi-species management, and integrating animals with other crops. The portability aspect is also important if you are leasing land: Who wants to invest in permanent fencing for land you don't own or don't have long-term tenure on?

The second reason is **cost**: Installing good permanent woven-wire fencing is expensive (labor and materials costs combined) and electric can be cheaper. Unless you are installing permanent high-tensile smooth wire fence, electric fencing doesn't require the fancy gates, H-braces, treated wood posts, or any of those sorts of things. A well-built barbed wire fence is expensive and requires more frequent posts, and they are not very effective, especially for sheep or goats and for keeping out predators. If you were just raising cattle then barbed wire perimeter fencing might be the way to go, but not for sheep or goats.

The third reason is **flexibility**. Electric fencing can be moved. It also allows you to experiment, try different animals, get to know your land better before you throw up something permanent. It allows you to use animal impact in certain areas and then fence them out of those areas later on. For example, you could try a short intensive grazing of a dry streambed in the summer and then get the animals out of that streambed when it runs again in fall. The possibilities are wide with electric fencing.

The fourth reason is for **predation**. Electric fences are not only more effective in keeping your livestock guard animals in with your flock or herd, but they also send a more powerful message to would-be ground predators.

There are five main components to an electric fence system. The first is the **energizer**, which generates the electric pulse that is sent through the **conductor**. The conductor is your fence wire and can be composed of wire, rope, tape, and twine that has metal filaments in it to conduct the electricity from the energizer. **Insulators** (usually made of plastic or fiberglass) hold the wire at a desired height and away from other conductors (metal posts, stray wires, wet grass, etc.) that will "ground out" your fence. The **posts** hold the insulators or can serve as both post and insulator if they are made of plastic or fiberglass. The **grounding system** acts as an antenna that completes the circuit from the conductor, through the animal and the earth, and back up through the grounding system to the energizer.

Energizer input can be plug-in electric (usually 110-volt), 12-volt battery powered, or AC/DC. Battery-powered energizers are often used with a solar panel to keep the battery charged: They can be used in remote sites with no access to electricity. If your only choice is to use a battery-powered energizer, then of course that's what you should use. If you have access to electricity, a plug-in charger is the easy choice. Plug-in chargers are cheaper, more powerful, more resistant to weed pressure, and they need less upkeep. AC/DC chargers offer the flexibility of using your energizer as both plug-in and 12-volt. Grazing consultant Jim Gerrish recommends that you size your energizer based on the length of fence line you plan to energize. He suggests 1 joule per mile of fencing (a joule is a unit of measurement for voltage x amps x time).

Wide-impedance energizers will perform well in dry or snow-covered conditions. For areas that have tall, green summer grass, a low-impedance unit is the best choice. The cost of the energizer is usually a small percentage of the cost of the entire fence, so dig deep and spend some money on an effective model from a respected manufacturer.

Electric fence **conductors** are rated by resistance (in ohms), durability, visibility, portability, strength, and price. Resistance is measured in ohms per 1,000 feet; a lower number is better. Don't buy cheap twine, tape, or rope conductors with high resistance. You want a low-resistance conductor that will lose very little energy over the length of the fence line. Choose a conductor that will be visible for the type of livestock you are going to raise.

Once you have chosen the type of conductor you will use to build your fence you need to decide how many "strands," or horizontal wires, you will need to contain the type of livestock you will be raising. For portable, multi-strand fencing for sheep and goats or when fencing poultry, use the type of electric net fencing that will best fit your needs. Netting has nonconductive posts built in, and is quick and easy to set up and move.

Insulators are made of nonconductive material and can be attached to wood and metal posts. Offset insulators hold the conductor away from the posts, and there are specially designed insulators for corner and line posts. Step-in posts and netting posts act as a post and an insulator in one. Insulators and conductors can be added to an existing permanent fence to make it more effective.

Improper **grounding** is the usual culprit when the fence isn't functioning properly, especially during dry conditions. The ground rods (made of copper or galvanized steel) need to be connected to the energizer using a heavy-gauge insulated wire. Use a non-corroding brass clamp to connect the ground wire to the ground rod. It is important to keep the soil around the ground rod moist. Again, Gerrish's rule of thumb is 3 feet of ground rod for every joule of energizer output. So a 3-joule energizer would need 9 feet of grounding rod. You could split that into three 3-foot lengths of grounding rod spaced around the fence line or one 6-foot-long grounding rod and another 3-foot-long rod. Gerrish recommends spacing your grounding rods throughout the network of fencing, not all clustered around the energizer.

Locating the ground rod under the eaves of an outbuilding will increase soil moisture around the rod. In very dry conditions, leave a hose dribbling next to the rod or place a 5-gallon bucket of water with a slow leak next to the rod. The moisture will increase the conductivity in the soil and will improve the effectiveness of your fence. In dry regions, long ground rods that can reach down to soil moisture work best. Multiple ground rods can also help in some situations. For very dry conditions a Pos/Neg fence, where some wires act as an extension of

> the ground, might be your only choice. For the animal to be shocked, it must touch both a ground and an energized wire at the same time when utilizing a Pos/Neg system.
>
> ### Training Your Animals to the Fence
>
> Livestock get trained to electric fencing by appearance, site location, and pain memory. If it's a good fence with a hot charge, they respect and avoid it. On the other hand, new animals just out of the livestock trailer may charge straight through your electric fences. That's why strong, tall, visible fences are essential for receiving corrals and feedlots. Temporary fences that are not physically strong pose the greatest risk of escape for newly acquired animals. It pays to train them to electric fencing inside a permanent fence. For example, you can put an electric offset wire on the inside of a woven wire fence or hog panels to train them to not mess with the fence. Once they get trained to that, they can be released into temporary paddocks with full electric fencing.
>
> We had a few crops of weaner pigs that we purchased that had never experienced electric fencing before. A few of them tried running right through the two or three strands of poly wire fencing that we had set up and gave themselves a great squealing shock in the process. When they got to the other side dazed and confused they looked around and noticed their mates weren't with them and then charged right back through the fence to their herd, giving themselves another great shock. That's all it took for them to "learn" the fence.

approximately 3 weeks and thrives in warm, humid conditions. Grazing animals pick up infective larvae on forages that are relatively short. Once in the rumen the larvae continue development, travel to the true stomach, and become adults. The adult female can lay thousands of eggs daily and can consume 200 microliters of blood daily. The eggs are deposited in the feces, hatch on pasture, and the life cycle begins again. Outbreaks are worst when warm summer rains break up the fecal pellet and create a moist environment for the hatched larvae.

During drought or very cold conditions, a majority of larvae become dormant or die and transmission to the animal is very low. Sheep are more susceptible because they graze so close to the ground, but goats will pick this worm up too if they are forced to graze (when there is no browse or live in confinement). The barber pole worm is associated with overgrazed pastures, because it thrives on short forages.

Brown stomach worm (*Ostertagia circumcincta*). This parasite feeds on secretory cells in the abomasum and causes loss of blood plasma. It is prevalent in areas with abundant winter rain. Symptoms include diarrhea, poor performance, reduced appetite, and bottle jaw.

Bankrupt worm (*Trichostrongylus axei*). This parasite feeds on mucous in the small intestine. FAMACHA scoring doesn't work for this worm. Symptoms include diarrhea, reduced appetite, and poor performance.

Tapeworms or cestodes (*Platyhelminthes* spp.). These mostly affect young animals in confinement systems. They can be identified as appearing like little pieces of rice in the animals' feces. They don't live off the animal, but they absorb food in the intestines and in this way compete for nutrition with the animal. Symptoms include slow growth, poor performance, and enterotoxemia. These worms are not considered as much of a problem as the other parasites. Usually if the animal is in severe distress it is because of the cumulative effects of tapeworms and barber pole worms together. In extreme cases they can cause intestinal blockage.

Lung worm (*Dictyocaulus filarial, Muellerius capillaris*). Low-lying pastures and wet areas in regions with high rainfall and cool weather are the breeding ground for this parasite. The animals ingest the eggs and the larvae travel to the lungs. Signs include fever, nasal discharge, coughing, rapid breathing, and poor performance. A secondary bacterial infection may cause death.

Liver flukes (*Fasciola hepatica*). Liver flukes can cause death or liver damage if not treated. They are prevalent in California, the Pacific Northwest, and the Gulf states. Snails are the intermediate host and infected animals are treated with flukicides.

Coccidia (*Imeria* spp.). These single-cell protozoa cause damage to the lining of the small intestine. They are host-specific and common in sheep, especially in young or growing lambs, and they thrive in warm and wet conditions. Keeping your animals out of areas with a lot of feces and out of wet areas will help. Fixing water line leaks or moving the animal waterers frequently to prevent wet, muddy conditions is also a great prevention tool. Coccidiostats in feed can also prevent coccidiosis, but there are some herbal remedies as well that show promise, such as garlic and black walnut. Symptoms of coccidiosis include diarrhea, weight loss, loss of appetite, blood in feces, and anemia. If left untreated, coccidiosis will eventually cause death.

Diagnosis and Control of Parasites

Selective treatment or deciding which animals to deworm can be determined by the use of FAMACHA, although it does not work for all nematodes. FAMACHA (FAMA for Dr. Faffa Malan, CHA for chart) was developed by a group of veterinarians and scientists in South Africa and was validated in the southern United States by members of the American Consortium for Small Ruminant Parasite Control (ACSRPC). A complete description of FAMACHA can be found on the ACSRPC website: www.acsrpc.org/Resources/famacha.html. Briefly, FAMACHA is a tool used by farmers that consists of examining the color of the lower eyelid and matching the color on a chart that ranges from red (healthy) to almost white (anemic). The lighter the color, the more anemic an animal is. Anemia occurs as a result of the adult parasite removing more blood than the animal can replace. There are other causes of anemia, such as poor diet, which should be considered before deworming the animal.

Whole flock deworming will only lead to parasitic resistance to the drugs. To avoid this, you can FAMACHA-score each animal in your herd, and then administer dewormers on a case-by-case basis. Individuals in your herd that show less resistance to parasites should be culled.

To prevent parasites in your goats and sheep, consider planting or grazing on *Sericea lespedeza*, which is a tannin-rich plant that likes to grow in the warmer climates. Where copper is limited in the soil, a copper oxide capsule inserted (forced down the throat) once a year in sheep and goats seems to be effective against barber pole worm, but not other worms. Don't use if you already have excess copper in your soils. Some research indicates that raw garlic juice, either administered orally or as a drench, can be effective for controlling many parasites in sheep (Masamha et al., 2010).

Encouraging your animals to browse instead of graze all the time will also help. You can accomplish this by not allowing your pastures to be overgrazed, allowing some taller weedy species to come in and get established, and allowing some access into brushy and forested areas at all times. Don't just confine your animals to pure grass, especially goats, as they have less natural resistance to parasites and will pick them up if forced to graze low on grass. In addition, keep your animals out of wet patches and marshy areas during the hottest, most humid part of the year. If you have leaky hoses or waterers, fix the problem, because wet patches host higher rates of parasites.

According to Janet McNally, grass needs to be managed in a way that keeps the caprids from eating large amounts of parasite larvae and coccidia oocysts that live in feces. Six to eight weeks of rest during warm humid weather is enough time for the majority of parasite larvae to die off. Clipping pastures, grazing other species (such as cattle or horses), or making hay are all ways of reducing parasite loads while also keeping forage species in a vegetative and palatable condition. Incorporating and encouraging the establishment of anthelmintic plants like birds-foot trefoil, chicory, and *Sericea*, which are natural wormers, can also reduce parasite loads in your flock. Grass that is grazed at a taller level will reduce parasite ingestion, as will waiting until later in the day to allow sheep into the pastures after the grass has dried off and the parasite larvae have climbed back down toward the soil. Fencing off standing water and providing a clean bed for the sheep to lie on at night will reduce the possibility of sheep being overcome by coccidia.

Sheep and Goat Illnesses and Diseases

These are some of the main illnesses and diseases that can affect sheep or goats.

- **Caprine arthritis encephalitus (CAE).** A viral disease in goats that produces arthritic-type symptoms. Newborn kids are infected by their dam when nursing colostrum or milk. Make sure you are bringing in CAE-free animals; you can do a blood test for antibodies.
- **Chlostridium.** Both goats and sheep are susceptible to this family of diseases, which includes tetanus and the enterotoxemias. Enterotoxemia causes scours, usually very early in life, but can affect animals of all ages. Prevention is far more successful than treatment. Ewes and does can be vaccinated 30 days prior to lambing or kidding.
- **Flystrike.** Various species of flies lay their eggs under the skin or in the wool of animals. When the eggs hatch, the larvae (maggots) eat their way under the skin. This can cause irritation, infection, and even death if left untreated. It is more common in humid, warm environments but is increasingly being seen in more northern locations. Heavy,

wrinkled wool breeds are more susceptible. Prevention includes choosing less wrinkled breeds, culling infected animals, shearing around the breech area, tail docking, and chemical dips.

Foot rot. Moist and warm conditions will cause this bacterial infection of the hooves of sheep and goats (see Hoof Trimming, below).

Ketosis. Pregnancy toxemia, or ketosis, occurs when the under-conditioned ewe's or doe's body starts metabolizing her own fat to produce enough energy for carrying twins or triplets. Overly conditioned females can also have ketosis, because the amount of fat in their gut doesn't allow for them to fill up with enough food. The ewe's or doe's breath will often have a sweet smell, and she will be lethargic and off her feed. Ketosis is often treated with propylene glycol drenches.

Mastitis. Bacterial inflammation of the udders, most often from rough treatment and unclean milking practices and close confinement. Conventional treatment calls for antibiotics. Organic producers report success with garlic tea when the infection is treated early. Keeping living areas clean and dry, as well as restricting ewes' and does' feed and water for a couple of days after weaning (to reduce milk production), will reduce the occurrence of mastitis. Iodine teat dips also kill mastitis-causing pathogens.

Scrapie. A degenerative nervous system disease that is always fatal in both sheep and goats. Prion proteins that cause scrapie are shed in birth fluid, feces, and other animal fluids. When you purchase sheep or goats, make sure they come from a scrapie-free flock and that they have the scrapie ear tags. Sheep can be tested for scrapie resistance.

White muscle disease. Selenium or vitamin E deficiency will cause this disease; symptoms include stiffness and pain in walking. Lambs may tremble in pain when standing. Test your soils and/or forages to find out if you have shortages of either of these minerals.

Other Body Modifications

HOOF TRIMMING

Goats and sheep need to be on the move in order to graze and browse, and unhealthy hooves will make it difficult for them to feed themselves. Overgrown hooves can lead to foot rot and other complications. Ensure that animals are current for tetanus before trimming hooves. Moisture softens hooves and makes trimming easier, so trimming them after a rain or heavy dew makes the job easier, as does trimming hooves on a regular basis. Try to time hoof trimming with other treatments and processing, and avoid trimming hooves during hot weather or late gestation.

Trimming hooves may not be necessary if you give sheep and goats access to rocky areas, or put down gravel in high-traffic areas such as around their waterers and feeders. Without rocky areas on the farm, it will most likely be necessary to trim the hooves of breeding animals that will be kept for several years. Black hooves are generally considered harder than lighter-colored hooves. Boer goats will likely need their hooves trimmed even in dry and rocky conditions.

CASTRATION

Banding is a very humane way to castrate bucklings and ram lambs. It's not always 100 percent effective; part of a testicle might be missed or the band or ring might break, but these cases are rare. Intact ram lambs and bucklings grow faster than castrated wethers, and the ethnic markets usually prefer intact males. On the other hand, intact males are more difficult to manage and may cause unwanted pregnancies. Some markets might discriminate against intact males. If ram lambs or bucks are left intact, they should be separated from their dams and female mates by the time they are four months of age.

DISBUDDING

Use a hot iron when the kid or lamb is very young to remove the horns. Horned animals can get their heads stuck in fencing; they can also be dangerous to humans and more dangerous with each other when fighting. Horned animals can cause bruising and tissue damage to other growing animals that may negatively affect their meat quality, especially when bunched in the livestock trailer on the way to slaughter. Disbudding is not really necessary with meat goats that don't get to live that long and whose horns won't develop that fully, but should be considered for bucks or rams that you will keep for several years. If you have a market for horned skulls, such as with four- or six-horned Jacob sheep, then leave the horns on the ramlings.

Handling and Slaughtering

When to Harvest?

You can view the optimal finish weights in the breed chart at the beginning of this chapter. Obviously, your particular animals may finish a bit differently and your markets may prefer a different-sized animal. In general, ethnic markets prefer smaller animals because they typically buy the whole animal, either live or as a fresh carcass. Restaurants like smaller animals too, so that the individual cuts are not too large (such as the chops) and the meat remains mild and tender. People of European ancestry sometimes like their animals a little larger and some even prefer older animals, such as mutton. Also, if you are selling by the whole or half, you should probably aim for a larger animal. Not many people want to order a half lamb and get back just 17 pounds of packaged meat or so.

Cabrito is goat meat from very young, milk-fed goats weighing between 15 and 25 pounds live. Typically cabrito is from kids that are harvested at 4 to 8 weeks of age and the meat is most often used for BBQ. **Chevon** is meat from young goats, approximately 6 to 9 months in age, often fed a forage and/or grain diet and weighing around 50 to 60 pounds live. Cabrito is typically more tender than chevon. Meat from goats that are larger than chevon can be of value as well and is typically sold to certain ethnic consumers. Goat meat is generally leaner than sheep meat and will appeal to a certain group of customers for that reason. It is increasingly seen on the menus of high-end restaurants too. Dairy breed kids will have a lower meat-to-bone ratio. If your does are a dairy breed, you might breed them to a meat-breed ram to produce more "meaty" kids.

Sheep will tend to accumulate exterior fat as they get older, which is something that some of your customers may not want. As they surpass about 100 pounds live weight, many breeds of sheep will accumulate only fat cells, especially if you are feeding grain. You can keep them leaner by using only forages. Goats will put on interior fat as they get older; this is not really a detriment, but it's not something you can capitalize on either. If your goal is to raise mature animals for meat, you will have to be careful not to have them eating high-energy feeds or forages, so they don't get overly fat as they get older.

Vanguard Ranch

Renard Turner, owner of the goat farm Vanguard Ranch Ltd. in central Virginia, uses mostly native forages in his goat pastures, although he overseeds his pastures with clover and *Sericea lespedeza* by frost-seeding in the winter. He recently cleared a dense stand of pines, which he also overseeded. He will allow his goats to access this area in about 3 years, when there is plenty of browse along with the overseeded species.

Renard's feeding plan changes with the seasons. In the summer the bulk of the animals' feed comes from native grasses and forbs: The native forage in Virginia is excellent according to Renard. In the fall the goats eat kale, turnips, and other forage brassicas which have perennialized and self-seed on his farm. In the winter Renard tries to feed hay for no more than 3 months and is always working on reducing this time frame. In the spring the pastures come alive with clover, chicory, plantain, native grasses, kudzu, honeysuckle, blackberries, and pigweed. What some might consider unsightly weeds are actually ideal goat browse and many are natural anthelmintics.

Renard feeds his lactating does a little cracked corn out of necessity, but would like to move away from the practice altogether. He asks himself, "Which came first, the goat or the feed store?" Renard is attempting to breed for does that don't need supplementation of concentrated feed at all.

Renard likes Myotonic goats because they are easier to handle and more resistant to parasites. They are shorter-legged and easier on fences. Their smaller size also makes them easier to handle for his wife and smaller people. They have the highest meat-to-bone ratio for goats. Preliminary studies have indicated that Myotonic goats may have a meat-to-bone ratio of up to 4 to 1, compared to ratio of 3 to 1 for most other

New Zealand sheep genetics have been selected to finish at 80 pounds live weight at around 8 months old. This is the "restaurant standard" for flavor, tenderness, and a carcass that isn't too fatty. It also makes for reasonable portion sizes, especially the chops. Jamison Farm of Pennsylvania harvests their lambs fairly young, because restaurant chefs have come to expect this level of tenderness and portion size. This is the classic "Spring Lamb" size—an animal born in the spring and slaughtered in the late fall around 8 months old. However, if you are harvesting for home cooks, such as sales at farmers markets or as halves and wholes, people may be looking for slightly larger portion sizes. We used to raise our Dorpers up to around 120 pounds live weight, around 12 to 14 months old, which seemed to be the perfect size for farmers market sales. Butcher Adam Danforth advocates for consuming older animals: He prefers sheep around 2 years old for their depth of flavor.

Renard Turner, the farmer profiled above, harvests his goats at a fairly young age of 6 or 7 months, when they are only 60 pounds or so. This is because he wants his goat meat to be tender, but also because goat growth stagnates a bit after 6 months anyway and his land can only support so breeds of goats. Renard crosses Myotonic bucks with either full Myotonic or Myotonic/Kiko does. They grow a little more slowly than the big Boer and Savannah goats, but require fewer inputs and have great cutting percentages.

Renard uses FAMACHA scorecards to determine which animals to deworm. He culls animals that always seem to be unhealthy, saying sometimes you just have to "be the wolf" by taking down weakened animals that should not be kept. Indeed, research shows that a small number of animals in each flock shed the most worm eggs (that is, 10 to 15 percent of goats in a flock might contribute to over 80 percent of the parasite loading). Renard's "take-no-prisoners" approach to improving his herd is rewarding him with hardier and hardier animals that are more naturally resistant to parasites. Renard's other poignant saying is, "My does are named Cash and my bucks are named Flow, and if they need help, then they need to go!" He does not overmedicate, grain-feed, or assist with birthing; all of his practices contribute toward a hardier flock and a more profitable business at the same time.

Renard has been very successful in marketing his goat meat and advises farmers to find the right crowd who will pay your price without blinking. He markets the majority of his goat meat via prepared meals that he sells through his mobile concession trailer at fairs and food festivals. Because he adds value to his goat meat by cooking it in rich, delicious dishes such as goat curry, the average price he earns for his goat meat is $44 per pound.

He says that according to the data, there is more demand for goats than supply in this country, but prices on live animals tend to not be very high, often in the $100 range. He can do much better than that and raise fewer animals by taking the goat meat all the way to the eaters.

many growing animals without diminishing his forage base.

In northern climates it is imperative to get lambs off to slaughter before winter weather sets in due to the high costs of holding lambs over winter. This creates a shorter time frame to get your animals to market weight, but is achievable with the excellent grass conditions in the north. Since the fall is the busiest time for most slaughterhouses, schedule your slaughter dates early.

Keep in mind that, regardless of the size of your animal, the slaughterhouse is going to charge the same kill fee. So you can pay $35 to kill an 80-pound lamb that yields 28 pounds of meat (that is $1.25 per pound) or $35 to kill a 120-pound lamb that yields 42 pounds of meat (that is $0.83 per pound). Forty-two cents a pound difference may seem insignificant, but it all adds up. You also have 14 more pounds of meat to sell from that animal, which is more like a difference of $122 (if you averaged $8 per pound for lamb meat). Of course, this is all dependent on the preferences of your market. If you can sell a smaller animal for more money that covers this difference in harvestable meat, then by all means go for it. This illustrates the powerful difference that harvest weights make to your bottom line.

Loading for Slaughter

After you have sorted your animals to determine which ones are going to slaughter and you have them in a sturdy holding pen, it's time to back the truck or trailer up to the holding pen gate. Ideally you have some sort of narrow enclosed ramp in which each animal walks up one at a time so that they don't try to climb on each other if they get spooked. Load gently and efficiently.

When you are just starting off with a small number of animals for slaughter, you could get by with a stake-sided truck or even a van, but you will quickly outgrow that. Look for a low-profile livestock trailer that is lower to the ground. Some even come with their own built-in ramps that fold up for transport.

It is important that all loading for slaughter be done as calmly as possible. Stress during the last couple days before slaughter will reduce the quality of the meat.

CHAPTER 4
Pig Production

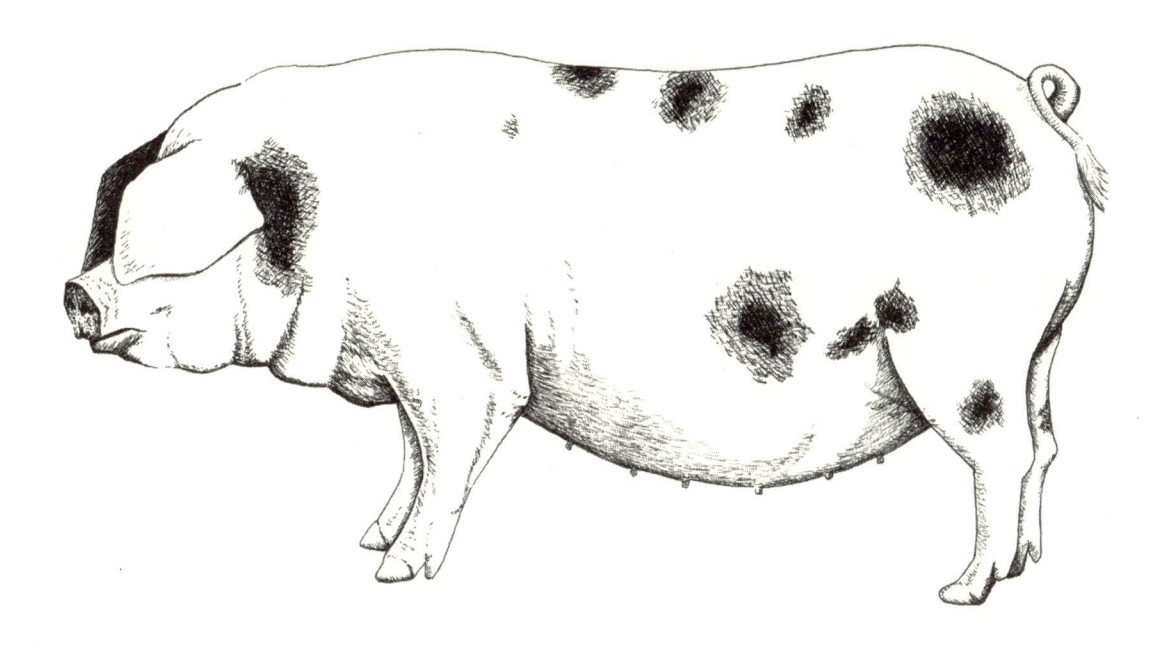

Pigs used to be creatures found on most diversified farms and were valued not only for their meat and fat but also for their roles as pasture renovators and food waste recyclers. Pigs thrive on dairy waste, cull vegetables, and other low-value feed, as well as forage and tree fruit and nuts. When we were raising large numbers of laying hens and pigs, our cracked eggs never went to waste, and we saved on feed costs by procuring cull vegetables from local farms and unsellable produce from natural grocery stores. Pigs are inquisitive, personable creatures that make a great addition to most farms.

In this chapter we will talk about a few different management systems for raising pigs, focusing on methods that are more humane and considered niche-type production. There are two ways to produce market-ready hogs: by breeding, farrowing, and finishing your own pigs; or by buying in partially grown pigs to finish on your farm. We will discuss some of the pros and cons for each of those

> **DEFINITIONS**
>
> **Barrow** a castrated male pig
>
> **Boar** an intact adult male pig
>
> **Colostrum** the first milk produced by a mammal; full of antibodies important for the newborn
>
> **Gilt** a female pig that has not yet given birth
>
> **Market hog** a pig that is ready for slaughter, usually weighing between 220 and 260 pounds and 5 to 7 months old; some people prefer to raise their hogs even bigger, between 275 and 325 pounds, so there is more intramuscular marbling and flavor to the meat
>
> **Sow** an adult female pig that has given birth before
>
> **Weaner** a pig that has been recently weaned, normally a minimum of 4 weeks old and up to 8 weeks old (15 to 30 pounds)

strategies, so you can decide which one would work best for you (or what blend of the two).

There are a lot of potential pitfalls in pig breeding and gestation both in terms of production and efficiency. We will give examples of what farmers are doing to minimize failures and maximize success. Farrowing is often the most labor-intensive time of the year for pig farmers and it is also the most important for the financial success of the producer. There is a lot to know about farrowing, so we will give this aspect of production a lot of attention and share what works for a variety of farrowing setups.

Pigs require shelter in most climates, either for shade or protection from the cold, so we'll discuss some options for housing, both stationary and moveable. We'll go over the watering and feeding concerns specific to pigs on pasture and in hoop buildings, and talk about some of the innovations that have been made by clever farmers. We'll also summarize the key points concerning fencing, feeding, and health management. Lastly, we will talk about individual breeds of pigs and their characteristics.

Management Systems

Deep Bedding or Hoophouse Production

Poly-covered hoop structures offer one of the best alternatives to confinement systems or full pasturing. This system consists of a barn (often a hoophouse) with a deep bedding pack that absorbs urine and manure. As the bedding composts it provides warmth. The bedding is refreshed as needed and removed between groups of pigs and composted for use as fertilizer, perhaps going back onto the same fields that grow the pig feed.

A hoophouse or barn can be sectioned off for farrowing stalls or you can bring your pasture farrowing huts inside. It can also be used to house pigs at all life stages, including grow-out. Barns can be further divided to allow for raising different-age pigs in groups, so they don't compete for feed. Some niche pig producers grow all or some of their market hogs in hoop structures because of the efficiencies in gain and production, especially during the winter months. They provide a lot of light and good air circulation and are easier to clean out than barn or stall-type setups too, making for a more comfortable environment for the pigs.

Many farmers use this type of system for growing and finishing pigs, but will farrow on pasture during the spring and summer. For example, Iowa organic farmer Tom Frantzen's sows farrow on pasture in the summer and then farrow in a barn in the winter, but all of his market hogs are raised in hoophouses. Innovative farmer and member of the Practical Farmers of Iowa Dan Wilson only uses a hoophouse for winter housing and keeps all of the pigs on pasture or dirt lots during the warmer months. Some farms, however, rely on hoophouses all year long and don't include any pasture in their production. If managed well, we believe even farms with 100 percent hoop production are a much more humane alternative to indoor confinement production (where typically pigs are on slatted or concrete floors with no bedding and are more densely stocked).

In the hoophouse model the farmer usually provides 100 percent of feed: The market pigs are not given access to pasture outside of the hoophouse. There may be a small outdoor yard (as required by organic rules for "outdoor access") but it usually doesn't include any feed sources. However, feed costs can be contained, because feed conversion rates are better in deep bedding systems than pasture-based systems. There is less feed waste and the animals exercise less, thus expending less energy. The variance in air temperatures is also less than outdoor production models, helping the pigs to convert feed more efficiently as well. Some farmers bring in green chop or alfalfa hay to give the pigs the benefit of eating some forage in their diet. As long as you manage the bedding well and the stocking rates are kept at reasonable levels, it is a low-stress environment.

The costs for this system include the building, bedding material, and a concrete pad for waterers with plumbing and electricity. If the house is going to be further subdivided or used for farrowing, you have to consider extra utilities and material to build the enclosures. Choose a design that has good ventilation for summer cooling and to reduce humidity levels that can reach up to 100 percent inside the hoop—pigs suffer with that kind of humidity. A well-designed sorting and loading system can be built in to allow for stress-free handling. A front loader is essential to clean out bedding.

Some benefits of this type of system include animal health and comfort; capture of nutrients; ease of cleaning with a tractor; good light and airflow; and ease of feeding, watering, and handling. Pastures can be rested when pigs are moved indoors, especially during wet weather when pig impact can be especially destructive. Less land is required in this system if you are not growing your own feed.

Disadvantages are the initial costly investment in the structure; eventually having to replace the poly cover; the cost of bedding material; and the need for increased manure management, labor, and equipment. Even when well managed, this system has a pretty high potential for odors and flies. If not managed well, it also has the potential for increased animal stress and disease pressure due to high stocking densities.

Pasture-Based Rotational System

In this system pigs are moved to a fresh paddock when they have impacted the pasture to a desired level or have run out of forage. While cattle and sheep are often moved on a daily basis when practicing rotational grazing, pigs on pasture are generally moved after a period of weeks or months, depending on the size of the enclosure.

Mob grazing, where pigs are moved more frequently with a multi-species herd including ruminants, would be an exception to this. Pigs thrive on pasture and certainly appear to enjoy themselves, as long as they are well managed. Pigs are great at renovating tired pastures and unproductive cropland. Co-grazing pigs with ruminants works well, either together or in a leader–follower situation; pigs disperse cow patties and aerate and fertilize the soil, while the grazers knock down the bulk of the grass. Usually electrified fencing is used to subdivide pastures.

Many Midwestern niche pork producers combine pig pasture in their field crop rotation. This can be done on any farm that raises pigs and also grows crops. The pigs fertilize the ground while breaking up disease cycles in the crop rotation. During the growing season, Iowa farmer Dan Wilson runs his sows on 16-acre strips in a 4-year rotation. Every year he rotates the production of each strip between corn, soy, small grains, and pigs. All four strips are fenced and have pig waterers even though pigs are raised in each strip only one season every 4 years.

Pigs can obtain 25 to 75 percent of their feed from high-quality pasture (and the various creatures that live in that pasture), but it requires a high ratio of land to pigs in order to satisfy more than half of their diet with forage. There is no formula for how many pigs an acre of land can support, and in most cases pasture-raised pigs are given supplemental feed, which further blurs the lines in terms of pigs per acre. For growing pigs, determine the best stocking rate by considering the type and size of pigs, the length of the grazing period, soil type and moisture, vegetation, and access to high-protein feed. Weaner pigs can be stocked at relatively higher rates, at least for the first couple of months.

The equipment needed to operate a pasture-based system includes moveable shelters, portable electric fencing, feeders, and portable or central waterers with lanes. All parts of the system must be easy to move or you won't move them often enough, and land will become degraded through continuous pig presence. If the pigs are on pasture during the winter in cold climates, frost-free waterers with buried water lines and more substantial shelters will be necessary.

Moveable pig shelters can be trailer-mounted or built on skids, making them towable with farm machinery or pickups. These handy structures enable the farmer to provide shelter in all the farm's paddocks. One of the simplest and most practical skid shelters is a three-sided, shed roof structure. All mobile structures must be built strong to withstand being towed over rough ground, as well as the rubbing and scratching of the largest sows. Skid structures can be carefully towed over a lowered two-strand poly wire fence, making moves between pastures a snap.

On pasture it is more difficult to feed, water, and handle the pigs, due to challenges associated with weather extremes, material handling, and site access. Maintaining the pastures in a green and growing state can be tough. Many farmers find it necessary to insert a nose ring in their sows in order to maintain pastures and limit rooting. Dan Wilson stopped ringing his sows in recent years and had to reduce stock density from twenty sows per acre to fifteen. He has found that orchard grass, because it develops a thick sod, helps to keep cover on the ground when dealing with unringed sows. Pigs will also flip over the soil, sometimes turning your once-flat pastures into an uneven mess, making life more difficult for other animals to walk around, for you to pull skid shelters over the ground, or for reseeding pasture. Wet areas can turn into giant mud wallows deep enough to lose your tractor in. These can be a pain in the rear to refill and repair, particularly if you are a low-machinery type of operation.

Upfront costs on pasture can be lower for housing, but the costs of fencing, water lines, transport, and feeding equipment can be relatively expensive. Housing may not be needed in warmer climates or where there is also access to the woods.

Pastured systems are more difficult to manage during the winter in cold regions, and pigs can easily turn the soil into a mud slurry that will take awhile to regrow forage. Often a seasonal sacrifice area is required to reduce the damage to the rest of the pastures, or seasonal hoophouse production can help relieve the pressure on the pastures in the winter. Pigs may be exposed to more environmental stresses such as cold, rain, wind, and heat in outdoor rearing systems. They may gain weight more slowly during inclement weather even while consuming more feed. There is a low to medium potential for odor and flies in a rotational pasture system. Appropriate stocking density and frequent rotations will keep odor, flies, and parasites to a minimum.

Woodlots

The forest or woodlot is the habitat that pigs prefer if given a choice because of the diversity of foods available to them and the protection from the elements. When managed well, pig impact can enhance the forest while producing a superior type of pork. By mixing and rooting around the forest floor the pigs may actually encourage understory growth by uncovering the latent seed bank in the soil. Woodlot-raised (or "forest-fed") pork is easy to market as a niche product. It is also a great way to utilize land that has limited value for other types of agriculture.

Raising pigs in woodlots requires fencing, waterers, and perhaps some winter shelters. This system might require a central watering system or development of one or more natural water sources. Pigs should be fenced out of natural water sources, and provided water in a way that isn't damaging to riparian areas. We do recommend utilizing natural water sources, but pipe the water to a tank or trough instead of allowing the pigs to degrade your valuable water supply. If you feed your pigs in the woods, care should be taken to spread the impact of this activity, or consider feeding them in adjacent pasture areas.

When high-quality mast (nuts, seeds, fruits) are available the pigs may hardly need any other feed (unless they are overstocked). Pigs can forage for a large percentage of their feed in the woods, but this is more seasonal than pasture and depends on the forest species. Hardwood forests have more mast-producing trees and support a richer understory than acidic, tannic evergreen forests. Mast-producing trees include walnuts, buckeyes, hickories, oaks, chestnuts, hazelnuts, hawthorns, beeches, crabapples, and more. Commercial fruit and nut orchards will also provide similar environments and diets for the pigs. If you have a chance to do some short-duration, high-intensity pig grazing in orchards, that can be a win-win situation for both the health of the orchard and your pigs. Fallen fruit and nuts can harbor pests—pigs can easily and happily vacuum them up.

As an experiment, we recently raised four American Guinea Hogs under a few acres of oaks in Oregon, and we didn't feed them anything for around 4 months. We even tried to get them to eat some melon, sweet corn, and other foods from our garden that pigs usually go crazy for, and they wouldn't touch it in favor of the acorns. That was the most affordable pork we ever produced (and it was tasty!). This is also the model used to produce the famous Spanish Iberian pigs, which are finished on acorns (aka Jamon Iberico de Bellota).

Woodlots can be very challenging to fence, especially when compared to pasture. Thick brush, trees, and uneven terrain make fence construction and maintenance difficult. Feed, water, and handling logistics need to be carefully considered, particularly if the terrain is hilly. There is a low to medium potential for odor and flies. Pigs may be exposed to environmental stress, although shade and protection from the wind is abundant.

Raising pigs in the woods can potentially be very destructive and should be monitored closely to avoid negative impacts, especially around the roots of the trees. Overuse of the forest can make it difficult for any tree regeneration to occur. If no new tree seedlings can get established, what will be the long-term future of that forest? We have seen several well-respected pig producers completely hammer their hardwood forests, resulting in extensive soil erosion, root compaction, physical damage to growing trees, and pigs eating all the new tree seedlings—thus preventing any future trees from getting started. Don't make the forest your sacrifice paddock: You must manage it well. Wet, rainy times are when pigs root the most; perhaps you can move them out of the woods during those wet times to keep damage to a minimum.

Dirt Lots

Raising pigs in dirt lots is a fairly common practice and usually consists of a permanently fenced area with a stationary shelter. This is not a system that we recommend, but since it is so ubiquitous, we feel that we need to include it in the discussion.

Keeping pigs in a system like this is relatively easy since all of the utilities can be set up permanently and aren't required to be portable. Handling and sorting infrastructure can easily be built into the system. This is the system with the lowest upfront costs, but also the worst animal health and the highest rates of stress of all the systems mentioned.

In the dirt lot model, the farmer provides 100 percent of the feed. Pigs raised in dirt lots will grow faster than those raised on pasture, but they will usually have more health problems and they don't spread their own manure like their pasture-raised counterparts. Disease can build up in the soil, and high internal and external parasite populations will require you to administer dewormers. It is difficult to manage and utilize the manure, because the pigs tread it into the ground. When the surface is scraped to remove the manure, it also removes a lot of dirt at the same time. This system has a high potential for odor, flies, and dust. It will become a mud lot when it rains.

Breeds

Of course it must be said that there are often more differences within a breed than between breeds, but we'll attempt to make some somewhat accurate generalizations about the different types of pigs. Instead of looking for a certain breed of pig, try to find pigs that will perform under the conditions on your farm and produce the type of meat preferred by your market. White pigs will sunburn in summer when raised outdoors. White pigs often don't have very much hair, which makes them less suited for outdoor production systems

TABLE 4.1. Characteristics of Different Pig Breeds

Breed	Weight of boars/sows (lbs)	Average litter size	Temperament	Use or market	Optimal harvest weight	Comments
Tamworth	600/500	approx. 10	active	• lean meat	260–275	• protective mothers
Gloucestershire Old Spot	600/500	6–10	docile	• lean meat • lard • cured products	260–280	• docile and friendly • good mothers
Large Black	750/650	8–12	docile	• lean meat • cured products	230–250	• docile • good mothers
Red Wattle	750/550	9–10	docile	• lean meat	260–280	• wide variability in breed • good foragers
American Guinea Hog	200/150	6–10	docile	• lean meat • lard • cured products	100–120	• friendly • great foragers

Adapted from the Livestock Conservancy website

but easier to scald and scrape on the processing end of things. Pigs from lines that have been raised in intensive confinement operations will generally be poor mothers on pasture and have a lot of their foraging drive bred out of them. Generally a heritage breed pig is going to be calmer than a lean and fast-growing "confinement-type" pig. Pigs can be really fun and friendly, or high-strung and aggressive—what kind of animals do you want to spend time with? Do you want pigs that will come up to you for a scratch or pigs that are more aloof?

If you are raising pigs for meat, you will likely want to cross-breed to capitalize on the heterosis (hybrid vigor) provided by your cross. If you plan on cross-breeding or buying in weaner pigs, you should have an understanding of the various pig breeds, their different temperaments, how big they get, their foraging ability, and their meat qualities. We will cover some of our favorite heritage and conventional pig breeds.

Heritage Breeds

Heritage breeds generally have more fat than conventional breeds and take longer to grow to market weight. They are suitable for niche markets and generally have more backfat and intramuscular fat, so you will have to be clever about using the whole animal and finding markets for the fat. Many of them are considered exceptional foragers, are hardy in a wide range of conditions, and possess excellent mothering abilities.

Berkshire. Medium-sized, long, black pigs with white points and prick ears. Although Berkshires are considered a heritage breed, the modern heavily muscled American

Berkshire looks nothing like its short-snouted British relatives. They have been bred to be large, heavily muscled, and fast-growing, just like other conventional breeds. The only difference is their ability to build considerable intramuscular fat marbling, making for tasty and juicy pork that is sometimes sold as "Kurubota pork." Good for using as terminal sire, lean meat production, cross-breeding, hardiness, and ability to produce in a variety of settings.

- **Tamworth.** Red pig with prick ears, long lean body, and long snout. Tamworths are more lean than other heritage breeds, and are known for producing high-quality bacon. They are good, protective mothers that often kneel and lie down gently. We have raised a lot these pigs and find them to be strong foragers and very hardy. We love the cross of Tamworths with Gloucestershire Old Spots, both for market hogs and replacement gilts.
- **Gloucestershire Old Spots.** Large, curvy white pig with a few black spots and big lop ears. Known as an orchard pig in England, they are a docile pig that produces a lot of fat. They are hesitant to cross an area with a removed electric fence because of limited visibility. Old Spots require patience when handling because ears reduce vision and their default is to stand still rather than run. Good for maternal lines, foraging ability, meat production, and lard.
- **American Guinea Hog.** Small black pig with prick ears and short snout. Guinea Hogs are easy to handle on small acreages and don't root as deeply. They are very fatty and easy to overfatten. We have raised Guinea Hogs and are amazed at their willingness to forage all day. They are the most docile pigs that we have raised, almost like pets. They love belly rubs. Good for small, easy to handle carcasses, lard, and making cured products. They have the smallest percentage of lean meat—roughly 40 percent of their carcass is lean meat.
- **Large Black.** Docile, medium-sized black pigs with big lop ears. Good for bacon, cured meats, and production in rough conditions. Large Blacks are good mothers with lots of milk, are good foragers, and can raise large litters outdoors.
- **Red Wattle.** Large red pig with wattles under chin. This is not a very improved or standardized breed, therefore there is considerable variation. Some may take 6 months to grow out, others 10. They are known to be good foragers and hardy in a wide range of conditions.
- **Cross-breeds.** Some favorite cross-breeds we have seen and farmers have told us about include: Tamworth (lean, bacon pig) crossed with Gloucestershire Old Spot (lard-type pig), Large Black with Tamworth, Berkshire with Tamworth, and Large Black with Red Wattle.

Conventional Breeds

Conventional pigs are generally leaner than heritage pigs and grow faster. Although many of these breeds have been around for a long time and could be considered "heritage" breeds, they are mainly being bred for intensive confinement production, heavy muscling, and lean carcasses, and lack many of their original traits such as mothering and foraging abilities. If you are going to use these breeds in your program, look for lines that have been raised on pasture for many generations. They may make a nice addition to a cross-breeding program.

American Yorkshire. Large white pig with prick ears. Good for bacon, ham, lean pork, maternal milk production, and large litters.

Hampshire. Black pig with white band around the middle. Good for hardiness, terminal sire, meat production, and foraging ability. High-energy, aggressive personality.

Chester White. Known for high conception rates and large litters with heavy weaning weights. Often used in cross-breeding programs. Produces heavily muscled offspring.

Duroc. Large, athletic-looking red pig with slightly drooping ears. Good for hardiness, foraging ability, meat, terminal sire, unequaled conversion rate of feed to meat, and protective mothers. They are most commonly used as a terminal sire because their fecundity rates are inferior (small litter sizes). Even the famous Iberian pigs are now using Duroc boars for 50 percent crosses because they grow bigger, meatier, and faster than 100 percent Iberian genetics.

Breeding

Keeping breeding stock is not for everybody. Indeed, many pig farmers just purchase weaners and raise them up to finish weight. Considerably more infrastructure and management is required when raising breeding animals. When we first got started, we just purchased weaner pigs each spring and grew the pigs out until the fall. This was a great way to learn how to manage pigs on pasture and refine our systems, while also not having to care for pigs over the winter. If you want a seasonal production model and maybe even a winter vacation, a weaner to finish model could work for you.

We eventually started breeding our own pigs for three main reasons. Most importantly, we wanted to raise heritage breed pigs that would perform well outdoors on pasture, and all we could purchase locally were conventional breeds that did not seem to forage that well. We also weren't able to buy large enough groups of pigs from one farm, and mixing groups of weaners from more than one farm was risky because of disease issues. And thirdly, our market was growing and we wanted to have more year-round production of a consistent quality of pork. Thus we purchased our first breeding stock and began expanding from there.

Systems of Breeding (applicable to all livestock species)

Farmers must choose a breeding system that will work best on their farm. This will most likely be determined by the genetics one is aiming for, the size of the herd, the quality of the animals, the infrastructure that the farmer has to work with, and the marketing of the animals. Farmers can utilize different systems of breeding based on their goals. For example, you could do cross-breeding for market animals (called *terminal breeding* because they don't rebreed; they are used for meat) and stick to pure breeding to produce higher-value breeding stock that you retain or sell.

There are a few different ways to choose the animals you will use for breeding. Adapted from a University of Florida publication (Walker, 2003), here are some of them:

Pure Breeding. A purebred animal is a member of a breed that possesses a common ancestry and distinctive characteristics and is either registered or eligible for registration in that breed. Pure breeding is the mating of two purebred animals of the same breed. Pure breeding can use line breeding and outcrossing for the selection of breeding stock.

Inbreeding. This is the system of breeding in which closely related animals are mated. This includes (1) sire to daughter, (2) son to dam (mother), and (3) brother to sister. Inbreeding is suggested for only highly qualified operators who are making an effort to stabilize important traits in a given set of animals. Intensive selection is needed to reduce the risk of producing undesirable traits in breeding stock when inbreeding is practiced. Inbreeding is usually fine for terminal market hogs since they will soon be slaughtered and not rebred.

Line Breeding. This is a system of breeding in which the degree of relationship is less intense than in inbreeding and is usually directed toward keeping the offspring related to some highly prized ancestor. The degree of relationship is not closer than half-brother/half-sister matings or cousin matings. Line breeding is practiced to conserve desirable traits of an outstanding boar or sow line.

Outcrossing. Outcrossing involves the mating of animals of the same breed that have no closer relationship than at least four to six generations back. This is the general system that is practiced by most purebred breeders and is classified as a safe system in the purebred business. However, for some rare heritage breeds it may be quite hard to find breeding stock that is not closely related. For example, whenever we tried to find new Gloucestershire Old Spot breeders, we found them to all be two to three generations distantly related to our own herd.

Cross-breeding. Cross-breeding is the mating of two animals that are members of different breeds. This system is being practiced by the majority of commercial livestock producers because of the resulting hybrid vigor, which makes possible improved production efficiency.

When to Breed

The length of a sow's heat period is two to three days. Breeding the sow on the first and second day is best, both in terms of conception rate and number of piglets. Two services in 24 hours will help to produce a bigger litter. Most farmers attempt to get two litters of pigs each year from their sows, which requires a high level of management and timely (and sometimes early) weaning in order to keep farrowing dates at the same time each year. Another option is to farrow three times in 2 years, which is less stressful for the pigs and farmers alike and allows the piglets to stay on their dams for up to 8 weeks, or even to self-wean. One litter or "all gilt" farrowing (described below) might be a good option for diversified farms and farms that don't have the facilities for year-round production.

By exposing the sow to the boar for up to a full 30 days after she weans her litter, she has two opportunities to breed. She will come into heat 4 to 7 days after weaning and again between 25 and 30 days after weaning. Continued access to the boar past 30 days will lead to a bigger spread in farrowing dates, and thus more potential management problems, if any of the sows are bred in their third heat (Practical Farmers of Iowa and Iowa State University, 2007). Sows that do not rebreed within the first two cycles will also have significantly smaller litters.

Gilts that are bred too young or too small can have trouble at farrowing. A piglet can get stuck in the birth canal (often called a "log jam"), blocking all the other piglets behind it. Usually this results in all the piglets behind the stuck one dying. This is a stressful event for the sow and a sad situation for the farmer, and points out the importance of not exposing your gilts to boars when they are too young. Sows that are overfed during the last third of their gestation can also experience "log jams" because their piglets get too big.

Hot weather during the breeding season can limit the reproductive ability of your boars. Boars that are heat-stressed have reduced fertility for 4 to 6 weeks after the event. This often manifests as decreased conception rate and reduced litter size in the sows that do conceive. Heat stress can be mitigated by providing wallows, sprinklers, shade, and other methods to keep boars cool during hot weather (Luce, Williams, and Huhnke, 2007).

ONE-LITTER FARROWING

An alternative to keeping and rebreeding sows twice a year is a one-litter farrowing, also called an "all-gilt system." One-litter systems sell or slaughter the young sows after the piglets are weaned. The growing pigs are then slaughtered at about 6 months of age and gilts are kept back from this group and bred at 8 months of age. These gilts will then farrow at 1 year of age and be sold or slaughtered at weaning time.

Advantages of this system include: Growing pigs are raised during the warmest and most productive time of the year. Pigs raised during the warmest months require very little shelter and simple watering systems (not frost-free hydrants). Pastures are rested during the winter when they are most prone to destruction. Winter housing is only needed for the gestating gilts. The farm has the fewest pigs during the time of the year that is most difficult to raise pigs. By not keeping sows the producer's cash flow is improved.

Disadvantages of this system include: Farm only produces one litter a year instead of two. Sows that are good mothers are not kept for rebreeding. Gilts may have smaller litters. The system also requires new boars every year (or artificial insemination).

ARTIFICIAL INSEMINATION (AI)

A teaser boar (or boars) is often used on bigger operations. Sows are introduced to the boar on the other side of a fence in order to get the sows to cycle together. This allows the farmer to artificially inseminate all of the sows in a tight time window.

By using AI you don't have to bring in new boars every couple years, effectively closing your herd and reducing the chance for disease introduction. AI allows you to breed for production traits or to breed for maternal traits for your breeding herd. Farmers must spend more time observing gilts and sows in this system to know when they are in standing heat. This type of breeding is obviously more labor-intensive, but it eliminates the cost of keeping boars (other than a teaser boar or two).

Gestation

The gestation period for a sow is 114 days. A traditional and easy way to remember the sow's gestation time is 3 months, 3 weeks, and 3 days.

Gestating sows can make the best use of pasture or harvested crop fields. They can live very well on pasture alone or pasture with supplements. Iowa farmer Tom Frantzen has found that one sow can be kept per acre of corn stubble for up to 6 weeks. Indiana farmer Greg Gunthorp also runs his sows on cropland and pasture and says that gestating sows can meet their nutritional needs on pasture alone, especially on clover using rotational grazing. We didn't learn this lesson of raising gestating sows mostly on pasture alone rather than giving them formulated feeds until we had been raising pigs for a couple of years. Had we known, we could have saved a lot of money on feed and had fewer farrowing complications due to overfed sows.

Some farmers keep their gestating sows indoors as a rule and others only move them inside during the winter. Tom Frantzen has a 30-foot by 60-foot hoop building that he uses for gestation in the winter. A concrete pad at the end of the house allows the sows to drink without making a wet mess. Down one side of the building he poured a 10-foot by 60-foot concrete pad with a walkway and a row of repurposed gestation crates. The crates now serve as feeding stalls that allow Tom to control the amount of feed that each sow receives and enable him to artificially inseminate his sows and administer health treatments. A box that holds AI supplies slides on rails mounted to the top of the row of crates. Most gestation crates are equipped with locking bars, so you can lock and unlock five stalls at a time with one lever mounted at the front of the stalls. Locking the sows in while they are feeding reduces fighting and ensures they each receive their proper amount of feed without stealing it from others. Tom cleans the sows' living area with a skid steer while they are eating and locked in their feeding stalls. With a little bit of metal fabrication, feeding stalls could be set up for sows on pasture too.

By utilizing feeding stalls you can more easily control variation in the sows' body conditions. Sow condition is scored between 1 and 5.

1. Emaciated: can see ribs
2. Thin: able to feel ribs
3. Ideal: in summer, can barely feel the ribs
4. Fat: can't feel the hips or backbone when pushing down with the palm of your hand
5. Overfat: can't feel the hips or backbone when pushing down with a single finger

A body condition of 3.5 to 4 might be better when going into winter on farms located in extremely cold conditions.

Two general rules of thumb are that large weight fluctuations are very hard on the health of a sow and that large sows have very high maintenance costs. Duane Reese of Oklahoma State University reports that overly thin sows have a delayed return to estrus, more lameness, and poor farrowing rates.

Conversely, Reese has found that overly fat sows eat less during lactation, have more lameness and morbidity, waste feed, and have poor feed efficiency. They produce more stillborn pigs, have a longer farrowing period, have summer infertility, and are more susceptible to heat stress. Fat gilts and sows also tend to lie on and crush more piglets. Sows should gain between 50 and 70 pounds and first-time gilts should gain between 70 and 100 pounds during pregnancy.

Reese says sows should be fed based on the temperature immediately around the sow for most of the day. For example, during winter the sow might eat, drink, and dung outside and then go into her warm hut for most of the day. Also, sows with more backfat have more insulation and are able to maintain their body temperature with less feed than sows with less backfat. A general rule is that a sow will need 0.4 pounds of extra feed for every 10 degrees under 55 degrees F (13 degrees C).

It is important to try to reduce heat stress on the sows during the early and late parts of gestation. Sows farrowing during the summer months often have smaller litters and lower weaning weights due to heat stress during gestation. Sows that experience heat stress during the first 15 days of gestation can have lower conception rates, fewer viable embryos, and lower embryo survival rates than sows that are not heat-stressed. Sows that are heat-stressed 2 to 3 weeks before farrowing produce more stillborn pigs and fewer live pigs per farrowing. Heat stress during the middle of gestation doesn't seem to produce results as negative as when experienced during early and late gestation (Luce et al., 2007).

We like to run our boars with our gestating sows in order to limit the amount of feed fed to both sexes. In fact, on good pasture or woodland you might be able to get away without feeding them much at all. Keeping both sows and boars in a fairly lean condition slows their growth and allows them to be productive breeders on your farm for a long time. Lean and fit boars have more stamina for breeding and limiting their size enables them to breed gilts that are smaller than your sows. Young gilts and small sows will find it hard to hold up the weight of a boar that is too big for them.

Sows and boars can also make do on a higher-fiber feed than growing pigs can. On top of all these other benefits, it is a more efficient use of feed and pasture to limit what you feed your sows and boars. Overfeeding animals until they become too fat to breed well is one of the biggest reasons we have seen farms fail at economical breeding.

Farrowing

Most producers consider a sow that can wean around eight pigs to be a good mother. She may start off with ten to twelve piglets—two could be born stillborn and two could be crushed—thus eight survive healthy to weaning. However, some breeds have smaller litter sizes, so keep that in mind.

Farrowing on pasture is the make-or-break time for pastured pig producers. Success is determined by a number of factors that can be controlled by the farmer and some that cannot. The prior selection of sows with genetics to farrow on pasture, and the situation that the farmer has created, including design and placement of huts, will all combine to make farrowing either a great success or a train wreck.

The sows teats and udder will be very swollen and might even drip some milk when the sow is within a day or two of giving birth. Her vulva will also be red and swollen. Your first clue that farrowing is imminent might be a sow that has bits of straw sticking to her leaky teats. Sows that are going to farrow within a couple of days will also begin to make a nest in the hut or location they plan to use. Straw should be provided for bedding, and it is generally better to have too little bedding in a hut than too much. Excess bedding makes it harder for the newborn piglets to maneuver themselves toward the sow's teats when they are born, and can also lead to more crushing, since in the deep bedding the piglets can't get out of the way of the sow, and it may be more difficult for the sow to locate the piglets.

In some regions sows can farrow big healthy litters with little or no shelter other than some shrubs or tree cover, but most likely you will want to provide a farrowing hut when sows are farrowing on pasture. Sows on open pasture will be attracted to the security of the hut and will (mostly) want to farrow inside it. The farrowing hut will protect the tiny piglets from predators and the weather and will also confine them for a week or so, until they are strong enough to run around with the rest of the pigs without getting trampled. In cold weather the hut will trap some of the sow's heat inside and keep the litter warm in the first couple of weeks.

Farrowing huts can be purchased or farm-built. Popular styles include Quonset, A-frame, and the E-hut (improved Illinois hut). The hut should be tall enough so that a large sow can't lift it with her back and heavy or secure enough that she can't move it with her snout. Sows can be really hard on their shelters and use them as scratching posts: Make them strong enough to handle that, yet light enough to be portable. Rails on one or both sides or simply a sloped roof provide the piglets a place to escape when the sow is up or about to lie down. Roller bars on the entrance keep the young piglets inside of the hut for the first couple of weeks, and allow the sow to enter and exit without painful scraping on her tender teats.

Farmer Greg Gunthorp tries to "think like a sow" when determining where to place farrowing huts on pasture. The sow's location preference can vary depending on the time of year and weather conditions. If the sow farrows outside of a hut, Gunthorp says it is likely your fault for not putting a hut where she wants it and not the sow's fault for choosing her own location. Putting out extra huts with bedding will give the sows more choices and increase the likelihood of the sow farrowing inside of a hut.

Sometimes, despite all of your efforts and especially in hot weather, a sow will choose to farrow in the open. In this situation, a hut can be moved to where the sow and litter are and the piglets can be put in the hut. A roller bar or a piece of lumber can usually keep the pigs inside for a week or so. To encourage sows to farrow in the huts during hot weather, all vents and doors should be opened to increase airflow, and huts should be placed in the shade if possible.

In wet weather, huts can be placed on berms or plowed ridges to keep them out of standing water. Hilly ground and well-drained soils are also good for farrowing. Sows should be provided with plenty of dry bedding material during wet weather.

Many experienced producers recommend farrowing in groups of ten sows per acre. Gunthorp farrows year-round in groups of between four and ten sows in 1- to 1.5-acre pastures. He says that you can farrow as many sows per group as you want, provided that they all farrow within 3 to 4 days of each other in warm weather and 7 to 14 days of each other in cold weather. In warm weather the older piglets will leave their huts sooner and will steal colostrum and milk from newborn litters. This is one of the biggest potential disasters when farrowing, since the newborn piglets may starve. To emphasize this point Greg states that he would rather farrow a group of one hundred sows over 3 or 4 days than a group of five sows over 2 weeks. He says that many folks give up on raising pigs because of the terrible repercussions of not farrowing in tight groups. This means you have to breed in tight groups as well.

Some farmers prefer to farrow indoors either year-round or during cold weather. Hoophouses and barns work well for this application. Farrowing huts can be moved indoors and installed with or without a pen attachment, or you can set up plywood partitions for sectioning off farrowing stalls. A hut with a pen attached allows for the individual feeding of the sow and protects the newborn piglets from older piglets intent on stealing milk and colostrum. Piglets can be kept in farrowing pens until they are weaned or the pens can be opened up to allow communal lactation once the piglets are strong. For indoor farrowing in winter, Dan Wilson keeps the barn temperature just above freezing, which encourages the sows to go into a hut to farrow.

Wilson culls his sows for good disposition and mothering abilities. A docile sow is more likely to farrow and stay lying down if she is not disturbed during the first crucial 24 hours. A sow that is high-strung and gets on her feet at the slightest provocation is more likely to step on and crush some of her newborn pigs. Every time a sow stands up is another time that she will have to lie down, and the newborns are at great risk of being crushed each time.

Greg says that most farrowing problems come from feeding too much or too little in the last 2 weeks before farrowing. If the sow is fed too much and the piglets are too big (4 pounds), they will get stuck (log jam) in the birth canal. If the piglets are too small, they will have viability issues. Somewhere in the high 2- to 3-pound range per piglet is a good size to ensure viability and ease of birthing. Greg says that ensuring the sows have access to forage and the ability to exercise will help keep them in good condition for farrowing.

Two hours or less is the normal amount of time it should take to farrow (Luce et al., 2007). If it seems to be taking longer than this, some human assistance may be necessary, although some producers contend that this is often a wasted effort because of the low likelihood of improving the outcome.

Lactation

Milk yield will often peak at 3 to 5 weeks of lactation. For this reason it can be economical to free-choice feed during the first 4 weeks of lactation and then limit feed the sows after 4 weeks, provided you are happy with their body condition. A sow needs to eat 4 pounds of feed per day plus 1 pound per day for every pig nursing. A sow with eight piglets would need 12 pounds of feed per day.

Communal lactation will be the default situation; it tends to occur naturally as both piglets and sows stop spending as much time in the farrowing huts. It is important that the piglets be close in age, so that they all get enough milk. The sows will often all lie down to lactate at the same time, which reduces the possibility of the bigger pigs stealing milk from the smaller ones, since they are all suckling at the same time.

Life-Stage Considerations

Pigs at different ages have varied needs in terms of amount and quality of feed, housing, and management. Efficient management of pigs at each life stage improves animal welfare and reduces costs to the producer.

Newborn Piglets

Processing of piglets within the first week might include ear notching or tagging, vaccinating, and castrating. Early processing of piglets seems to be less stressful for them. Dan Wilson castrates his piglets within the first 12 hours and says that there is less squealing if he does it early. This makes the process less stressful for the pig, the farmer, and the sows. Most pasture pig farmers don't dock tails, file teeth, or give iron shots. Iron is usually not necessary when pigs have access to pasture and soil to root in, because they consume it from the soil.

Castration

On Gunthorp Farms they castrate boar piglets at 1 to 3 days old. They usually wait until the sow is out of the hut and eating. Then they will grab the male pigs and take them outside to a processing box mounted to the three-point hitch of a tractor. He recommends keeping a couple of plastic sorting boards handy in order to keep any overly aggressive sows at bay.

Farmers castrate boar pigs for a couple of reasons: Boar meat can have an off flavor or smell called boar taint, and castrated barrows are easier to manage. Boar taint is caused by androstenone, a pheromone produced in the testes, and skatole, a bacteria-produced compound found in the large intestine. About 25 percent of humans are unable to detect the smell in cooked pork, but for folks who can it is very unpleasant. Boar taint is affected by age, breed, and environment. Castrated boars are also less aggressive and can be run with the gilts, which allows farmers to manage one group of pigs instead of two.

On our farm, we didn't like to castrate pigs, so we didn't. At first we developed a market for young boars as suckling pigs at high-end restaurants in and around San Francisco. This wasn't very profitable for us because of the processing costs for such a small animal, and all the logistics involved in restaurant sales. We stopped doing suckling pigs and instead separated the young pigs at about 4 months and ran two groups instead of one. The boars were leaner than the gilts (they are also leaner than barrows), but we didn't run into any issues with boar taint. In fact, at one of our markets we had a lot of Chinese and Korean clients who actually preferred the leaner pork from the relatively young boars. We slaughtered the boars at about 6 months of age.

The only commercially viable alternative to physical castration is immunization with a protein product that is legal in sixty-two countries including the United States and has been used extensively in Australia and New Zealand for over 10 years. This product, called Improvest, creates what is called temporary immunological castration because the boar will not produce viable sperm for a period of time. In Europe, physical castration can't be performed legally without analgesics or anesthesia, and there is a plan to end physical castration by 2018. We believe the trend against castrating pigs will extend to the United States as well. Interestingly, Whole Foods Market's Global Animal Partnership program gives the highest rating (Step 5) to farms that don't castrate their boars. Yet they don't buy boar meat for their meat case. Obviously finding a market for boar meat is imperative should you decide to avoid castration.

Weaners

Weaning usually takes place at about 6 weeks of age or once the pigs reach a certain weight. Piglets are best weaned by leaving them where they are and removing the sows. If the piglets are moved to another area, they will be stressed by the new surroundings and will often try to escape to get back to the sows and their former shelter.

By this time they should already be used to eating solid feed and the transition should be fairly easy. It's important to keep an eye on the recently weaned pigs for the first week or so to ensure they are making the transition from their mother's milk to pig feed. Weaners that are stressed will often get diarrhea. After a couple of days the weaners should be over any stress from being separated from the sows.

Greg Gunthorp weans his pigs at between 6 and 8 weeks of age. At this size they can be stocked at up to one hundred pigs per acre, but by the time they are market size it is hard to go much above twenty pigs per acre.

On smaller farms, some producers will allow the pigs to self-wean. This works for farmers who aren't in a rush to rebreed their sows and should only be done as the sow's body condition allows. Be sure that the sows' teats aren't damaged by overly aggressive larger pigs. After a while the sows will allow less and less access to their udder and the piglets will, at the same time, be eating more feed for their caloric needs. The sow will dry up, but not as fast as if the pigs were weaned at one time. Keep in mind that self-weaning will likely lead to a longer period before the sows are rebred and might increase the length of the next farrowing period, necessitating a higher level of management. This also impacts the economics of raising sows.

Growers and Finishers

Many larger producers that farrow on pasture will move their weaned pigs into a hoophouse or drylot for finishing, even during the summer. The pigs gain faster and more efficiently when they are more confined. Growing pigs on pasture gain weight more slowly and are less efficient feed converters, but they are usually more healthy and able to forage for some of their own feed.

Unlike non-lactating sows, who can usually maintain their body condition just fine on good pasture or in the woods, growing pigs need a lot of nutrition in the form of grains, milk products, nuts, and other high energy and protein feeds. Considering the fact that this is the largest group of pigs on the farm and they eat the most feed, it is easy to understand why a farmer with a large number of pigs would want to keep them more closely confined. For example, in a hoop building with deep bedding the pigs can stay warm and dry, and the feed system can easily be mechanized to reduce labor. Water and electricity are also more easily supplied than out on pasture.

Gilts

Gilts can be bred as early as 8 months old, but 10 to 12 months is better. More ideal though is to use the weight of the animal as an indicator of when she is ready. A gilt's weight at breeding is critical to her longevity. There is general agreement that the ideal weight range of breeding for full-sized pigs is 300 to 350 pounds. For smaller pig breeds, you can use a smaller weight range, essentially whatever weight you would raise those animals up to harvest. If too light, gilts will be challenged at their first farrowing and a high percentage will experience a second birthing number dip caused by poor body condition, an inability to breed back, and, subsequently, a small litter size. If the gilts are too heavy, it may have an adverse effect on the milk production of their first birth (parity) and these gilts are often culled because they get too big for effective conception. If you don't have a scale to weigh your gilts, a flank-to-flank measurement can offer guidance. If the tape measures over 34 inches, her weight is probably over 300 pounds.

You can improve your genetics by selecting gilts within your herd. One of the most important heritable traits is the mothering instinct of the sow. A good sow with strong maternal traits will raise more pigs. These traits are passed on to the gilts, so select gilts from your best sows. Strong maternal traits will improve "pigs out the door" quicker than selecting for production traits. Select the fastest-growing gilts in your herd, as the rate of growth is highly heritable compared to the number born alive. Gilt selection should begin in the first week of life. Ear notch the gilts from your best sows and then track their rate of growth. Remember that the largest gilts in a group might not be the fastest growing in the group, but might be older (Practical Farmers of Iowa and Iowa State University, 2007). Ear notching will help you keep track of age and what litter they came from.

Boars

Boars that are selected for breeding should be physiologically correct and fast-growing, and they should come from the big litters of the best mothers, too. The number of teats on the boar is a heritable trait as well. If you buy replacement boars, try to get ones that are raised in conditions that are similar to your own.

One boar can reasonably service up to twenty sows using natural mating. Unless you have at least five sows, it is not recommended to keep your own boar; instead, borrow one or consider artificial insemination. Boars should be at least 7 to 8 months old before used for servicing. They should be kept in thin, thrifty condition so they are able to breed both gilts and larger sows. Boars can reasonably be kept for around 36 months before you should consider replacement. If you keep them lean and healthy, they may be kept for longer.

Boars of the same size and age can be run together during the off-season. Boars of different ages should not be run together because they can fight and even kill each other. Holding lots for boars should be constructed out of strong material that will restrain the animal adequately. Build pens narrow and long. To encourage exercise, place feed at one end and water at the other. Furnish adequate shade and shelter for inclement weather. We rotated our boars on pasture, just like the sows, using only two strands of electric fence, with no problems.

Housing

Most of the year pigs can live on pasture without too much protection from the elements. Greg Gunthorp notes that pigs are the livestock least suited to temperature extremes and should have some type of shelter year-round. Pigs—especially lighter-colored breeds—like to have a shady spot to get out of the sun on hot days. Light-colored pigs can actually get sunburned; you will see irritated, sometimes bloody skin, especially around the backs of the ears. Pigs are tolerant of cold to a point, as long as they are not also wet. Pigs like to have a deep layer of bedding that they can bury themselves under to stay warm. Large groups of pigs can generate enough heat to keep themselves warm on even the coldest days. Although they may be able to suffer through the cold with no shelter, they will not gain weight, as most of their calories are expended keeping themselves warm. Why not provide them a little shelter?

Dan Wilson uses sheds that are on skids and can be moved in two parts for large groups of growing pigs to stay warm together. They are 18 feet by 20 feet and each half is 9 feet wide. Greg makes his shelters from corrugated steel sheets meant for building feed tanks.

Shade over wet sand or concrete is one way of dealing with heat events. Sand must be wetted several times a day and can be set up on a timer. Many pig farmers create, or allow the pigs to create, a wallow. Wallows work very well to keep pigs cool, but some farmers find them to be unsightly and destructive to the pasture. Some of our pig wallows got so large you could have lost a tractor in them. You can even lose small piglets in them if they are too deep and the sides become too steep for the pigs to get out. If you ever want to hay or reseed that ground, you are going to need to fill in those holes. Prevent giant wallows by frequently rotating your pastures and limiting the amount of water leaking in different places on your farm. Your pigs will quickly turn a wet area near a leaky pipe or valve into a giant mud pit if you aren't careful.

Feeding

Selecting feed for different life stages of your animals is the most economically efficient strategy. For example, protein needs for pigs are highest when they are young and reduce as they mature (from 18 to 16 to 14 percent). Diets for pigs should be based on the amount of available lysine, not just the levels of crude protein, since lysine is the limiting factor. High-fiber diets have lower digestible energy and nutrients. You may be tempted to give whole or milled beans and other legumes to your pigs, but they all should be roasted, steamed, or extruded first to reduce the trypsin inhibitors present in legumes (with the exception of peas).

Peas are high in lysine, energy, and protein and have considerably less trypsin inhibitors than other legumes. Wheat can be a great option too, particularly if you have access to cheaper wheat middlings and shorts. They are similar to corn in digestibility and protein levels, with slightly lower energy.

Much more than with other livestock species, swine meat quality and flavor is highly influenced by the variety of feeds given to them. People have been experimenting with feeding pigs a wide range of foods for hundreds of years.

Feeding a large percentage of barley, corn, and other high-energy feeds late in the finishing phase will increase fat deposition, sometimes to an unmarketable level. These grains also produce more saturated fat on the animal, which makes it less effective for curing because it goes rancid more quickly. Barley produces a firmer backfat than corn. Sprouting your grains will reduce the energy levels in them and make them more digestible (Mattocks, 2014). Feeds higher in polyunsaturated fats, like peanuts and other nuts, will produce a softer, creamier fat that is less likely to go rancid, so it's better for curing purposes. High-protein oilseeds and fish products fed late in finishing may give the pork an off or fishy flavor. If you use any fish- or crabmeal-type products, limit them at the end of life. We'll say more below about pork meat quality and what affects it.

Since feed costs are the most expensive input in pig production, farmers are experimenting with using free or low-cost feeds. One feed that is increasingly becoming available as craft breweries sprout up across the nation is brewers' grains (BG), often to be had for free. BG has lower digestible energy levels than corn or barley and higher protein and fiber levels. This might be useful if you want to limit the backfat deposition of fatty-type pigs. BG is also higher in lysine than corn or barley, an amino acid often in short supply in other grains. Don't feed moldy BG to pigs and don't allow access to it after it gets moldy in one or two days.

Milk products can be a great source of energy, protein, and amino acids for pigs, but they are often logistically hard to handle. Unless you are getting them for free or very cheaply, it is probably not worth the added handling and storage required. Using liquid milk or dairy products will increase pig manure production by two to three times, so be careful about that.

You can feed some hay to your pigs, but they will waste and soil much of it. It is low in energy and high in fiber, so not suitable for young, growing pigs that don't have a fully developed hindgut and have much higher energy needs. We would feed some hay in the last couple of months if we were trying to raise older animals with more intramuscular fat, without giving them lots of energy that just lays on excessive cavity and backfat. In general though, hay is better suited to non-lactating breeding animals.

According to our research, it is not recommended to feed alfalfa hay to growing pigs, due to several anti-nutritional factors, such as saponins and tannins that reduce the growth rate of young pigs. However, you can feed it to a sow because her hindgut is fully developed and colonized by microbes that help her digest the fiber. This will add bulk to her diet (but not calories), which is useful in keeping her in good breeding shape.

Pasture

Use whatever happens to be growing in your pastures for the easiest, quickest results. Your pigs will improve those pastures over time and will often revive the latent seed bank hiding in the soil. If you have the equipment and tillable soils, you could try seeding and improving your pastures for your pigs or you could rotate pigs into a grain or legume crop rotation. Dan Wilson likes the durable cover he gets from a dense turf of orchard grass. Greg Gunthorp likes quackgrass pasture, but it is controversial in his region, because row crop farmers despise it.

Pigs can even seed their own crops by allowing them to till a field and tread in the broadcasted seed for a couple of days, then removing them for 3 to 4 months while the crop grows. Then you can put the pigs back in to self-harvest the crop. Good seeds to sow include rape, forage brassicas, mangel beets, forage carrots, squashes of all types, wheat, oats, barley, sorghum, corn, field peas, and rhizomatous grasses.

Other forages that are seeded on Gunthorp Farms include white clover, red clover, dwarf Essex rape, rye, and forage turnips. Greg overseeds rape and rye into corn during the last cultivation. He likes Berseem clover, but finds it hard to manage in a vegetative state, which is the most palatable for the pigs. Once it goes to flower he finds himself questioning why he spent the money on the seed. Greg has tried broadcasting rape in the woodland paddocks, but finds that it is outcompeted by rye. He likes to frost-seed red and white clover in his pastures, but in his experience the rape does better with cultivation.

Feeders

Self-feeders have flip-up lids that a pig can lift with its snout to access feed. The lids keep out vermin and moisture and conserve feed. The feeders are manufactured in many different sizes with up to twelve lids and capacities of over a ton. Feeders are checked daily to ensure that feed is flowing and that there is plenty of feed. This type of feeder can be filled with an auger or a tractor front-loader with supersacs of feed. We have even filled this type of feeder with barrels of feed tipped over from the back of our pickup truck.

Greg uses big wheel-style self-feeders for his pigs, turkeys, and ducks. He mixes the feed on the farm and hauls it to the 1- and 2-ton self-feeders. Greg recommends feeding once a week for smaller groups and feeding big groups at the same time that you are checking on them. Dan Wilson says that feeder adjustment is very important for reducing feed costs. He puts a couple of hundred pounds of oats in his feeders to slow the flow of feed but still allow it to flow.

Greg's market hogs eat between 6 and 12 pounds a day depending on the weather and the availability of pasture forage. They eat the most feed during the winter in order to maintain their body heat. They eat less feed in the summer because of the warmth and the availability of pasture. He provides a non-GMO feed to supplement his pigs on pasture. Greg raises big pigs (325 to 350 pounds) for flavorful, well-marbled quality pork, keeping in mind that feed efficiency declines as the pigs get this large.

Watering

Clean, fresh water should be provided for the pigs at all times. Pigs need 2 to 5 gallons of water a day per 100 pounds of body weight. Pigs should not be allowed continuous access to creeks or ponds because they will foul the water with their own excrement and then be drinking that same water. Unless you want to become dependent on antiparisitics and antibiotics, letting your pigs drink from the same place they poop is a bad idea.

If a pig can flip over its waterer, it will. Pigs will also try to climb into water troughs to cool off. When they do this they introduce fecal contamination to the water, which will raise the risk of parasite infection. Small pigs can use small water troughs, but a different system should be employed for growing pigs and breeding animals.

Areas around waterers experience a lot of impact and will often become wallows. Nipple waterers are often the cleanest and least wasteful type of waterer. They can be mounted to a moveable platform or a trailer in order to spread the impact. A moveable water system for pasture-based farmers is a huge advantage. A water system that is easy to move allows you to utilize more land and makes pasture moves easy. A trailer with water storage and gravity-fed nipple waterers works well and is very portable. An alternative would be a central waterer mounted on concrete or wooden platforms with lanes for access.

During the growing season Iowa farmer Dan Wilson runs a black plastic pipe on top of the ground for a supply line to his pig paddocks. Along the line are valves that he can tee off and run to his wooden pig waterer platforms. The platforms are along the fence line and give the pigs a mud-free area to drink while at the same time protecting the watering infrastructure. Dan says that the weeds grow over the pipe and the water stays cool this way in the heat of the summer. An aboveground pipe without some type of shading would make the water too hot for the pigs to drink.

Winter watering can be a challenge and in cold climates often consists of a stationary frost-free waterer (e.g., Mirafont). Greg Gunthorp strongly advises installing underground water lines for the winter paddocks and says that setting up automatic waterers should be one of the highest priorities for improving farm infrastructure. By locating waterer pads between the two paddocks you can supply water to each paddock with one water point. These improvements are relatively inexpensive when compared to the amount of labor they free up and for the peace of mind they bring. Obviously in warmer climates you don't have to worry about frost-free watering. But you will have to pay attention to providing plentiful cool water during heat events.

Keeping Healthy

It is far easier to manage your pigs in a way that will keep them healthy, rather than skimp on the management and treat the symptoms that will surely develop. Appropriate shelter and clean bedding, as well as appropriate stocking rates, will go a long way toward keeping your pigs healthy. The biggest risk you take in terms of jeopardizing herd health is bringing new pigs onto your farm. Follow the biosecurity guidance found in appendix A to reduce the possibility of disease transmission onto your farm.

Practical Farmers of Iowa (PFI) warns that pigs coming on to your farm are a risk, such as new boars, gilts, and feeder pigs. A closed herd is one way to reduce introduction of disease to your farm. New blood in this type of system is introduced by the use of artificial insemination.

"All-in all-out" systems give groups of pigs isolation in time. Market all the pigs at the same time in order to clean and disinfect their housing and then allow it to sit empty for a couple of weeks to further reduce pathogens. Don't hold back runts and introduce them into the next group of pigs, because you will expose the young pigs to the sickest animals of the previous group.

If you do bring in outside animals, house the new pigs away from other pigs for a minimum of 30 days and do the chores for these pigs after you have taken care of your existing herd. Some farmers will run new pigs across the fence from cull sows in order to inoculate the new pigs. If the cull sows get sick, they will be leaving your herd soon anyway, and won't infect the rest of your pigs.

PFI also recommends keeping your new and main herd gilts separated during the first month of gestation in order to spot any problems and to allow for the embryos to firmly attach to the uterus.

You can also separate pigs by age. Piglets don't have a fully developed immune system and should be isolated from older groups of pigs. In order to do this effectively, you must breed sows in tight groups. This is explained in the breeding section of this chapter.

You want to avoid a situation in which newborn pigs are born and then are immediately faced with a large parasite load from their mother or the environment. For this reason farmers will often worm the sows or gilts before farrowing. Sows can be dewormed only in the first trimester for certified organic producers. Farmers that have success without worming their pigs at all often have a tightly managed rotational grazing system with low stock densities. As pigs grow and their immune systems strengthen, they can gradually be exposed to greater parasite loads. In all our years of raising pigs, we only had to deworm one group of weaner pigs we naively purchased from a not-so-clean farm. You can prevent parasite loading through good rotational grazing practices.

New Stock

When purchasing pigs be sure to visit the farm where they are raised. This gives you a chance to see the facilities and the parents of the pigs you might purchase. Always give yourself the option of not purchasing any pigs. If there are any "red flags" it is best to look elsewhere. You should visit a farm to look at some pigs that you *might* buy, rather than visiting a farm to buy some pigs. This slight mental distinction will allow you to more easily walk away instead of buying the best of the poor stock that is available. This is true for all livestock species.

The pigs should be active and curious, not slow and lethargic. They should have a smooth gait and not be too far up on their toes. Their tails should be curly and they should appear healthy. Look at the pigs breathing when they are lying down. They should breathe in and out smoothly without thumping or pausing in their exhalation. Thumping may indicate the presence of pneumonia. We have purchased pigs in the past that were from a herd that was treated for pneumonia and then they completely fell apart on our farm because it came back. Pigs that are "pulled-apart" or wide-stanced have a larger chest cavity and are more resistant to pulmonary diseases. Put simply, they can get pneumonia and possibly live through it.

For organic and natural producers, the best way to deal with disease is to promote conditions that reduce the chances of disease. Don't purchase diseased stock. Quarantine new stock and slowly introduce them to your existing stock (for instance, across a fence line). Reduce environmental stress by providing warm and dry shelter. In cold weather provide deep bedding. Feed the pigs on clean ground or from a feeder. Provide abundant clean water. Provide shade and a wallow in hot climates. Rotating and resting pastures is a great way to reduce parasite loads. Allowing chickens to access former pig pastures is a natural way to reduce parasites.

Parasites

Pigs become infected with worms by eating the eggs that hatch and become the larvae. The eggs are thick-shelled and difficult to destroy. Roundworm eggs can remain viable in the soil for up to 10 years, which makes them difficult to control. Worms may be diagnosed by the presence of eggs in the manure. When processing pigs, worms may be found in the intestines, or "milk spots" on the liver may indicate their presence.

- **Adult roundworms** live in the pig's small intestine, but immature phases pass through the liver and lungs. All of the life-cycle stages cause a reduction in pig growth rate because the worms consume nutrients. Infected pigs may cough and have diarrhea. Roundworms are the biggest parasite problem on Greg Gunthorp's farm. Greg says it is worse without good forage. If he could, he would like to rest his pastures more, in order to really break up the roundworm life cycle, but he doesn't feel that he has enough acreage for that to be practical. He tries a little bit of everything for parasite control, including feeding diatomaceous earth and charcoal.
- **Whipworms** are smaller and live in the pig's large intestine. Symptoms include lack of appetite and diarrhea. The manure of infected pigs often contains blood and mucous.
- **Mange** is a parasitic disease of the skin caused by one of two mites—either *Sarcoptes scabiei* or *Demodex phylloides*. Sarcoptic mange (sometimes called scabies) is by far the most common; it is a skin irritant and uncomfortable for the pig, causing it to rub and damage the skin. It significantly depresses growth rate and feed efficiency.

As a last resort, using an antibiotic or medication is often the most humane way of treating livestock diseases. Simply letting an animal suffer and die is inhumane and far worse than treating or even euthanizing it. Treat the animal and then cull it from your program as you see fit.

Fencing

Physical barriers alone will work for pigs, but they will often suffer from wear and tear. Pigs love to rub and scratch on fences and a large sow or boar will bend even the strongest hog panel. We once watched as a 500-pound sow put her snout through the bottom of a hog panel and lifted the entire fence—T-posts and all—out of the ground and over her back. It takes a very strong physical barrier to stop a sow, which generally means an expensive and labor-intensive fence. The same sow can easily be contained with a single strand of electrified poly wire.

Permanent fence is best for perimeter fencing, fencing pigs out of sensitive areas (e.g., waterways, gardens, cropland), and for delineating large paddocks that will be cross-fenced using temporary fence. Pigs respond well to a woven wire fence with an offset electrified wire placed near snout height. This combination creates a barrier that pigs can't dig under or run through. The offset wire prevents the pigs from damaging the woven wire, and also becomes a power supply to any cross fence that you want to install. For a long-lasting offset conductor, many farmers will use high-tensile smooth wire, which will last longer than poly wire. Another wire option that we recently installed on a perimeter fence is maxi-twine from Premier 1 Supplies. It is braided so it isn't springy like high-tensile wire and it can be easily cut and tied to splice it. It can also be unrolled without special tools, unlike high-tensile.

For temporary fences we like to use electrified poly twine. Poly twine is plastic twine with strands of copper and stainless steel. It is lightweight and can be cut with a knife or scissors and is easy to wind up on a reel. It also has some built-in elasticity, which enables it to be tightened without tools or hardware. With older pigs and in low liability situations we use one strand of poly twine. With young pigs or pigs of varying sizes we use two strands. Two strands are usually enough to contain even the hardest-to-fence pigs. We have set up some three-strand fences in the past in order to contain sows and piglets in a high-liability situation and to contain cattle and pigs in the same paddock.

Energize cross fences by connecting them to the offset wire located on the perimeter fence. We end the cross fence where it meets with a T-post or a wood post on the perimeter fence by adding another T-post insulator or wood post insulator at the connection point.

Some people think that lop-eared pigs need to be fenced with a more visible conductor like rope or tape, but we have found this to be unnecessary. In fact, the farms that we have visited that have the worst, saggy fences are ones that use rope or tape. It is nearly impossible to keep the conductors tightly clamped on the rope/tape, plus the material starts to stretch with exposure to sun, snow, and wind.

We've used many different types of temporary posts for pigs, but have settled on fiber rods with one or two screw-on plastic insulators. The screw-on insulator allows the height of the fence wire to be customized along the length of the fence. This is important for fencing over varied terrain and for lowering the fence wire to drive over it. The benefit of being able to adjust this type of fence while it is energized cannot be overstated. It is also handy for raising the wire (while it is still hot) to "round up" an escaped pig. Simply raise the hot wire to the top of the post on one or two consecutive posts and then walk the pig past that section of fence. The pig will notice the opening under the wire and will scoot under it in order to rejoin his buddies in the paddock. You can just as easily lower the fence if you want to walk over it without the risk of getting zapped.

We have visited a lot of farms that use plastic step-in posts with built-in supports to support low- or high-tensile metal wire. Sooner or later the plastic posts break under the strain of this wire. With poly twine the wire stretches, but with metal wire the posts take the brunt of the force and they break, ending up as more landfill fodder. The metal wire is also less visible, so pigs are more likely to run through it. Some folks try to make it more visible by tying plastic flagging strips to it, but in our opinion this just looks junky. Metal wire is also more difficult to cut, splice, and roll up on a reel for temporary fences. The ease of moving your fence will directly correlate with how often you will be willing to move it. Save yourself the trouble and make your cross fences with poly wire on fiberglass rods with adjustable insulators.

The corners and any changes of direction to the fence line require extra support. Beginning farmers often try to use lightweight posts for this purpose and end up with saggy fences that often need attention. Rebar or T-posts work well in this situation to provide a solid post to pull to or to hold a change of direction. We have a bunch of recycled cut-off T-posts that we use for pig fence corners.

Pigs must be trained to electric fence in order for it to be effective. For some reason, when a pig is first shocked by an electric fence, its first reaction is to run right through it. A group of young pigs exposed to a double-strand fence for the first time will always result in numerous escapees. In order to prevent this situation the pigs should be introduced to electric fencing as a single strand offset from a physical barrier fence that they cannot run through. That way, when they touch the fence, they will learn to avoid the hot wire. After a couple of days they will all be trained to avoid the electric fence and can be moved to a pasture that only has electric fence.

Keep vegetation off of your temporary electric fence by string trimming the fence line prior to installation. Farmer Dan Wilson used to use herbicide to clean his permanent fence line, but now uses a string trimmer. He says that it takes half a day to trim the whole fence line and he only has to do it twice a year. Pigs love to root up clods of soil along the fence line, so it is important to regularly remove any dirt clods or adjust the fence line up (another advantage of the adjustable insulators). Direct soil contact will ground out your electric fence more than weed contact.

Handling and Sorting

Tom Frantzen has a Temple Grandin–designed sorting and loading corral on his farm. The corral features crowd gates and pressure release points that allow Tom to load a semi-truck in about 20 minutes, with very little stress experienced by the pigs. A double loading ramp allows two pigs to load side by side, separated by a see-through divider, which is much less stressful for the pigs than a single loading chute.

Farmer Dan Wilson has a portable corral that can be pulled by a tractor. He uses this corral to sort and process sixteen sows and their litters at one time. He sets it up in the corner of two lots so that he can work pigs from two different lots with one corral setup. It takes three people and a dog to round up the pigs. Dan says that a well-trained dog and a human could do all the work on the farm without outside help. Accordingly, Dan believes that up to $5,000 is not too much to pay for a well-trained dog that will save a lot of money in labor costs. He likes Border Collies and English Shepherds for that purpose. He wants a dog that will not be too aggressive with the pigs and will not bite and draw blood.

When we were farming on leased land we didn't have any handling facilities for our pigs, so we improvised. When we wanted to load and sort pigs we would use our livestock trailer. Often the pigs were already used to the trailer from being moved between properties and even on the same farm. If the pigs weren't used to the trailer we would park it so the pigs could access the trailer along the fence line. By feeding them a couple of times in the trailer, they got used to it. It's important to park the trailer in a way that the pigs can't tear up your trailer lights or wiring. By skipping a feeding and then "chumming" (baiting with food) the pigs onto the trailer, we could then sort the ones we wanted to the front compartment. This was time-consuming, but it worked. A variation of this method is to presort all the pigs that you want to haul and then stage them in a separate paddock or corral. Loading them at a later time doesn't require tedious sorting. Ideally a corral with a set of gates and sorting pens is the best and least stressful way to handle the pigs.

Meat Quality

Restricting dietary intake a few weeks prior to slaughter has detrimental effects on pork flavor (Stringer, 1970). Other feeds that negatively affect meat flavor include too much fish meal, spoiled meat scraps (this is illegal anyway), horse manure, and cooked garbage (Stringer, 1970). Don't feed more than 5 percent fish scraps or fish meal during the pig's growing phase (Davies, 1939) and withhold all fish products from the pig's diet at least 2 weeks before slaughter. Restricted feeding leads to leaner carcasses compared to free-choice feeding (Ellis et al., 1996). Intramuscular fat deposition is also reduced by up to 25 percent in the loin muscle of restricted fed pigs compared to free-choice-fed animals (Lebret et al., 2001). As a result, eating quality of pork can be negatively affected with lower tenderness and juiciness (Ellis et al., 1996). Restricted feeding makes sense for breeding stock, but not for market hogs.

However, if you are raising a breed of pigs that has a genetic tendency to put on a lot of intramuscular fat or backfat, perhaps too much for your customer preferences, you could consider free-choice feeding with lower energy feeds. According to Lebret (2008), a gradual reduction in lysine/energy ratio together with limited energy intake seems to be a more efficient strategy to modify rate and composition of growth at both carcass and muscle levels for improved pork quality, rather than free-choice distribution of a protein-deficient diet or, worse, feed restriction alone.

Fatty acid composition of pork can be easily manipulated through the feeding regime as a consequence of the well-known influence of the dietary fatty acids on fatty acid deposition in both intramuscular fat (IMF) and backfat (Wood et al., 2004). Grazing of grasses and consumption of oil-rich nuts such as peanuts, acorns, chestnuts, and hazelnuts has been shown to increase the levels of polyunsaturated fats (PUFAs), the omega 3/omega 6 ratio, and the levels of tocopherols (vitamin E). These all contribute toward a healthier pork for consumers while also making the meat less likely to go rancid and better for dry-curing (less lipid oxidation during storage due to the increased presence of tocopherols) (Lebret, 2008). Consumer taste tests have also shown they prefer the flavor of the fats and that the pigs taste less greasy and more oily. Overall, the level of juiciness increases as well with nut finishing, due to the increased IMF deposition (Lebret, 2008).

Other studies have shown that pigs raised outdoors exhibit less stress hormones and fighting at the slaughterhouse, and this leads to improvements in pork quality. The outdoor pigs thus had less skin damage, higher pre- and post-mortem muscle glycogen levels, and lower pH after slaughter (Barton-Gade, 2008). These all equate

Pig Production

Gunthorp Farm

Greg Gunthorp is a fourth-generation pastured pig farmer from LaGrange, Indiana. Greg's family had always raised pigs on pasture and sold them on the live animal market. In the region of Indiana around Gunthorp Farm outdoor farrowing and pasturing of swine was fairly commonplace until recently, with some farmers farrowing over one thousand gilts in "gilt-only" operations. In 1998 Greg was paid less for his pigs than his grandfather received during the Great Depression. He didn't want to be the last Gunthorp to raise pigs, so he decided to try direct-marketing his pork to get out of the economic slump. Greg noticed an increasing demand for local, humanely raised meat. He began calling a lot of restaurants to try to market his pork, but didn't get a warm reception at first. Greg says that he "really got lucky" when he found out, while chatting with some farmers after a farm conference, that the famed Charlie Trotter's restaurant in Chicago was losing their supply of milk-fed pork. Greg followed up on this lead and became their new pork supplier. This opened up the many restaurant doors that followed.

Currently Gunthorp Farm raises pigs, ducks, chickens, and turkeys, and processes them on the farm in their USDA-inspected abattoir. Greg sells his products in the Chicago, Detroit, and Indianapolis metro regions. He mostly sells to high-end restaurants including Rick Bayliss's Frontera Grill. Greg believes that you make your own luck (that it doesn't fall into your lap) and that marketing is a lot about networking and who you know.

Gunthorp Farm's pork production is a farrow-to-finish outdoor operation, with three groups of eighty sows each, finishing around three thousand pigs per year, all on a little over 200 acres also used by other animal species. Greg tries to have his sows farrow in the spring and fall, but some farrow in winter and summer. He mostly raises Durocs and Duroc crosses. His challenges include struggling to get an average of fifteen pigs per sow per year—sometimes he can't get two litters per sow in a year. He also contends with maintaining the right sow body condition for optimal breeding and keeping sows comfortable during the hot, humid summers.

Greg buys in boars each year for breeding. When he is done with them he sells the boars at auction or at the buying station. Some of his cull sows go to auction and some are used to add to the sausage trim. He gets a decent price at auction because his sows are pretty fat when compared to conventional sows that are often skin and bones. Because his buyers are not interested in highly fatty pork, he doesn't really have a market for sow meat himself.

to a better eating experience. Pig carcasses with high pH after slaughter often result in pink, soft, exudative (PSE) pork, which is usually discarded. We have yet to hear of a pastured pig producer having any problems with PSE pork. PSE is also a function of genetics: Highly muscled, highly stressed indoor-raised pigs are much more likely to have PSE pork.

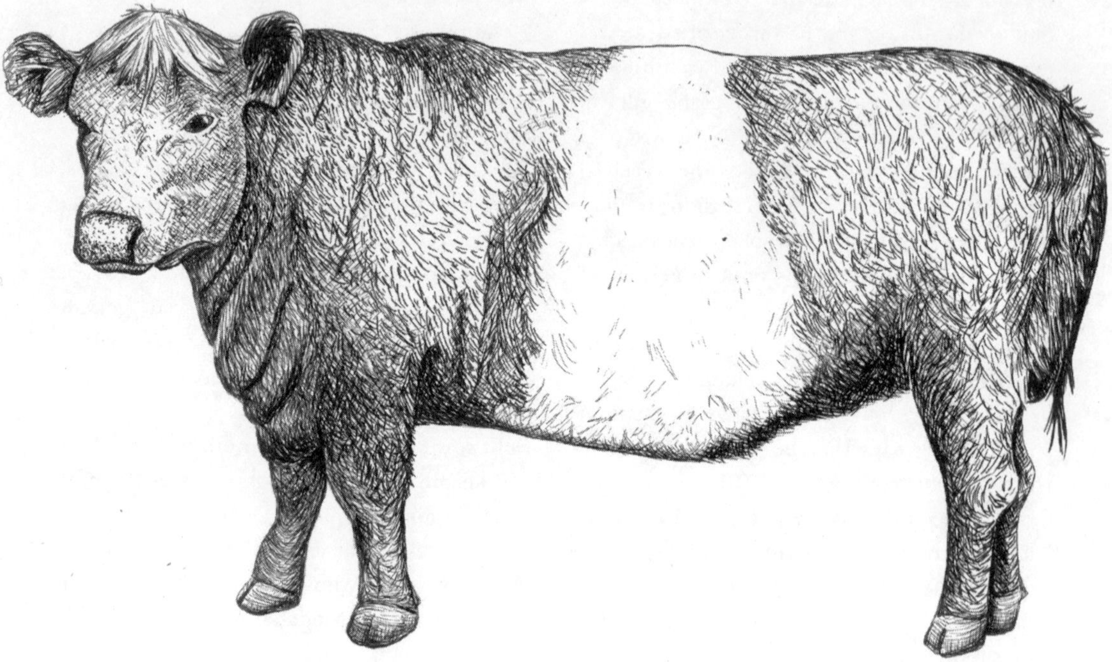

CHAPTER 5
Cattle Production

Production Systems

You probably already know that cattle are ruminants and are designed to eat vegetation. Accordingly, grassfed production systems will be covered in depth below. We will also describe cattle production systems that utilize limited grain feeding (but not feedlot production) during the finishing phase of production. You will need to take into consideration several factors, such as climate, land base, animal genetics, customer preferences, and others in order to determine the best system for you.

Producing good grassfed beef is both an art and a science; it is not easy or simple. Even though cattle have evolved to eat vegetation, there are a lot of factors at play, such as the animal's genetics, rainfall patterns and soil types, vegetative communities, and human management. In terms of genetics, keep in mind that for the last 70 years or so, we have been breeding for an animal that will gain muscle but won't become overly fat in the feedlot. Cattle that won't fatten easily on grain will be very difficult to finish on grass and stored forages. Animals that have the ability to put on fat while consuming forage alone are a minority of the animals that are available for sale. The art of

DEFINITIONS

Bull an intact male bovine used for breeding purposes

Calf a young bovine, 0–12 months or so, which has not been weaned

Colostrum the first milk produced by a mammal; full of antibodies important for the newborn

Cow a female bovine that has had a calf

EQIP Environmental Qualities Incentive Program, a program of the USDA Natural Resources Conservation Service to cost-share various on-farm conservation practices

Heifer a female that has not had a calf

Open cow a cow that is not pregnant; either she didn't take or has not been bred yet

Steer a castrated male used for finishing for beef

Yearling a young bovine that has been weaned and is not yet in finishing phase grassfed production really comes through experience, observation, and adaptation. The science is becoming more refined over time: Some of it will be described in this chapter.

Kit Pharro of the grassfed cattle seedstock company Pharo Cattle Company describes three types of management practices that will make the most efficient use of your available forage. The first is to use some type of rotational grazing. By moving the animals regularly to new paddocks you give the forage species time to grow back during the same growing season. It also allows you to stockpile forage for grazing during the dormant season. The second practice is to calve in sync with nature. This means that the land is producing the best and most abundant forage at the same time that the mother cow is experiencing her highest nutritional requirements while nursing her calf. The third practice is to produce cows that can do well on the forage available with little or no inputs. The cow must be highly adapted to her environment and produce the type of calves that will be profitable in the existing market.

Each region has a unique climate, soil type, and forage base. There are also cattle that are better adapted to different regional conditions. We will describe some of the regional models that have been developed to take advantage of these differences.

Southeast and Year-Round Grazing Climates

The humid Southeast allows for year-round grazing with a minimal need for forage stockpiling and very limited stored forage feeding if pastures are well managed. Seasonal irrigation may be required during drought conditions. Shade can be a limiting factor for pasture usage in the long, humid summer season in the Southeast. Overheating can reduce weight gains and fertility and cause other health problems in your herd. Parasites thrive in this region, and parasite prevention and management is a high priority for livestock producers in this part of the country. Shelters are typically not needed unless they are for shade or protection from monsoon-type rains.

During heavy rainfall times of the year, muddy conditions and soil pugging can become a problem, particularly on clay soils. Sandier, coastal plain soils don't have this challenge. Farmers in this region often do some type of mechanical pasture clipping seasonally, in order to keep the grass in a vegetative state when the cow herd can't keep up with forage growth. Some of the forages that thrive here, like Bermuda and Bahia grass, are not the best in terms of producing the average daily gains that you want to see in finishing cattle, but are fine for maintenance conditions. Farmers have tried fertilization, rotational grazing, and introducing other annual grasses and legumes to improve cattle fattening.

Southwest and California

The dry Southwest is characterized by short-season, range finishing (often no more than three months) that is sometimes extended with irrigated pasture or periodic summer rainstorms. Cowherds on extensive rangeland live on stockpiled forage in the summer and early fall, when the rains start again. Ranchers will often feed hay during the two or three coldest months of winter when forage production also slows down.

Generally ranchers in this region, without irrigation, have a short period of time during the spring and early summer to finish cattle. Many

Southwest ranchers will move their cattle to mountain pastures during the long summer dry season and down to lowland or floodplain pastures during the cold winters. For example, cattle rancher Joe Morris of San Juan Bautista, California, moves his later finishing cattle to mountain pastures to finish by June and July, after a first group has finished on the best forage of his home pastures in late April. His cows and calves stay behind to be maintained on stockpiled forage. Forage can be stockpiled throughout the summer since the grass dries while it is still standing. While this type of stockpiled forage won't be good enough for finishing cattle, it will provide feed for bulls, cows, and calves.

Midwinter is often the leanest time for grazing in this region. At this time of year the leftover summer-dried grass has been broken down by the moisture from the rains and the new spring grass has yet to start growing in earnest. Because the ground doesn't freeze up for very long in this part of the country, mud can be a big problem in the winter, and pastures can easily be damaged.

Many farmers at lower elevations do not need to provide winter shelter in this region, but higher elevations or snowy areas normally do call for shelter or at least some type of natural protection from the wind and snowdrifts. Access to trees or shade structures in the summer is important due to the hot, dry conditions. Again, cattle don't gain well in high heat. It always amazes us how many dark black cattle we see in desert country sweltering under the sun. Researchers from the USDA (Brown-Brandl et al., 2006) found that Angus cattle in a Nebraska feedlot had average hide temperatures in the summer of 101.6 degrees F (38.7 degrees C), while tan Gelbvieh were at 96.2 degrees F (35.7 degrees C) and white Charolais were at 91.5 degrees F (33 degrees C). Panting and other heat stress indicators were also higher for the Angus. Sadly, the demand for black-hided cattle in the commercial marketplace has influenced many ranchers to produce Angus cattle in regions that they aren't suited to. This is something to keep in mind when selecting the breeds you plan to work with.

Because of the short finishing window in this region, there can be a bottleneck at the processing facilities, which creates a problem for the rancher and the butcher. The ranchers struggle to get processing slots, and the butchers work their tails off and can't handle all the business during that time of the year. It is important to have a good way of storing commercial quantities of frozen meat if your business needs to sell meat over the course of the year. If the beef is going directly to the end customer at one time then this does not apply.

Northern and Seasonal

Cattle production in this region is characterized by a short forage-growing season of 3 to 5 months with stockpiled, stored, or brought-in forage for the rest of the year. Frozen ground allows access to pasture during months of the winter that would be muddy and prone to pugging in more southern climates. Cattle are often given some type of winter shelter or are moved to pastures that have protection from the wind, rain, and snow, such as forested areas.

Maintaining body condition in the winter can be challenging; many cattle producers try to finish their animals in 18 to 20 months so that they only have to get through one winter—and that first winter is on mothers' milk—in order to reduce feed costs. Summer pastures are lush (sometimes too lush) and are normally rainfed.

Breeds

Marbling Breeds or British Breeds

British cattle breeds include Angus (Red and Black), Hereford (polled and horned), Galloway, Devon, Dexter, Jersey, Shorthorn, Scottish Highlander, and Holstein. Some of these are considered dairy breeds (Jersey, Holstein, some Shorthorn) but can be used in beef programs for certain characteristics such as ease of fattening, smaller frames, calving ease (smaller calves), mothering abilities, and beef tenderness.

Some producers choose to raise dairy bull calves because they are easy to obtain, usually for a very low price (depending on the strength of the beef and veal industry). One must factor in the cost of milk replacer or come up with another source of milk such as dairy goats or a nonconforming nurse cow for it to be economical. When sourcing dairy bull calves, try to find a farmer who allows the bull calves colostrum from their dam. Dairy bull calves are delicate enough even with colostrum, and even more so without it (scours is more prevalent in calves that do not receive colostrum).

Continental Breeds or Exotics

These large-framed cattle breeds include Tarantaise, Salers, Limousin, Maine-Anjou, Gelbvieh, Charolais, and Simmental. Many of these breeds were used as oxen centuries ago and to a lesser extent are still used as oxen today. They are large-framed animals with a lot of fast-twitch muscling. They produce large, meaty carcasses, but generally have leaner and potentially less tender meat. Some grassfed producers have been able to cross these animals with British breeds or select for smaller-framed continental breeds for grass-finishing quality beef. (Gelbvieh and Maine-Anjou are a couple that seem to do well on grass.) The downsides of these animals are the difficulty to adequately finish them on forage alone and the potential for calving difficulties due to their large size.

Composite Breeds and Cross-Breeding

Included in this group are Murray Grey, Red Poll, Brangus, Black and White Baldies, Balancer, and many others. Composite breed cattle have the advantages of a cross-bred animal in a stabilized breed.

There is a science to cross-breeding that we can't possibly cover here, but there are a lot of advantages to mixed-breed cattle. Hybrid vigor is one reason to produce cross-breeds; hybrid vigor is the production advantage that can be obtained from crossing breeds, or strains, which are genetically diverse. The new combinations of genetic material can lead to production advantages over and above the average of the two parent breeds or strains. If you are breeding for meat production, then cross-breeding will likely be to your advantage in developing a mixed-breed animal that is best adapted to your regional situation.

However, some customers will be looking for a purebred animal. For example, there seems to be growing consumer awareness about the eating qualities of Scottish Highlander beef, and of course everyone goes goo-goo for so-called Angus beef, even though that is mainly marketing hype, since most "certified Angus" today is not even 100 percent Angus cattle—they just have black hides. However, a black-hided animal will always fetch a premium at auction, so it's a good fallback plan for producers to sell off extra animals or ones that don't make your finishing program. Even if you

TABLE 5.1. Characteristics of Different Cattle Breeds

Breed	Attitude and personality strengths	Maintenance level	Marbling and tenderness	Good crosses?
Red Angus and Black Angus	• can be high-spirited • not long-lived • known for quality meat	medium-high, large-framed (except Lowline Angus)	high	Angus × Hereford Angus × Shorthorn (for better mothering and milk production) Angus × Galloway (for smaller frames and hardiness)
Hereford (polled and horned)	• docile • good mothers • longevity	medium-high, large-framed	medium	Hereford × Angus cross (called Baldies) with Hereford bulls and Angus cows
Shorthorn	• docile • good mothers • good milk production (can be dual-purpose) • long lasting bulls	medium, moderate-framed	high	Shorthorn × Angus
Gelbvieh	• docile • good mothers	medium, moderate-framed	medium	Gelbvieh × Angus (called Balancer sometimes)
Jersey	• moderately docile • inquisitive • good mothers	medium, small-framed	high but more yellow fat	Jersey bulls × Angus or Hereford cows produce smaller calves, their heifer calves will produce more milk, plus add more marbling to the steers

have a goal of protecting a rare breed of cattle, you can still do terminal cross-breeding for beef animals while keeping another line pure for breeding purposes. More on the ins and outs of cross-breeding can be found in chapter 4.

Bos indicus *Breeds*

These include Zebu and Brahma. These animals are best for hot and sometimes humid climates. Although not noted for their eating qualities, some farmers choose to breed in one-eighth to one-fourth *Bos indicus* types for hot climate conditions. Most branded grass-finishing programs prohibit *indicus* cattle genetics in their protocols. That said, many independent producers in the South (-west and -east) have incorporated some *indicus* genetics into their herds for better adaptability to hot climates. The Beefmaster breed developed in Texas and Colorado is one such example of a composite breed that incorporated some Brahma genetics for hardiness, mothering abilities, and other desired characteristics.

Genetics for Grass-Based Finishing

The genetics for grass-based cattle production are not breed-specific, but rather bloodline-specific. It is your job to identify that bloodline. That's the hard part.

According to a paper written on the genetic aspects of beef marbling by Dr. A. D. Herring of Texas A&M University (2006) there are substantial differences in marbling ability across breeds of cattle, and within breeds of cattle (bloodlines). Heritability estimates of marbling ability have ranged from 13 to 88 percent in particular groups, with a mean value of approximately 45 percent. As a result, marbling will respond to selection in all breeds, but the amount of genetic variation is not constant within all breeds, and the relationship of marbling with other traits is probably not constant across all breeds. This means that marbling ability is *kind of* heritable.

Relationships between important cowherd traits and end product traits need to be considered in beef production systems. In the attempt to increase genetic ability of marbling, for example, producers need to be careful not to ignore, and thus possibly sacrifice, desirable cow functionality and reproduction traits. In plain English, this means don't sacrifice important mothering, longevity, and other traits in your quest for maximum marbling and size.

Kit Pharo asserts that many of the measurable traits and expected progeny differences (EPDs) do not directly affect ranch profits. Some breed associations are coming up with ways to measure and compare economically relevant traits (ERTs) instead of EPDs. Two rather new measurements that Pharo Cattle Company is excited about predict the amount of energy required by mature cows to maintain their body weight. Kit says that almost all of the bulls promoted as sires by artificial insemination and seedstock companies score poorly for maintenance energy. This means that the cows they will sire are too big and produce too much milk to be efficient in a grass-based, low-input system. That is the definition of a high-maintenance cow.

Pharo Cattle Company has also devised a measurement formula to predict longevity in cows sired by the bulls it sells. Longevity is the best predictor of a profitable and efficient cow. The old cow in your herd is one that produces a calf every year without fail, because you cull the ones that don't. Longevity requires the cow to possess a combination of fertility, structural soundness, mothering ability, correct frame size, productivity, efficiency, and good disposition. Kit concludes that these are time-proven animals that can seldom be improved upon.

Another trait that Kit keeps scores for in his bulls is for healthy hair coat. Bulls with a soft, healthy winter coat that is easily shed for a shiny short summer coat are scored 5 on a scale of 1 to 5. Kit learned from famed South African animal scientist Dr. Jan Bonsma that, for functional efficiency, there is no single factor that can give such positive results as early hair shedding, and that acclimated, well-fed, and hormonally balanced cattle share this trait. Kit recommends a high hair coat score for bulls destined to live in the South and Southeast because this indicates that they will be fully shed and less likely to overheat in the hot, humid summers found there.

One of the most important economic traits that Kit selects for is fleshing ability. This is a measure of the animal's ability to gain and maintain body condition. Cows with good fleshing ability will generally breed back sooner and have better longevity in the herd. Easy-fleshing cows also appear to have a higher level of fertility, which makes this trait even more important.

Disposition, or personality, is a highly heritable trait in cattle. Bulls with a bad disposition will pass this characteristic on to their offspring. Cattle that are high-strung are hard to work with and their stress instigates unruly behavior in the rest of your herd. Stressed cattle will also have lower meat quality due to the stress hormones. Stressed cattle may also fight during transport or while penned at the slaughterhouse, which can damage meat with exterior bruising or even puncture wounds if horned. Cull these animals out of your herd.

Breeding

There are two ways to conduct breeding: through natural mating or through artificial insemination (AI). Natural breeding is simpler, but it requires that you purchase, borrow, or rent a bull. AI is more complicated, but it allows you to get more specific about the genetics you want and to maintain a more disease-free, closed herd. Either way, you need to do some planning. At the minimum, follow these steps:

1. First figure out the best calving season: either in the spring when the grass is the best or in the fall if you have decent winter pasture, so you can take advantage of the peak market value for calves, which is usually in March through May. If you are keeping your calves to grow out yourself, then market value for calves is not your driver.
2. Make sure your calving season is no longer than 80 days, because you need to breed back your cows in the same year and gestation is around 280 days (280 + 80 = 360). Try to get all your cows bred during the same window of time. You could also split your cows into a couple of different breeding groups so they calve together but the groups calve at two different times of the year. This is helpful if you are moving toward more of a year-round beef production model. Of course, you have to have the pasture base to support this.
3. Plan and implement an estrus (heat) synchronization and AI program, or release bulls into the pasture with females. To get the cows cycling around the same time, run your bull in the adjacent pasture for at least 30 days. Even if you are going to use AI, you should use a teaser bull to get the cows into heat at the same time.
4. After each calving, cows should be rebred within 80 days. Do not rebreed a cow in poor health. Cows with a low body weight or other health conditions may lack the energy reserves needed to support simultaneous lactation and conception, so it is best to allow these cows to recover before rebreeding. The best production usually occurs in cows with a body condition score (BCS) of 5 or 6. Too fat (over 6) will also result in reproductive problems.

Life-Stage Considerations

Cattle have different needs at different points in their lives, and mother cows, naturally, have their own unique needs. It's important to look at the different nutritional needs during those life stages and between bulls and cows.

Cows and Their Calves

Gestation length varies to a small degree between the age of your cows and their breeds, but averages around 280 days. Make sure you mark breeding dates down on a calendar so you know when to expect calves. You may want to ultrasound your cows to make sure they got pregnant and cull those that did not take (these are called "open cows"). Peak lactation requirements are one month after calving, so it makes sense to have cows with one-month-old calves on your highest-protein pastures.

Moving the cattle at this time might be more difficult and stressful, and some precautions need to be taken in order to keep cows and calves together. Long moves and moves over terrain with limited line of sight might make it more difficult, due to cows leaving their calves behind in the old paddock as they move forward with the herd. Eliminating the back fence of the paddock in this situation will make it easier for the cows and calves to stay together but will allow the cows and calves access to already grazed paddocks and the flies and parasites that they contain. Cows and calves can be left on one big paddock, but they will be at a higher risk for disease, especially in wet weather.

Calving

Most beef cows give birth in the pasture. If this is the case, merely observe the birthing stages from afar and only intervene if the appropriate amount of progress is not being made. Pay special attention to first-calf heifers. Move your animals into a well-drained, grassy pasture for your cows to give birth, ideally with some tree breaks or woods for protection from the elements. The biggest enemy of the calf is wetness. Make sure the calf starts suckling right away, which will ensure that it receives its share of colostrum, and step in to assist if this does not happen naturally. Use caution when approaching newborn calves. Mother cows can be protective of their newborn and may be dangerous, so have a head gate or cattle chute ready to go just in case you need to milk the colostrum and bottle-feed the newborn.

To keep the pathogen load low in your pastures, limit the number of cows giving birth in one area and rotate pastures regularly.

In the southern states, both western and eastern, spring calving, when the grass is often at its best, means the lactating cow is getting her best nutrition. In northern latitudes, however, grasses are just getting going in the spring and lack the protein needed for your cows, thus you may need to supplement with some stored forages. Supporting your lactating cow with the best nutrition will translate into better gains for the calf. It also may mean that the calf will only have to go through one winter before finishing, which reduces feed costs (unless you are raising slower-growing breeds like Scottish Highlanders or Jerseys). A calf born in April, for example, could be finished by October of the following year, depending on the breed. That would be an 18-month-old animal. There are other factors to consider though, such as getting processing dates at your slaughterhouse. Many slaughterhouses are at their busiest in the fall with a combination of livestock and hunted animals filling their slots. Keep the market in mind as you build your calving schedule.

While visiting the Lasater Ranch in the short-grass prairie of Colorado, we learned that some producers are turning to nature to determine when to do their calving. The Lasaters' cows used to calve earlier in the spring, but the grass growth just was not good enough to support the needs of their lactating cows. They noticed that the deer and antelope that live on their ranch would give birth around June when grass growth was peaking with warmer temperatures. Now the Lasaters mimic nature by having their cows calve in June as well.

Some producers with more year-round markets opt for two calving seasons a year—one in the spring and one in the fall. The heifers and the steers grow at different rates, so this means you have animals finishing almost year-round.

Weaning

Calves should be close to having a fully functioning rumen when they are weaned. This usually happens around 120 days. Early weaning can be done if for some reason the pasture conditions are not fully supporting the lactating cow, and her body condition is dropping. We would recommend waiting until at least 8 weeks before early weaning. On the flipside, if there is no hurry to wean the calves, many producers wait up to a full 10 months before they wean the calves, giving the cows about a 2-month break to recover from the work of lactation before they calve once again.

Calves can be left with the cows for up to a month and a half before the next calving if the cows have the appropriate body condition. Sometimes the cows will self-wean the calves over the winter as their fat reserves are depleted. This makes the whole process a lot less traumatic and makes it easier to separate the cow and the calf at spring calving time.

Calves can be weaned from the cows over the winter if the BCS of the cow drops. Cows can be separated from the herd when their BCS drops to a predetermined level. The calf can be left behind with the familiarity of the herd where it can steal milk, if needed, from other cows. The cows removed earliest can then be fed supplements before calving, and might be candidates for culling due to their inability to maintain condition with the available forage. Joe Morris will wean his calves when all the cows get down to a 4.5 to 5 in body condition. The health and longevity of his cows are most important.

Weaning rings can be used to wean the calf while keeping the herd intact. Run the calves through a squeeze chute and apply the weaning ring. The cow will kick off the calf due to the pain from the weaning ring. After a week or so with no suckling the cow will start to dry up. At this point the calves are weaned and can be more easily removed from the herd. They should also at this point be used to consuming forage and other feedstuffs available. Remove the weaning ring within 7 to 10 days.

Natural weaning is one option where the cow kicks off the calf right before calving time, but this can be rough on the cows, as they are not allowed to rest between lactation and calving. In some open-range conditions or more laissez-faire type ranches this may be the default option.

Corral weaning is when calves are sorted out from cows and are left behind in the corral to eat hay or grain, but this can cause undue stress on the calves because of the lack of fresh forage and the possibility of higher parasite populations in the corral.

Calves can be weaned across a secure fence line with electric fence, called nose-to-nose weaning, which some consider the least stressful on the calves. Leave the cows in the adjacent field for a few days until you notice that they aren't paying much attention to each other. The animals will be making a lot of noise during this process, so don't do this next to your house or near your neighbors. Cows can then be moved to another part of the farm with little or no stress on the cows and calves.

First-calf heifers will have a stronger drive to get back to their calves. Sometimes open cows will have the same drive. It is important to have strong fences in order to keep them from trying to get back with their calves, as they will be crawling the fence line. In this situation it might be advantageous to put the cows in an escape-proof corral for a few days following weaning. Regularly check the condition and weight gain of weaned calves to ensure that they are healthy. Put the calves on your best grass and not in wet or dusty dirt conditions.

Castrating, dehorning, and other calf processing should not be done at the same time as weaning in order to reduce the stress on the animal. Wean the calves by removing the cows and allowing the calves to stay in a familiar paddock. Do not provide supplemental feed to the cows at this time; doing so will prevent the cow from drying up and will inhibit weaning. Yearling heifers or open cows can be left with the calves for a while in order to transition them with less stress.

Cows

Dr. Allen Williams recommends that ranchers start with the phenotype (what the cow looks like and its behavior) that they are looking for and the genotype (genetics) will follow. Dr. Williams says that the most important trait for profit is longevity. The average beef cow only has 4.2 to 4.6 calves before it is culled, but it needs to have at least 5 calves before it breaks even on the investment. Obviously, the national average is not adequate for profitability—we can do better than this.

Cows that are going to produce on grass need to exhibit a high level of fertility, low to moderate milk production, sound feet, and good legs. The cow's teats should be well formed and small enough that the calf can easily grab on to them. Selecting for high milk production has been shown to frame up cattle, which is not the desired phenotype for grass-based production. The stock that you acquire should be highly adapted to their environment. The cattle should be of moderate size with a frame score of 3 to 4.

Cows that produce well in a grassfed operation have frames that are deep in the midsection and midrib. The teats and udders should be feminine in appearance. The cow should have a clean and lean head, neck, and shoulders. The udder structure should be tightly suspended and the teat should be small. The cow should be short-legged and from above should have the appearance of a wedge that explodes toward the rear of the animal. The widest part of the cow should be the lower midrib and she should show a lot of side depth. For good examples of this kind of body type, see the photographs of grassfed cows in the color insert.

Heifers

Cattle producer Miller Livestock Co. in Ohio primarily uses heifer calves for their grassfed beef program. They do this for a few reasons: (1) the heifers marble more easily than steers; (2) their customers prefer the small size of the cuts, especially when they are buying sides; and (3) they can be purchased for cheaper than steer calves. Finding quality heifer stockers may be difficult though, since most cattle producers keep them as replacement stock.

Bulls

The bulls you select should be thick, deep, and meaty-type individuals, says Dr. Williams. Bulls should have a very masculine appearance. They should be thick in the neck and shoulders, deep in the heart girth and flank, and muscled in the hindquarters. Bulls from behind should have a lot of width in the rump, but the widest part of the body should be in the lower to midrib section. The bull should have a large capacity. From the front of the bull you should see the rib expression. Breeding bulls should not be overly fat, just like all breeding stock. Keep them in good breeding condition.

Small-acreage producers probably don't want to keep their own bulls due to the necessary space and pasture requirements. You can either contract a stud bull to come in, borrow your neighbor's for a trade, or consider artificial insemination. With AI, you can better control the genetics that you are breeding in as well as the timing of insemination to stay in line with your preferred calving schedule. However, AI can be expensive and does not always take, requiring multiple semen sticks and inseminations. Hire a trained AI professional or consider taking an AI training course before you try to do it yourself.

Finishers

You can finish bulls, steers, or heifers for meat. Bulls will grow faster and larger with less feed than a castrated steer, but will require stronger fences and more dedicated management (along with being separated from steers, heifers, and cows). Castrated steers are most commonly used for finishing animals, but heifers that you don't want or need for breeding can also be adequately finished. When your animal reaches 60 percent of mature body weight it will be at the point where it grows intramuscular fat cells or connective tissue (gristle cells). If a steer is gaining well from that point forward it will marble, if it is not gaining it will produce more connective tissue. According to famed Argentine beef expert Anibal Pordomingo, above 1.8 pounds a day weight gain is preferred in the finishing stage of grassfed beef. How to grow good grass to get those sorts of gains will be discussed later in this chapter.

In one study in northern California, feeding steers through two winters resulted in a higher death loss and higher feeding costs (Alexandre, 2013). Other regions with winter forage, such as the southern states, or farms that have ready access to stored feeds don't have this same problem. But it may point out the need to time your calving so that the steers don't go through two winters. Finishing a beef animal in the winter is challenging and potentially very expensive. However, it is hard to get grassfed steer to size in less than 24 months, so they may have to go through two winters.

Shelter

In some of the hotter regions of the country shade may be a limiting factor in terms of pasture utilization. Pastures without shade or with limited shade won't be ideal for subdivision if there is no way for the cattle to escape the unrelenting heat of the direct sun in a particular paddock. Hot cattle have all sorts of reproductive and health issues. Ranchers like Will Harris of Georgia are actively planting trees in order to create more shady spots that will allow for further subdivision of pastures during the hottest part of the year. Will likes to plant trees along his farm roads so that his pastures stay open for haying. Other ranchers are using shade structures on wheels or skids to provide a moveable shady spot, which also aids in spreading the animal impact and fertility around the pasture.

Feeding

Veteran rancher Walt Davis says, "When a cow is out grazing, she is working for you; when you are feeding a cow, you are working for her." He goes on to say that if you are regularly feeding large amounts of hay, then you should consider ways of adjusting to your environment. He recommends adjusting your stocking rate as your first course of action.

Ranchers in even the warmest non-brittle (subtropical) regions feed hay or other stored forage to their cattle during some parts of the year. It may be nearly impossible on even the most efficient operations to avoid feeding some stored forage, but ranching experts suggest trying to reduce the amount you feed and to find ways to get more grazing days out of your pastures. When you are feeding stored forages you are decreasing your profits. Every time you feed hay you are reducing your profitability, even if the hay came from your own land. Ohio State University researchers are finding that producers that feed more than two large bales (1,100 pounds each) of hay per cow per year are not profitable. The one benefit of buying in forage is that it adds fertility and biomass to your pastures, but there may be cheaper ways to improve your soil fertility. We once encouraged a grassfed beef producer who was cutting his own marginal hay and feeding it to his cattle during the dry season to actually stop cutting hay, rotate his animals more effectively, and buy in high-quality hay from elsewhere if he needed to. That would result in a net gain of fertility and water-holding capacity in his pastures, plus he could sell off all his expensive haying equipment that was tying up his capital. Once his pastures recovered, he could reduce his reliance on purchased forages and improve his overall profitability.

Options for winter feeding include providing stored forages or allowing the animals to self-harvest stockpiled forage. Stockpiling forage is accomplished by removing grazing animals from a pasture at some time during the growing season and allowing forage accumulation for later grazing. Stockpiling forage can also be achieved on a field cut for hay and allowed to regrow.

Ideally, forage is stockpiled after it has gone to seed, or been clipped by grazing animals or haying equipment, and then left to regrow until it is dormant. When growing forage for stockpiling, or any time forage is grown, the goal should be to provide the greatest amount of leaf material possible to grazing animals.

Kind and class of livestock is an important factor when considering whether stockpiling forage will be beneficial on your farm. Generally, stockpiled forage is of moderate to poor quality, so it may not meet nutrient requirements for finishing or lactating animals. However, it can economically maintain dry cows, bulls, or other breeding stock.

Any forage can be stockpiled, but some do better than others through the winter. Endophyte-free tall fescue, bromegrass, orchardgrass, even alfalfa and red clover to some extent are good options. Cattle will dig through soft snow up to a couple feet deep or so to reach grass, but a cover of ice will make that pretty hard. Keep some stored forages on hand in case the snow or ice are too severe. Choose pastures to stockpile that are nearest to your animal shelters and frost-free waterers to make logistics easier.

Some ranchers prefer to feed their cattle hay that they bring to their herd. This can be done by simply tossing out flakes by hand from the back of a truck, or more mechanized options may be used. Round bales can be delivered and fed to cattle with a round bale feeder attached to the back of a pickup truck. Big bales are often fed using a big bale feeding truck in larger operations.

Allowing the cattle access to the front of a haystack and controlling that access with an electric fence reduces the amount of handling involved with feeding hay. You could even stack the hay you baled in the field and control access to stacks with portable poly wire electric fencing. Electric strands may be moved up or down in order to

Grazing Management

All ruminants thrive in well-managed grazing systems. Although this section is included in the cattle chapter, this information applies to the production of sheep, goats, bison, elk, and deer, as well as any other ruminant animal.

Continuous grazing, where you set up a large fenced pasture and put all your animals in it to eat for the rest of the season, is not what we are going to talk about, because it is an all-around inferior system. Controlled, rotational, or management-intensive grazing (MIG)—all different names for the same concept of moving your animals around—can reduce soil erosion and compaction, increase forage production, improve fertility, reduce parasite loads, and usually increase the number of animals you can raise in a given area, sometimes doubling your yields. Improved forage production and diversity of forages contributes to healthier animals and more flavorful meat as well. This topic could take up an entire book, and indeed there are many great ones out there devoted to the topic (by Greg Judy, Jim Gerrish, Allan Nation, Joel Salatin, Alan Savory, *Stockman Grassfarmer* magazine, etc.). We will cover just a few salient points to get started; these are adapted from several ATTRA publications on pastures. Do these five steps, adapt, then repeat.

1. **Assess the current status of your pastures, including your soils.** A soil test or a forage test or both will take the guesswork out of choosing which type of amendments your pastures need. You may need to do some major nutrient, micronutrient, or pH corrections before you get started. Many pasture problems such as compaction, sparse plant growth, and noxious weed growth are caused by poor grazing management. Before you proceed to the next step, figure out why those problems exist in the first place.

2. **Decide whether to renovate existing pastures or establish new ones.** Planting a new pasture offers the opportunity to choose plant species best suited to the soils, climate, and livestock you want to raise. It also allows you to plant forage diversity that can extend your grazing season. However, it is expensive to reseed pastures and it requires copious tillage, something you may not want to do on certain soils or where erosion potential is high. It can burn up a lot of the carbon stored in your soil, and unless you have irrigation, it can be difficult to get reseeded pastures properly established.

   Renovating existing pastures might be a better strategy, particularly if you are low on equipment and capital. You could try overseeding more desirable species onto existing pastures using a no-till drill or by frost-seeding. Or you can feed seedy hay that is composed of the species you would like to see established. The livestock will eat the seeds and spread them around your land in their manure. Make sure you time the planting close to rainfall and keep your animals off of it while the plants are getting established.

3. **Design your grazing system.** This is not as simple as just turning the animals out to

pasture. A controlled grazing system is based on three parameters: grazing period, rest period, and understanding forage growth rates. Some people do high-density, short-duration rotations where they put a large number of animals in a small area for a very short time, sometimes 12 to 24 hours. Others use a lower stocking number and a slightly longer duration of 1 to 3 days. When grass production is at its highest for your region, moves are often more frequent and rest periods shorter. When it takes the grass longer to recover, the rest period is extended. Some people graze several livestock species together to more evenly graze the existing forage, such as cattle with goats or sheep with goats or goats with horses.

4. **Determine your grazing periods and rest periods.** In general, the shorter your grazing period, the more forage the pasture will produce. Cattle prefer green leaves to stems, so manage your forage to stay in leafing mode more so than maturing and getting "stemmy" (except when you are stockpiling forage for the winter). To minimize the potential for animals just selecting their favorite species, the grazing period should be no more than 2 consecutive days. This also limits the potential ingestion of parasites—many worm eggs don't develop into larvae for 4 to 6 days. If you have moved your animals on, the larvae will hatch and have nothing to eat, so they will eventually die. In areas with high parasite loads (or when raising parasite-prone species like sheep), a 30- to 35-day rest period seems to limit larval survival more than just a 20-day rest.

Don't let your animals graze below the growing points of the forage or you will lose a week or more of production. This also makes the plant less drought-proof because a short plant means a shorter root system (this is called the root/shoot ratio). Roots die off when the top of the plant is clipped. This can increase carbon levels in the soil due to increased root turnover (a good thing) but reduce the drought tolerance of the plant because it has less root mass and the roots are concentrated in the top layers of the soil where it dries out more quickly (a bad thing). Therefore, it is really a balance of the aboveground with the belowground biomass of the forage.

As for the rest periods, this is largely determined by how heavily your animals grazed in the first place and what species of forages you have, plus your climate. For example, orchardgrass and meadow bromes need at least 20 days of rest, while perennial ryegrasses need about 2 weeks. Legumes need longer rest periods, more like 20 to 35 days. Cool-season grasses also need longer rest periods. Seasonality will also affect rest periods; plants will always grow more slowly during the cold and will often go dormant during extreme heat. Shorter rest periods will require more nutrients, which you may have to supplement. They can also increase parasite loading.

Rest periods that are too long can have negative effects as well. Plants can become senescent and lignified, less palatable, less nutritive, and they won't bounce back as quickly. However, if your plan is to stockpile forage for the winter, you need to let some pastures go for that long rest period.

5. **Monitor and record.** Measure your pasture plant heights when animals enter and exit a paddock, along with how many days it takes to regrow to the same height. Also keep track of average daily weight gains by sampling a subset of your herd or flock. Estimate your average daily dry matter production per acre. Then figure out how many paddocks and what kind of stocking rate you can get away with. Test it, tweak, and repeat. Over time you should see an increase in the number of pounds of meat you can produce per acre. As 35-year pasture specialist Dr. Darrel Emmick says, "Keep one eye monitoring your animals, one eye monitoring your pasture, and both eyes monitoring your overall land health and your wallet."

If you are going to be a grass farmer, you owe it to yourself to invest in a course (or two) on the subject of planned grazing. A few good ones include Holistic Management courses, the Savory Institute, Ranching for Profit, and others that may be available in your region.

make the cattle clean up the bottom or top of the stack. The stack should only be one level high. The sites that you choose for feeding should not be prone to mud, and many individual sites should be set up in order to spread manure and animal impact. Choose locations that have protection from the elements and plan on using these sites during the months with the harshest weather. Only the front face of the round bales are accessible to the cattle, which means that they can't ruin the rest of the hay, and very little hay gets wasted in this way. The width across the face of the pile should be enough for all of the cattle in the pasture to eat at the same time. Setting up these self-feeding stacks should be done during the summer and fall when the ground is less muddy and conditions are better than in the middle of a snowy winter. Wrapped bales of hay can be left in the field where they are cut and access can be allowed using an electric fence. A silage clamp is a silage pile that is covered and fenced in a manner that allows controlled access using electric fence.

Wisconsin grazer and manager of the Wisconsin Grassfed Beef Producers Coop, Rod Ofte, says it is important to plan for harvested forage. Plan for consumption and have hay in storage, so you don't have to buy hay at high prices in the winter. You don't need a fancy shed to store and feed hay. Rod built a simple hayrack out of scrap wood that keeps the hay protected from the rain and prevents the cattle from wasting hay. Build a round bale feeder that's strong and easy to move, and has a bottom, so that the cattle can clean it all up. Another method is to feed with a bale feeder wagon. Make sure the feeder you choose is going to last and work for you. Don't spend too much money, and be creative. Whatever method you choose, it must be moveable in order to spread out animal impact.

Many ranchers in colder climates choose to use a feeding shed in a pole barn. Cattle are easily fed in the winter in this setup and can access feed through head gates, while the barn can also provide some shelter from the wind and rain, which may improve gains. However, this type of centralized indoor feeding system will require more manure management than multiple pasture feeding areas.

Watering

Average cattle water needs in gallons per day are the following: bulls 7 to 19, dry cows 6 to 15, lactating cows 11 to 18, and growing and finishing cattle 1 to 2.5 per 100 pounds of body weight.

After you have perimeter-fenced your pasture, it is important to determine where and how many water points you will have. Choose strategic locations for your water points in order to minimize the distance that the cattle need to travel. The location and amount of water points will determine how you can subdivide your pasture. Water in every paddock is ideal, but lanes are okay as long as the animals don't have to walk too far. Rod Ofte warns that if your finishers have to regularly walk over 500 feet to drink, they will lose weight or they will get lazy and not consume enough water for proper gains. For cows and calves, walking far for water is not as big of an issue.

There are many methods of providing water to your stock. Ranchers use solar- or wind-powered pumps, float valves, bubblers, black plastic pipe, quick couplings, and portable tanks and troughs. Ranchers can minimize negative impacts to pasture and reduce cattle health issues by moving or changing water points.

Effort should be made to keep water cool by locating it in the shade or providing some sort of shade structure over the water. Cattle will drink less water if it is too hot. If you utilize a natural watercourse such as a stream or spring, care should be taken to protect the water quality and the vegetation around the water body. Cattle should not have direct access to a creek or spring, but rather to a trough or tank that is fed off of it. Likewise, if you water your animals from a well, keep the animals out of the area directly around the wellhead. Draw at least a 100-foot radius out from your wellhead and fence animals out of it. Also make sure that runoff from animal pastures does not flow toward your wellhead. These good water quality practices are true for all animal production, not just cattle.

Soil, Nutrients, and Plants

Dr. Allen Williams says that all good beef production starts with the soil. Healthy soil should have an incredible array of life. Ninety percent of soil function is mediated by microbes. Microbes are dependent on plants. The fungi-to-bacteria ratio and the predator-to-prey ratio are important indicators of soil health. What we find is that most soils in pastures are bacteria-centric and prey-centric. They contain very few predators and very little fungi. Predators are important because they consume prey bacteria and release nutrients. Fungi are important in making nutrients available to plants. Healthy pasture soils should also show ample evidence of earthworms.

High levels of organic matter in the soil benefit the soil in many ways. A 1 percent increase in soil organic matter equates to an increase in water-holding capacity of 25,000 gallons per acre. Soil with 5 percent organic matter can hold about 53 percent of a rainfall, and at 8 percent it can absorb 85 percent of a rainfall. High soil organic matter reduces erosion and nutrient runoff. Ranchers can increase soil organic matter with high plant diversity and managed livestock impact.

Short grasses have short roots, which mine the nutrients in the top strata of soil. Short roots also run out of water quickly. Diversity of forages allows for the selection of other species that are still palatable to the animal. With diverse plant species we expand the root systems to expand the territory for microbes and increase mineral uptake. Ranchers using animal impact and forage diversity can see a 0.5 to 1 percent annual organic matter increase each year, until a threshold is reached. Without cover on the soil the temperature gets too hot for plant growth, because water is unavailable.

A Brix measurement, which measures the amount of plant solids to water in a plant, will give the true nutrient density in a particular forage, notes Dr. Williams. Solids in the plant include sugars, minerals, lipids, pectins, amino acids, and proteins. To measure Brix in your forages you can use a $120 refractometer to get a reading. The degree of finish on cattle is dependent on the Brix of the plants they consume. Young grass is not a finishing diet and will cause the cattle to have runny manure from the high percentage of protein. High-protein pasture, when used for finishing, can also give the beef a liver-like off-flavor. The sugar-to-protein ratio found in pastures slightly beyond mid-stage maturity is good for finishing cattle. High Brix makes plants more resistant to pests, freezes, diseases, and drought.

Higher microbial numbers in the soil lead to an increase in Brix. An increase in the Brix of your pasture forage directly increases the amount of gain per day that your cattle can achieve. For example, a Brix reading of 5 degrees or less will only allow your cattle to gain in the low 1s (pounds per day). A Brix level of 12 degrees will enable gains in the low to mid 2s. Brix levels of 12 to 15 degrees can allow for gains in the high 2s, and a reading above 15 degrees Brix can yield gains in the high 2s and up to 3 pounds per day. For every 1 degree increase in Brix you can add 0.1 to 0.3 pound of average daily gain. Brix levels are always highest during the middle of the day. For this reason, it might be advantageous to move your finishers to new pastures during this time of day. A Brix reading is raised by a plant's ability to access more nutrients in the soil.

Keeping Healthy (Adapted from Heather S. Thomas, *Cattle Health Handbook*)

One of the most effective ways to keep contagious diseases off your farm is to keep as closed a herd as possible. This means avoiding bringing in new cattle unless you are sure they are disease-free. It's often safer to raise your own heifers than to purchase new cows, and usually cheaper too. If you buy cows, heifers, or a new bull, buy from a reputable breeder, where you know the animals' genetics and health history. You can also use AI so that you don't have to bring in new bulls all the time.

If you purchase new animals, keep them isolated for a couple of weeks to make sure they are not incubating a disease; this is a good practice to use for all livestock and poultry species. If an animal becomes sick while in the isolation paddock, you have an opportunity to clean that area and remove all the manure, rather than having the sick animal infect your herd. Depending on the illness you may elect to cull that animal or seek veterinary care.

Diet as Cornerstone

An adequate and balanced diet is the cornerstone of all livestock health. Each class of animal—young calves, lactating cows, dry cows, bulls, or yearlings being finished for meat—needs a sufficient amount and balance of feed and minerals to meet its particular needs. Poor nutrition is the underlying cause of diseases such as scours, respiratory illnesses, and foot rot, and it causes infertility in adults and slow growth in young animals. Pregnant cows with insufficient protein levels don't produce enough colostrum for their newborn calves, making them more vulnerable to disease during their first weeks of life.

Trace mineral deficiencies can be corrected by adding them to your soils, using supplements added to salt mixes, or giving them to each animal orally or by injection. Consult your vet, extension agent, or a cattle nutritionist to examine your feeds and help you make necessary adjustments. One farm we visited in Texas provided a trace mineral and kelp bar to their cattle, where the animals could pick which minerals they needed most. Also, by providing the minerals in the diet of the animal, it helps spread those nutrients around the farm in their dung. Over time, this can help partially correct soil deficiencies.

Always provide adequate sources of clean water. Dirty water may spread disease. If cattle are short on water, they suffer from dehydration or impaction, and steers may develop urinary stones if they don't drink enough during cold weather, causing their urine to become too concentrated.

Parasites

Even if cattle look healthy, internal and external parasites may be robbing them of nutrients. This can result in lower weaning weights in calves, cows producing less milk, lowered immune system activity, and lower fertility rates.

External parasites such as flies, mosquitoes, and ticks carry disease and can steal an animal's nutrients by sucking blood. Solutions include herbal sprays, fly parasites, insecticide applicators such as back rubbers or dusters, or insecticide ear tags. Good manure management underpins any fly control program. Either don't let your manure build up in any one spot through good rotational grazing or consolidate your manure and manage it with hot composting that kills off most insect eggs.

If cattle are spread out on large pastures or moved frequently, internal parasites are not as great of a problem. In small areas, however, cattle continually graze where they defecate, picking up worm larvae that hatch from eggs passed in manure. Because most of the larvae are on the lowest part of the plant, overgrazed pastures—where plants are eaten off close to the ground—are most risky for re-infecting the cattle.

Break the worm's life cycle and keep parasitism to a minimum by seasonal worming of only infected animals (instead of whole-herd worming), avoiding overgrazing, and moving cattle to a new pasture after they are dewormed. There are also natural herbal wormers and anthelmintic plants, many which were mentioned in chapter 3. Rotating poultry with cattle will also reduce external and internal parasites, because the birds will eat many of the larvae and they will scratch up the cow patties, exposing them to air and sun, which kills many parasite eggs and larvae.

Tetany and Bloat

In early spring, fast-growing cereal grains or certain grasses may cause a metabolic disorder called grass tetany, a magnesium deficiency that particularly affects lactating cows. Avoid the problem by using these succulent pastures for weanlings or dry cows during risky times of year, or make sure every animal gets an adequate magnesium supplement.

Some legumes, especially young, lush alfalfa, can cause bloat. Try mowing the legume pastures a few hours ahead of turning your cows in, so the plants are starting to wilt and dry before being eaten. Some farmers say that bloat is reduced if you turn cattle onto lush and leguminous pastures in the afternoon instead of the morning. If pastures are botanically diverse and animals have sufficient selection, there is less chance of bloat. Additionally, if animals are full when they are let into the new pasture there is less chance of bloat.

Scours

Calf scours, or diarrhea, causes more financial loss to cow-calf producers than any other disease-related problem they encounter. However, calf scours is not a disease but a symptom, which can have many causes. Since a calf is approximately 70 percent water at birth, loss of body fluids through diarrhea can produce rapid dehydration. Dehydration and the loss of certain electrolytes produce a change in body chemistry in the calf. Although infectious agents may be the cause of primary damage to the intestine, death from scours is usually due to loss of electrolytes, changes in body chemistry, dehydration, and change in acid-base balance rather than by invasion of an infectious agent. The infectious agent that causes scours can be a virus (bovine virus diarrhea or BVD, rotavirus, coronavirus), bacteria (*E. coli*, *Salmonella*, enterotoxemia), or protozoa (coccidiosis, *Cryptosporidium*).

Treatment for scours is very similar regardless of the cause, and should be directed toward correcting the dehydration, acidosis, and electrolyte loss. Dehydration can be overcome with simple fluids mixed with electrolytes given by mouth early in the course of the disease. If the scours progress, antibiotic treatment can be given. The age of the calf when scours begins is an important consideration in its survival. The younger the calf, the greater the chance of death.

Recent research has indicated that many scour cases can be directly related to lack of colostrum intake by the newborn calf. A calf that is well mothered and consumes 1 to 2 quarts of colostrum in the first few hours after birth absorbs a higher level of antibodies and is far less susceptible to scours and other calfhood diseases. Most beef calves will have received colostrum, but sometimes dairy bull calves won't because they are not needed by the dairies. If you use dairy bull calves for beef production, try to find ones that at least received colostrum and mother's milk for a few days or buy colostrum to administer yourself.

Stress

Stress contributes to higher incidence and severity of disease. Stressed cattle do not eat well, so stress interferes with proper weight gain, reproduction, and disease resistance. If susceptible animals are highly stressed at the same time they are exposed to disease, they generally become sick.

Inadequate nutrition, as mentioned, can physically stress animals, as can bad weather. Provide shelters to minimize heat or cold stress. In a hot climate, use pastures with shade or create shade structures tall enough to allow good air movement above the cattle. Winter pastures need windbreaks—either a grove of trees or manmade structures will be adequate.

Cattle become stressed when overcrowded, during weaning, when normal social interactions are disrupted, or during improper handling. Stressed animals produce more cortisol, a hormone that helps them cope with short-term stress by changing body metabolism to help it function better. But over a longer period of time, the extra cortisol hinders the immune system. Lungs are especially vulnerable, since some pathogens are always present in the respiratory tract, waiting for an opportunity to invade the tissues.

Common stressors in cattle handling include moving, sorting, vaccinating, branding, dehorning, tagging, castrating, weaning, and transporting. Avoid doubling up on stresses if at all possible. For example, don't dehorn, castrate, and brand calves at the same time you wean them. It's best to perform some of these procedures when they are small and still suckling.

Develop a quiet and conscientious way of handling cattle. When they are handled gently and with patience, rather than enduring chasing dogs, cattle prods, and manhandling, they learn that coming into the corral is not frightening. Ultimately you will have healthier animals and better meat quality as a result of less-stressed animals.

Fencing and Pasture

Permanent fences for cattle are often used for pasture perimeters and for large pasture subdivisions. These fences can be post-and-rail, woven wire, barbed wire, high-tensile (electrified or not), or multi-strand electrified poly wire. Of these options, the electric fence is the cheapest in terms of labor, as well as the easiest to build.

Single strands of portable electric fencing can be attached to the permanent fence to further subdivide the pastures for maximum grazing efficiency. Cattle need about 2,000 volts to stop them. Cattle must be trained to respect electric fencing starting at a young age, so that they don't just go tearing through it on their first encounter. Train them by confining them in a corral or secure pen with a hot wire or two attached to the physical barrier fence. After getting shocked a few times over the course of a couple days, they will be ready to be released into a pasture fenced only with electric fence. (For more on electric fencing see chapter 3.)

Cows waste less grass by trampling when there are more pasture subdivisions. Portable electric fence allows you to be flexible with your grazing. When there is a lot of grass production, paddocks can be made small. Conversely, when there is less grass (or grass production has slowed down) the paddocks can be larger. With portable fencing the rancher is able to seasonally customize the paddocks. In winter you can ration grass across a pasture with one all-weather water site by eliminating the back fence and allowing the cattle to walk back across already grazed areas to reach the water point. Simply move the front fence daily to ration the grass.

Cows become easy to move once they get used to the daily pasture moves. Using the same easy-to-recognize call will alert the cows that it is time to move to a fresh pasture. Cows will herd up and compete more for the grass, while being less selective about the type of forage they are consuming. This herd or mob action spreads manure and urine more evenly over pastures and will also trample down most of the inedible forage. The trampling of the plants and the manure combine to fertilize and lightly till the soil while bringing much of the standing biomass into contact with the soil organisms, which hasten the breakdown of nutrients.

Rod Ofte advises not to fight Mother Nature when building your fences. Don't fence around every little curve and tree, but instead keep fences streamlined. Every little deviation from a straight line takes more support structures, which cost time and money.

USDA EQIP financing has cost-share money available to help defray the cost of building fences on your farm, especially if the fencing helps keep animals permanently or seasonally out of sensitive areas such as wetlands and stream courses. Some farmers may receive up to 90 percent of cost-share help in implementing their fencing plan.

Rod also has some good advice when it comes to what kind of fence you should build. Some folks like a five-strand barbed-wire fence and others prefer a four- or five-strand high-tensile fence. Fences that are up against the woods, where cattle won't push as hard, can be four strands. Cows and calves are easy on fences, but yearling steers and heifers will push, and will need a more substantial barrier. Internal fences don't need to be as substantial. Fence more strongly around your liabilities. If you are located near a busy road or your neighbor's corn field, then the liabilities are pretty high, and that needs to be taken into account when designing your fence. It is important to build strong corners and supports so your fence won't sag.

Handling and Slaughtering

Handling facilities, such as those described in Temple Grandin's book *Humane Livestock Handling* (Storey Publishing, 2008) will become essential as you scale up your beef operation and regularly load animals for slaughter. You need to have a safe and sturdy infrastructure for gathering, sorting, and selecting the animals that are ready for harvest.

Traveling to slaughter on a livestock trailer can be a potentially stressful event for your cattle. The animal's genetics and how often it was gently handled throughout its life will play a big role in how calm it remains on the trip to the slaughterhouse. Cattle that have been rotationally grazed using low-stress stockmanship methods, and have had regular contact with humans, will be much calmer when transported than a range steer that has not. We've already mentioned how calm disposition is very heritable in cattle. Ranchers like Joe Morris, who has been culling "high-headed" cattle from his herds for years, experience few problems in transport. Even one agitated animal can cause the stress to spread to all the animals on the trailer. This can lead to injuries and the release of adrenaline in the animal's body, both of which will lead to diminished meat quality.

Beef Quality

A grassfed animal that is ready for slaughter should have a proper degree of finish on the top line, tail-head fat, cover over the ribs, depth and fill in heart girth and in the flank, and depth and fat deposit in the brisket. The animal should have a heavy, distended brisket that is not held up tight, round or even indented over the top, no ribs showing, and pones of fat on the tail head. The tail head is the last place to deposit fat (called "poning").

Weight is a better indicator of readiness for harvest than age. Although feedlot cattle are now being finished in as few as 14 months, grassfed animals need longer, usually 18 to 24 months or so, depending on the genetics, gender, and forage season. A finished steer should weigh around 100 pounds more than its mother and a finished heifer should weigh 25 to 50 pounds more than its mother, according to Dr. Allen Williams. Keep in mind that if you take longer than 30 months to finish a beef cow, the USDA will require that you bone out the entire spinal column due to bovine spongiform encephalopathy (BSE) concerns. This means you won't have those pretty bone-in rib eye chops that your customers may be looking for. However, some breeds, like Scottish Highlander, simply take longer and some producers may opt for longer-maturing animals for more depth of flavor.

Always harvest beeves when they are "on the gain," which means when they are actively gaining weight and not losing body condition. Brookshire Farm in Blanchett, Louisiana, follows the advice of Brazilian grassfed beef researcher Dr. Anibal Pordomingo, opting to harvest their beeves after they have been gaining at least 2 pounds a day for at least 6 weeks. Brookshire Farm believes this provides the tenderness that their grassfed beef customers desire. It also follows the work of Dr. Guillermo Scaglia of Louisiana State University about improving meat quality through diversifying pastures.

Dr. Scaglia headed up an experiment with three different pasture systems for grass finishing in the South. He found that pastures with the most diverse forages finished beef that resulted in the highest taste scores, moisture content (which translates into tenderness), color, and appearance (Bhandari et al., 2013).

Beef breeds gain more weight at a lower cost; however, dairy breeds (such as Holstein and Jersey) will grade higher and typically have higher intramuscular fat, leading to more tenderness of the meat (Rust et al., 2005). Eating quality of dairy breeds has to be balanced with the economics of raising them, because they produce smaller carcasses. If you have a ready supply of dairy bull calves and a way to cheaply feed them (buying milk replacer is *not* the cheap way), then they may be a nice addition to your beef program. Dairy cull cows, while not tender, provide an excellent ground beef product. You can also cross in some dairy genetics for improving the marbling abilities of your beef animals. Likewise, you could raise up the dairy bull calves for rose veal (veal that has color because the calves are allowed to move around rather than live in a crate), opting to slaughter them at around 5 to 7 months instead, avoiding the costs of raising them up longer. The markets for rose veal are strong and growing.

Finishing on grain can help cover up bad management and breeding, but grassfed beef quality is more clearly obvious: You can't hide behind the flaked corn. There is a lot of poor-quality grassfed beef out there, so much that it has given grassfed beef a bad name and a poor taste experience for many people. Mark Schatzker, while writing the book *Steak*, tasted beef from around the world. He had both good and bad eating experiences with grainfed and grassfed beef. One grassfed beef out of Texas he likened to the taste of "swampwater," while other grassfed beef he tried was tough and gamey with a seriously "off" flavor. On the other hand, slow-raised grassfed Scottish Highlander beef in Scotland was one of his favorites (adapted from Schatzker, 2010).

In traveling the world and talking to experts, including Dr. Allen Williams who is mentioned so often in this chapter, Mark found that the following issues contributed to poor beef quality. Clearly, the opposite of these would contribute to better beef quality. These issues included:

- Cattle were not ready to be harvested, not fleshed out enough.

- Cattle were not "on the gain" when they were harvested, meaning they were losing weight or holding steady instead of consistently gaining weight.
- Cattle were not grazing the right kind of grass during the finishing period.
- Cattle were grazing spring grass that had too much protein and not enough carbohydrates, leading to off flavors.
- Cattle finished on fescue that was infected with fungus, giving it an off flavor (this can happen with grain-finished animals as well, if the grains have ergot or other fungi).
- The beef came from breeds or breeding lines that don't finish well on grass (or they were designed for grain finishing). These are often taller, larger-framed animals that are considerably higher maintenance.
- The carcass had been dry-aged for too long, particularly if it was a lean carcass without much fat cover to protect the meat during aging; molds could get into the meat, giving it an off flavor.
- The carcass was "cold-shortened"—the hot carcass was chilled too rapidly in a very cold cooler (resulting in tough beef) instead of transitioned more slowly to cold storage.

Dr. Williams developed a method to assess live cattle to ascertain if their meat would be of high quality. It involves using an ultrasound machine. He uses four different measurements to give the animal a score. First is measuring the marbling in the longisimmus muscle (the rib eye section) using ultrasound on the live animal. Next he examines the connective tissue: how thick it is and how it runs with or across the muscle grain. This will indicate how much gristle, and hence lack of tenderness, the beef will have. Williams then has a way of analyzing the amount of stress the animal has been enduring, because a highly stressed animal will burn up much of its fat to handle the stress. If marbling is missing from the top of the longisimmus, he can tell that the animal has been stressed. He then uses the ultrasound to look at a rib eye cross-section, just like a meat grader would on a beef carcass, to see how big the rib eye area is and how much fat cap is on it. Live animals that don't make the grade can be separated out and finished for different purposes, or perhaps sold to a feedlot for fattening.

Of course, not all cattle ranchers can afford an ultrasound machine. However, it may make sense to pay for an expert to test any breeding stock you are considering purchasing or to come out once to assess how your pasture management and feeding regime seem to be working. Even if you can't do this all the time, it will give you a sense of how well your system is functioning with regard to meat quality. Eating the meat is obviously a good way to assess meat quality too, but you can't eat your way through every animal to tell you if you are doing a good job!

Post-Harvest Handling

Beef tenderness is determined by genetics; by how the animal was raised, killed, cooled, and aged; and by how the meat was cooked. Longer aging will help tenderize meat, which is especially important for typically leaner grassfed beef. However, the longer you age the more you have to trim because a hard crust forms on the outside layer of meat and fat. If the body is covered with more fat, like a yield grade of 3 or 4, that hard fat can be trimmed off while the meat underneath is fine. A leaner animal will have more exposed meat with less fat covering, and therefore more meat will be exposed to the

Morris Grassfed Beef

Morris Grassfed Beef, run by husband and wife team Joe and Julie Morris, is located in the oak-studded coastal hills of central California. Until the drought started rearing its ugly head, this was ideal cattle country with a long growing season, a coastal influence to moderate temperatures and keep grass production going, and large oak trees for shade. The Morrises have been raising cattle here for six generations now. Their cattle are a fixture on the hilly countryside on the outskirts of Watsonville, San Juan Bautista, and Hollister.

When we were living and farming on California's central coast we used to sell Morris beef along with our meat and eggs. Our customers loved Morris beef, and their ethic for taking care of the land was also fairly well-known in the area, especially among the foodie crowd. In the past few years California ranchers like Joe have been experiencing a severe drought. When we spoke to Joe at the end of March 2014, he told us that they had only received an inch and a half of rain since the first of the year. Late winter and early spring are generally the wettest months in central California, with more like 15 to 20 inches during those months.

We were hoping to hear Joe's perspective on the drought and what strategies he had implemented on his ranch in order to make it through the dry weather. What Joe said left us with a much better understanding of the severity of the drought and the hard decisions that ranchers were taking to stay economically viable. The Morrises sent their 168 mother cows to Nebraska to graze, with hopes of bringing them back when the rains came, but the rains had not come. Now they were getting ready to sell the herd that they had spent decades building. Joe had also sold all of his weaned calves in order to have enough grass for his finishing animals. At the time we spoke to him Joe was only running his finishers and considering getting out of the cow-calf business altogether and instead concentrating on finishing stockers. He said that the only silver lining was that he was able to sell his animals at a good price, and that maybe it was forcing a change on him that might lead to a more sustainable type of operation in the future in terms of quality of life. Reflecting on this, Joe suggested that all ranchers develop a "drought plan," for drought will happen, whether short-term or long-term. Selling cows is a difficult decision, but degrading the land is not an acceptable solution, which would have been the result had the Morrises tried to "feed their way out of the drought." One can learn and be flexible given the manifold challenges of working with nature, and one needs to do both if one is to be sustainable.

Joe tries hard to manage his animals on pasture so that he doesn't have to buy feed for them. He provides them with the minerals that are lacking in his soils and only feeds them hay in special situations. Sometimes Joe will get grazing contracts on state land. Some of this land hasn't been grazed for years and has a lot of old lignified feed. In these situations he will feed a couple of pounds of alfalfa per day in order to bump up their protein consumption.

Joe likes his cattle to always be gaining at least a little and not be going backwards. He manages them to be gaining 2 plus pounds per day from early March until they are finished. Generally his cattle will finish in 18 to 24 months.

Joe doesn't have the facilities to weigh his cattle often, but he will sometimes get an average when his animals are transported by weighing the truck. He uses visual cues to determine finish, relying mostly on the presence of brisket fullness and fat deposits on the tail head. Joe's cattle regularly grade high Select to low Choice, which he says works fine for his customers, most of whom are individual families. Joe's beef is plenty tender, but the thing that sets it apart is its delicious complex flavor due to his diverse pastures and rotational grazing.

Most of Joe's cattle are naturally polled and he will dehorn the rare individual that has horns, stemming from breeding some Hereford into his herd years ago. Joe doesn't mind working with horned animals, but is most concerned with the possibility of bruising on the way to slaughter. His animals are a mix of Angus (Red and Black) and Hereford genetics.

His cows and bulls stay on the hilly rangeland around the central coast. His bulls are from Colorado ranchers Kit Pharo and Leachman Cattle Co. and have been selected to produce animals that will perform well on grass.

Joe markets his beef as split halves to individual customers, as discussed in chapter 8. He also makes his beef available to customers in a CSA-like scheme. Customers without freezer space can pick up their beef in manageable amounts at predetermined drop-off points at predetermined times. This is made possible by the availability of a nearby frozen storage business to warehouse the meat.

drying process. That meat will have to be trimmed off, which represents a loss of money. You could turn that hardened meat trim into dog food to gain some value off of it, which is what some butchers do. Can you charge a higher price for longer-aged meat? Only very high-end restaurants will care about this. It seems that many producers and processors compromise on the aging process by going for a 10- to 14-day dry-age and no longer to reduce excess shrink (not only weight loss from drying, but more trim that has to come off). We'll discuss aging further in chapter 9.

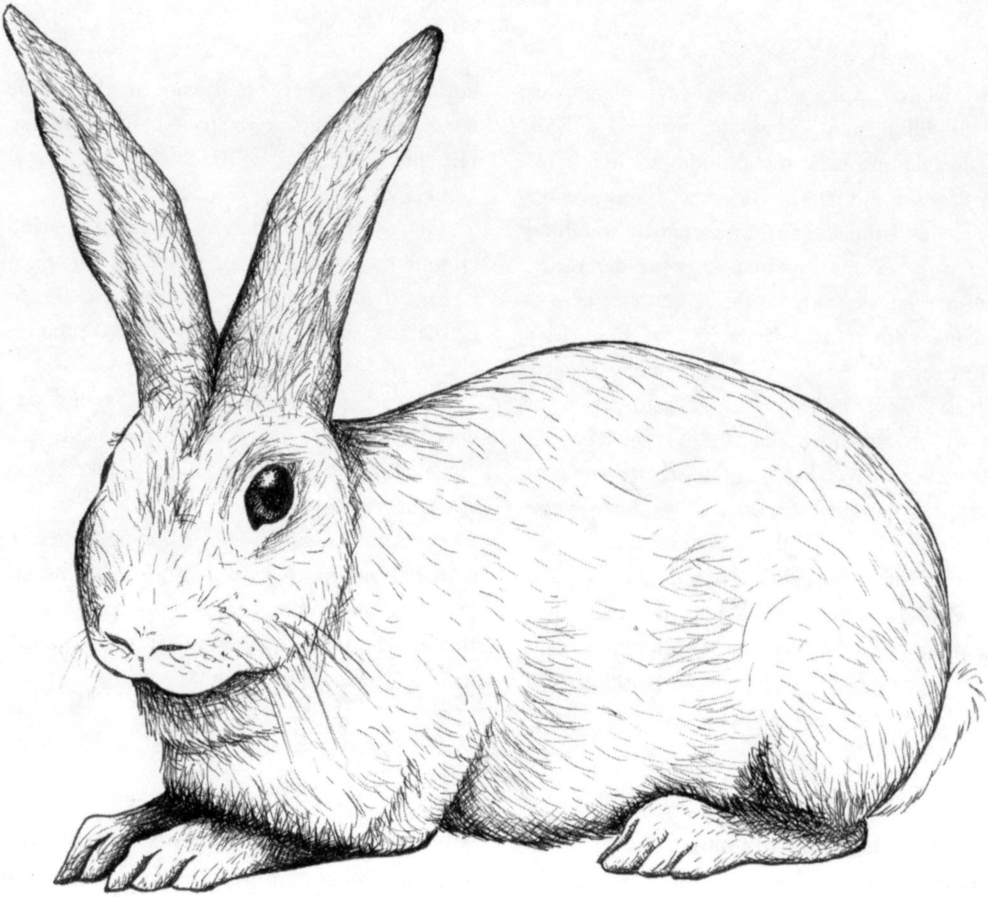

CHAPTER 6
Exotics: Rabbits, Bison, Elk, and Deer

RABBITS

Some consider the smallest commercial rabbit-raising unit to be twenty does with two bucks. A small rabbit business can provide supplemental income as well as fresh meat for the farm family. The amount of space needed to produce rabbits is relatively small, especially when compared to most other livestock enterprises. A small rabbitry can even be set up in a typical suburban backyard. Rabbits are quiet, so neighbors are unlikely to have any objection to them. The care of rabbits, especially in cage systems, can be done under cover and is less physically demanding than other types of livestock. Rabbits are also highly efficient meat animals: A single doe can produce 1,000 percent of her body weight in meat in a single year! Rabbits can easily be slaughtered onsite and don't require any special equipment to do so (see chapter 7 for more information on slaughter regulations). Rabbit meat appeals to those seeking low-fat, low-cholesterol meat; it should be an easy crossover for chicken lovers (if they can get over the "cute" factor).

A part-time rabbit operation might consist of fifty to one hundred does. A full-time rabbit operation might be closer to six hundred does and sixty bucks. Each doe can produce twenty-five to fifty live rabbits annually, which will yield between 125 and 250 pounds of meat. You could envision a full-time operation producing between fifteen thousand and thirty thousand fryers per year; at a conservative sales price of $10 a fryer you could earn in the range of $150,000 to $300,000 in gross annual income.

Rabbits provide opportunities to produce usable products other than meat. Rabbit pelts can be preserved and processed to add further value to the carcass. Rabbit manure is loved by many vegetable gardeners, because it is ready to apply without composting. Some farmers, including the one profiled in this chapter, sell the manure and also raise earthworms under the cages.

The initial investment in a rabbitry can be high if you don't already have an existing outbuilding that can be adapted for rabbit housing and you have to start from scratch. However, the rabbits themselves are inexpensive in comparison to other livestock,

Definitions

Buck a male rabbit

Doe a female rabbit

Fryer a rabbit ready for harvest between 8 and 11 weeks old and weighing 3.75 to 4.5 pounds

Kindling birthing process for a female doe

Kit a baby rabbit

Rabbitry a rabbit-raising enterprise; a place where rabbits are kept

Stewer a rabbit that is 3 months or older and weighs about 6 pounds

Warren where rabbits are kept together as a group (for breeding and grow-out) so they can exhibit some of their natural behaviors

Breeds

There are many rabbit breeds, but here are some of those commonly used for meat and to a lesser extent for their pelts. Starting with the largest to the smallest:

- Giant Chinchilla: large, 12 to 16 pounds live weight, medium-boned
- New Zealand: medium, 9 to 12 pounds live weight, fine-boned
- Champagne or Creme D'Argent: medium, 9 to 12 pounds live weight, fine-boned
- American Chinchilla: medium, 9 to 12 pounds live weight, fine-boned
- Californian: medium, 9 to 11 pounds live weight, fine-boned
- Florida White: small, 4 to 6 pounds live weight, fine-boned

A variety of other meat and dual-purpose breeds of rabbits are not on this list; you will have to experiment early on to find your ideal breed or crossbreed. When you buy rabbit breeding stock, make sure you buy from producers who are selecting for meat production traits and not showing or pet traits. If your goal is to produce meat, then who cares about things like cute coloring patterns on the pelt? Instead, be observant for things like blocky, meaty appearance, and ask about the bone structure (fine or thick) and the meat/bone ratio. Maybe you can buy one of the fryers from the potential breeding stock you are considering and dress it out yourself to see what the meat/bone ratio looks like.

Many of these breeds are used for crossbreeding programs in order to produce superior meat animals. For example, Flemish Giants are huge rabbits, but they have a poor meat/bone ratio. They can be bred to a rabbit with finer

and used cages and equipment are ubiquitous on Craigslist and other classifieds advertising sources.

There aren't many developed marketing streams for rabbits, which means that you will probably be on your own in terms of securing markets for your rabbits. One specialty meats distributor that distributes rabbit to chefs, Nicky USA, is profiled at the end of this chapter. If you are looking for distribution, there may be a wild game or specialty meats distributor in your nearest major metropolitan city. Marketing to chefs and restaurants will likely be your strongest niche. As with all animal husbandry there is a steep learning curve in order to achieve efficient production. Unlike some livestock, rabbits generally require daily care, but the labor can be greatly reduced by installing automated or more efficient systems.

bones, such as New Zealand or Californian. Just make sure the doe is the Flemish, not the buck. You don't want to use a large buck on a smaller doe because that could result in kits that are too large for her to easily birth.

If you have a market for pelts, you may want to consider the French Argent breeds with their gorgeous fur or the Chinchillas or American Blue strain. Truthfully all rabbit pelts are beautiful and may have a market if you treat them well during slaughter. One rabbit farmer we spoke to in preparing this chapter did not currently have a market for pelts, so she froze them laid out flat in stacks with the hope that she would eventually find somebody interested in buying them. If worst comes to worst, you could use the pelts and fur as mulch around fruit trees. If you have the time or inclination, you could start processing them yourself, perhaps in the slow season, to turn into beautiful blankets, rugs, scarves, shawls, hats, and other creations.

Because rabbits breed so copiously, it does not take long to create your own cross-breed that works best for your climate and your market. For example, the Salatin family at Polyface Farm in Virginia have been crossing New Zealands, Californians, and a little Dutch genetics for a couple of decades now and have created their own unique breed that thrives on forage and works well in their climate.

Does

Select does that come from maternal lines known for good milk production, large litter sizes, large weaning weights, and that don't cannibalize their young. Ideally pick does that have a docile nature; however, frequent gentle handling will make your does more friendly with humans. If any of your does cannibalizes their young, cull them immediately. If she doesn't raise at least 80 percent of her kits to weaning age, cull her. Keep the best and eat the rest. This is not hard with rabbits because the investment per animal is not high.

On commercial operations, a doe can produce up to six litters per year successfully and should be able to produce good-sized litters for 2 to 3 years. You can choose to slow this down a bit in the winter so they are not kindling when it is the coldest and mortality rates are higher.

Bucks

When selecting a buck, first make sure he is a male. Choose an animal that has good physical conformation for his breed and one that has good, meaty muscle tone, and who grew out quickly to both his weaning weight and breeding age (6 or 7 months). One buck may service up to ten does but can only be used two to three times per week.

Breeding

Inbreeding of rabbits is an acceptable practice. A father can be bred to a daughter, a mother can be bred to a son, or two cousins can be bred together. Breeding a brother to a sister, however, should not be done. Although inbreeding can be done, it doesn't necessarily mean that it will work with your particular pair and produce the results you are looking for. If it doesn't work, you can always eat your mistake.

Medium-weight breeds are ready to breed at 6 to 7 months of age, with the males maturing a month later than the females.

Keep a very tight breeding schedule, because it is often difficult to tell from outward signs that the doe is in heat. Put the doe in the hutch with the buck and they should mate right away. Never bring the buck to the doe because she will aggressively defend her territory. The doe should be placed back in her hutch after a couple of days with the buck.

Gestation will take 31 or 32 days, but at around 28 days you should place a nest box with clean nesting material in with the doe just to be on the safe side.

Commercial does should produce six to eight kits per litter for medium-weight breeds and eight to ten kits per litter for heavyweight breeds, and an average of thirty to thirty-six successful fryers per year. Unfortunately, kit mortality can reach as high as 25 percent in commercial operations: You'll need to try and understand what the causes of the mortality are if you want this enterprise to pay. Also, low-quality feeds and lack of proper nutrition will mean your does won't wean as many kits. So you may save money on feed, but you will lose the same amount if not more on lost babies (fewer kits equals less meat to sell).

At 48 hours after birth check the kits and remove any dead ones. When the kits are 5 to 21 days of age, remove the nest box. Wean the rabbits at 30 days of age. Remove the young rabbits and keep the doe in her same space; she is very territorial and wants to stay in her space, whereas the young are less so.

Housing

There are several different styles of rabbit housing and each can be further modified according to the life stage of the animals. Breeding animals will have different space requirements than the weaned fryers that are growing to harvest size. Typical rabbit housing is in wire cages, but there are many people experimenting with open-bottom "rabbit tractors" on pasture, and more communal warren-style housing in barns, outbuildings, or even greenhouses.

Cages

Cages can be stacked two high in order to maximize space, but make sure the manure from the top layer is not falling on the animals underneath. One tier is less economical and more than two tiers can lead to management problems. It is generally better to install the cages on legs as opposed to hanging them from the ceiling. The legs make for stable cage supports and don't require any extra reinforcement for the ceiling. It also means they don't swing around during high winds.

It is necessary to have a drainage system for urine and leaky water systems or to use bedding or some type of carbonaceous material under the cages to absorb liquids. Manure will have to be removed when it builds up, but this chore can be delayed somewhat if you are raising earthworms under the cages and also have a bedding pack. Without a source of carbon under the cages the amount of ammonia gases can reach levels that are unhealthy. Respiratory illnesses can occur from excess ammonia gas and dust inside a poorly managed cage system or rabbit house.

If you are building a barn from scratch, it is important to calculate the optimum size in order to efficiently hold the amount of cages you plan to install. It will most likely be necessary to provide for some type of ventilation for at least the hottest months of the year. Rabbits are very susceptible to heat stress. Insulating the roof and walls is important to reduce heat buildup in the summer and to keep the rabbits comfortable in the winter.

Raising worms under the rabbit cages can create another income stream. The worms can take advantage of the manure and any feed that spills through the wire bottom of the cages. Raising worms can be done indoors or, in milder climates, outdoors. The rabbit manure and the worms can both be sold from the farm or used in other farm ventures. Rabbit pellets make a well-balanced, low-nitrogen manure that can be applied directly to your crops without causing nitrogen burn.

Cages can be hard on the feet of rabbits, particularly the large breeds like Flemish Giant and Giant Chinchilla. If you add a little piece of plywood in part of the cage so they can periodically get off the wire, you can easily pull that piece of wood out for washing.

Rabbit Tractors

"Rabbit tractors" come in a wide variety of shapes, but they are usually designed to be pulled around by hand to different areas so the rabbits can have access to fresh pasture and they can get off of their manure. This is a great system, particularly if you are trying to fertilize a garden, orchard, or other crops with rabbit manure. However, if the bottoms are fully open, rabbits will easily dig their way out, and chasing rabbits is not fun. Some people have used chicken wire for the bottoms; others use small wooden slats spaced no more than 2 inches apart; still others build a small lip that goes in 6 or so inches on the inside of the tractor so the rabbit is less able to dig out. A picture of the slatted floor design is shown on the top left of page 12 in the color insert.

Building rabbit tractors is a fairly new endeavor, and farmers are still working on perfecting their designs. You will likely build and rebuild yours several times to get it "just right."

Warren

Another style of housing is called a warren, in which rabbits are raised in more communal groups. You can have a warren-style cage system, but we will explain warren-style production in a building of some sort. With a warren, does have their own nest boxes to kindle and the kits stay in those boxes for around a week or so. The does can come and go to get to the group feeder and waterer. The key to this system is trying to have all the does kindle at the same time so they are not harassing each other and the kits don't try to find other does to rob their milk. It's similar to group farrowing with pigs. The kits are all weaned by removing them to a different area where they can grow out to harvest size. It could be in the same building, just separated by plywood or some other material. Or the kits could be moved to outdoor rabbit tractor–style housing to finish on pasture. In addition to feed pellets and water, green chop or hay is often brought in for dietary diversity and for something to keep the rabbits occupied. We have seen people use old dairy barns and hoophouses to create warren-style rabbitries. Just make sure predators, especially rats, can't get inside or set up shop living in the walls. In addition to stealing rabbit feed, they will eat the kits.

Feeding and Watering

Active growing and producing meat rabbits need about 1,100 calories a day. A doe and her litter will usually eat 100 to 120 pounds of feed during the 8-week grow-out period that is typical for fryers. Bucks and non-lactating does need 6 to 8 ounces of feed per day, and pregnant does and does with litters should be given all the feed that they can eat. A variety of hay, straw, and green chop should be provided all the time so that your rabbits get all the fiber and roughage they need. Be sure to also include plenty of twigs for them to gnaw on; this keeps their teeth filed down. Rabbits can be fed twigs from apple, pear, fir, hazel, hawthorn, maple, spruce, and willow trees, or from blackberry and raspberry patches.

There are essentially three types of feeds you can give rabbits: hay and grains, formulated pellet feeds, and pasture or green chop; or you can do a mixture of all three. Pellets are easier than formulating a hay and grain diet because the micronutrients have already been provided in the right amounts for rabbits. However, it can be expensive to furnish an all-pellet diet and it will easily be your biggest input cost. If your goal is to make money from raising rabbits, you will have to experiment with some of the alternatives described in this section. Fryer rabbits will take an extra couple of weeks to grow out if you are using more forage-based feeds. However, the meat will probably be more nutritious (particularly in vitamins A and E, and the omega 3 fatty acids that come from grass), and you may be able to save money.

Commercial rabbit producer Graeme Harris of Tasmania reports that he was spending $760 every 3 weeks on rabbit pellets when he decided to add a fodder system to his operation. At the time he was feeding fifty to sixty does with kits and sixty to seventy young fryer rabbits. Now he is producing 30 kilograms of barley fodder per day with the same feed costs, but he is able to feed ninety does and their kits along with double the number of young rabbits. He has also noted a decline in scours in his weaned rabbits. A fodder system is where you sprout trays of grain seeds under lights, such as wheat or barley, to produce small mats of short, succulent grass.

Some rabbit farmers have reported using the leaves, seedpods, and small branches of mesquite to make up to 25 percent of their rabbits' feed. Mesquite is ubiquitous in many areas of the Southwest and contains high levels of protein.

Mulberry is another tree or shrub whose leaves, twigs, and bark can be used as rabbit feed. One study done in Nigeria showed that there was no change in growth rate versus traditional diets at levels of up to 50 percent mulberry in the ration. Of course, mulberry is not as commonly grown in the United States, but there may be other similar shrubs or trees that are high in energy or protein.

The leaves, twigs, and bark of the sycamore tree are also readily eaten by rabbits. Fresh leaves and twigs can be fed during the growing season, and the sycamore bark, which readily peels and sheds, can be stored for use throughout the year.

Raspberry and blackberry leaves can be used for rabbit feed. They are high in tannins, which can relieve diarrhea, and they also contain high levels of calcium, manganese, and vitamins A and C. Chicory, which can be planted as a forage crop, contains essential oils, which are good for eliminating parasites. Likewise, the leaves and small branches of willow and poplar of all varieties can be fed to rabbits as a good source of protein and other nutrients. Willow may also act as a natural anthelmintic (which kills parasites) and a coccidiostat.

Many weeds and forbs are also great rabbit feed. These include plantain, lambsquarters, pigweed, dandelion, chickweed, clovers, alfalfa, vetch, and many others.

If your rabbit is leaving a lot of food behind from one feeding to the next, cut back on feed. On the other hand, if your meat rabbit seems to be hungry all the time, give it more food. Just remember that overfeeding can lead to poorly producing meat rabbits, especially does. Does start to build up fat, which leads to breeding complications and makes it more difficult for them to give birth. If your meat rabbit suddenly loses its appetite or has no interest in food, it could be a sign of health problems. Also, spilled feed is wasted feed, which is wasted money. Watch for this and nip the problem in the bud early. It may be a particular animal that loves to scatter its feed everywhere, or it may be a sign of overfeeding.

The easiest way to tell if you are giving your rabbit the right amount of food is to stroke its backbone regularly. If the ridge of the backbone is present but feels rounded, your meat rabbit is receiving the right amount of food. If the ridges of the backbone feel pointed and sharp, start giving your rabbit more food. If you can't feel the backbone, decrease your rabbit's intake.

Water

A doe with her litter can consume a gallon of water per day. Automatic waterers are the easiest way to provide water while reducing contamination and waste.

Winter weather and freezing temperatures can greatly increase the amount of time and effort needed in order to care for the rabbits. Waterer or water system freezing is one of the biggest challenges. Frozen water bottles can be thawed and replaced or swapped out with fresh bottles. This takes time and effort as well as the added expense of having twice as many water bottles. Nipples for these bottles may also be damaged when the frozen water expands inside the nipple. Automatic water systems may also freeze during cold weather, requiring the whole system to be defrosted. This may prove to be time-consuming or nearly impossible.

Freeze-proof water systems use a heated water source that continuously cycles heated water through the water lines. This might be a good option for rabbitries that are located in areas with cold winters. It is important for young rabbits and lactating does to have plenty of water, but during freezing events they can go without water for up to 8 hours.

Keeping Healthy

There are many illnesses, parasites, and diseases that affect rabbits. We will go into the most common below. For overall biosecurity and disease prevention, do not allow casual visitors into the rabbitry, or at least make them walk through a boot wash and perhaps put on gloves. When purchasing stock be aware of the cleanliness and condition of the rabbitry and ask to see health records. See appendix A for more biosecurity ideas.

Malocclusion

Rabbits' teeth are constantly growing and they need to chew on hay and twigs so that they can keep their teeth at a comfortable length. Overgrown teeth can cut into the mouth and cause infection. Rabbits may also stop eating and become emaciated because it is too painful for them to eat. Provide your rabbits with a constant supply of long and fibrous hay or grass, rabbit-safe twigs, and green chop to chew on to help grind down their teeth. When you cannot keep the malocclusion at bay, you can have the teeth trimmed by a vet, but this is costly and not practical for most commercial rabbitries. Rabbits with malocclusion should not be used for breeding purposes because it is a heritable trait.

Pododdematitis or "Sore Hocks"

Unlike dogs or cats, rabbits do not have pads on their feet. Cages without a solid bottom are usually the cause of damaged feet, and constant moisture, prolonged confinement, obesity, and overgrown nails can also exacerbate the situation. Rabbits with sore hocks have raw, inflamed feet with callouses or abscesses. This can lead to them having seriously infected ulcers when not treated immediately. To treat this form of rabbit illness, you have to fix their housing situation and provide them with proper flooring. Also, exercise and nail-trimming can keep the situation at bay. Topical ointments can help.

Respiratory Infections

This form of rabbit illness can be transmitted from one rabbit to another and is usually a result of stress, malnutrition, or bacterial infection. Allergies and environmental irritants can also be causes. Of all the rabbit illnesses that come in the form of bacterial infection, *Pasturella* is the most frightening one as it is very contagious. Symptoms such as sneezing, wheezing, runny nose or eyes, and fever indicate infection. Some rabbits develop ear infections and exhibit head-shaking and head tilt. *Pasturella*-free breeding stock can be purchased. Quarantine all new animals and disinfect the quarantine hutch. Identify all animals with a tattoo or ear tag and keep a health and breeding card attached to each hutch. If you suspect *Pasturella* you can have your vet do a culture test to determine the kind of bacteria causing the respiratory infection.

Heat Stress

Rabbits do not tolerate hot temperatures very well and can even die from heat stress. A rabbit's optimal air temperature is between 50 and 70 degrees F (10 and 21 degrees C), a range not always achievable in the summertime.

Rabbits with thick or long coats of hair, ones that are overweight, and young or old individuals are at an even greater risk for heat stress. Temperature, humidity, and air ventilation are all factors that contribute to heatstroke in a rabbit. Rabbits are individuals and could respond to these conditions somewhat differently. It is important to check your rabbits constantly to ensure they are comfortable and not overheated. Early detection of heatstroke and proper corrective steps could mean the difference between life and death for your rabbits.

When you are observing your animals daily, and multiple times a day during heat events, here are some symptoms that will help you recognize that your rabbit has or is beginning to get heat stroke.

- The rabbit is fully stretched out and the feet are sprawled apart with tail limp.
- There is wetness around the nose area.
- The rabbit's eyes are half closed. The rabbit has a sleepy or dazed appearance.
- The rabbit's tongue is hanging out. Its breathing is rapid and possibly labored.
- The rabbit has fast, shallow breathing.
- The rabbit is reluctant to move.
- The rabbit refuses to eat or drink.
- The rabbit has hot ears.

Prevention includes providing plenty of ventilation, shade, and frozen ice cubes in a dish; making sure there is plenty of space for them to stretch out; providing plentiful cool, clean water; placing frozen water bottles for them to lie against; and even using misters. If you house your rabbits in a greenhouse or hoophouse in the winter, it's best to take them out of those structures, or cover the structure with dark shade cloth in the summer, because their function is to magnify heat.

Manure Handling

A medium-sized doe will produce between 350 and 400 pounds of manure and 50 gallons of urine in a year. Dealing with this waste in an efficient manner should be part of the design of the rabbit housing. Make it easy to access the manure pile from both sides if possible or make worm bins where worms digest and break down the manure into fine worm compost. Turn your rabbit waste into an additional income stream (compost, worms) or fertilizer for your own garden.

Handling and Slaughtering

Rabbit fryers are typically ready to harvest after 8 weeks for confinement pellet-feeding-style production and up to around 11 weeks for more forage-based production. Animals older than this can be harvested as roasters and stewers.

If rabbits are slaughtered for sale in commercial establishments (food stores, restaurants, etc.), they must be processed in a manner that meets local or state health codes. These codes are usually established and enforced by state agencies, although community agencies may also have regulatory control. In most instances, meeting state requirements and slaughtering with a "state license" is sufficient for sale of rabbit carcasses. Check with your local county health department or state meat inspection agency for the policies that govern the processing and sale of rabbits in your community. In many states, rabbits are lumped in with poultry and allowed to be processed on-farm without USDA inspection for up to either one thousand or twenty thousand animals per year (see chapter 7). Strangely, according to the USDA, rabbits are not considered an "amenable species" or a normal meat species that requires USDA inspection and are therefore regulated by the FDA as "food." Thus you typically only need a food-processing license and a commercial kitchen to process them, just like other foods.

Rabbits can be stunned by a couple of different methods. One method is to dislocate the neck. Hold the rabbit by the base of the head and the rear legs and stretch the rabbit to full length. The neck is dislocated by bending the rabbit's head backward with a hard sharp pull. The rabbit can also be stunned with a hard blow to the back of the head with the side of the hand or with a blunt club.

After stunning, hang the rabbit by the hock and cut off the head to bleed out the rabbit. Then cut off the front feet. Next cut the skin around the hocks and make a cut on the lower part of the body between these two cuts. Remove the tail. Pull the skin down over the carcass. The skin is easier to remove on younger rabbits and harder to remove on older rabbits.

To eviscerate, make a cut on the abdomen from near the anus to the ribcage. Remove the intestinal tract and the lungs. Remove the rabbit from the hook and cut off the rear feet at the hock. Wash the rabbit to remove any dirt or hair that may be present on the carcass. Do not allow rabbits to soak in water, as their carcass may take up water. Chill the carcass at 35 degrees F (2 degrees C) and don't let it get above 40 degrees F (4 degrees C). Rabbits may be sold either whole or cut up. Liver, heart, and kidneys are usually left with the carcass and considered in the dressed weight.

Animals with an average body condition should dress out at about 55 percent, while animals with good body condition will dress out at 60 percent or more. In general, dressing percentage increases with age until the animal is almost mature, but the meat also becomes tougher.

Northwest Red Worms

The name Northwest Red Worms may not sound like a rabbit farm, but if you think about the diverse uses that rabbits can play on an integrated farm, then it makes sense. Owner Doug Knippel raises rabbits, chickens, pigs, and red worms on a small farm in Camas, Washington. He also makes wooden compost and worm bins and chicken coops. Doug sells most of his products on Craigslist, which works out well for him since he is located very close to the Portland–Vancouver metropolitan region of over two million people. He also makes educational farm videos in which he documents the happenings on his farm and teaches about various innovative farming techniques.

Doug started in rabbits when his daughter bought thirty does. Doug had a steep learning curve, but is now comfortable with raising and breeding rabbits. He has cut his doe numbers down to around eighteen and is happy with this size of rabbitry in his diverse operation.

Most of Doug's rabbit sales are of live animals. He says that the overwhelming majority of his rabbit customers are young couples that want a buck and a couple of does for backyard breeding. He also sells already bred does or offers to service does with his bucks for a fee. He says a lot of customers appreciate the breeding service since it doesn't really make sense for most hobbyists to keep their own buck when they have just couple does. The rest of Doug's customers for breeding rabbits are Mexican and Russian immigrants and urban homesteaders.

Doug says that many newbies want to purchase young rabbits to raise for breeders, but that he recommends against this. He says that it is difficult to determine the sex of the young rabbits and sometimes the young rabbits will die, since they are not as hearty as older stock. He suggests buying older, proven breeding stock instead.

He says that the number one mistake that people make is overfeeding their breeding rabbits. Overfed bucks are less likely to breed and overfed does have smaller litter sizes. Other problems that new rabbit breeders experience include death losses of young kits due to the cold, and sometimes new moms will pull kits out of the nest boxes and drop them on the wire or they will just birth them outside of the box. Cull any does that refuse to use the nest boxes.

Doug's farm doesn't experience a lot of long-term freezing events, so he is able to get away without having a heated water system. During freezing events that last more than a couple of days, he will provide the rabbits with apples and carrots to provide water for their needs. He says the rabbits do fine with just the fruit and vegetables for up to 3 days in a row without water.

The rabbits' diet consists of a variety of feeds, including around 25 percent pellets, 25 percent orchardgrass hay, and 50 percent fruits and vegetables. The produce consists of cabbage, broccoli, cauliflower, carrots, corn on the cob, and apples. Doug saves a lot of money by feeding the waste produce that he collects anyway for other enterprises on his farm. He originally started collecting waste produce for his composting and red worm operation, but is able to feed any excess to his pigs, chickens, and rabbits. The biggest problem with the produce collection is contamination in the form of packaging and other plastic trash. Doug has to spend valuable time picking trash from the produce of some of his accounts. He works hard on maintaining relationships with his produce sources in order to maintain the supply and keep the waste stream as clean as possible.

Doug's cages are stacked two high and in four rows inside a greenhouse. He uses plastic or galvanized roofing sheets between the top and bottom tiers to shed the manure and urine from the top cages. He says that the plastic roofing works better because the galvanized will rust and make it more difficult to shed the manure. If he had to build it over again he would only use three rows in order to provide more space in the aisles. During the hottest days of summer Doug will use a fan to circulate the air in his rabbitry and he will mist the rabbits with a garden hose. He has also used frozen water bottles to keep the rabbits cool, but only on the hottest days.

Doug has built some of his cages, but now he prefers to buy them pre-cut and assemble them himself. He says the cost is about the same or maybe a little more because of shipping. The cages are better because the floors are flat. It's hard to get the floors flat when you build your own from rolls of wire. He says that building his own cages was really hard on his hands. He now buys his cages from KW Cages in California.

Doug has experimented with some communal rabbit housing. He doesn't plan to do any communal housing for his breeding animals, but thinks it might be a good match for his growing rabbits. His main concern is mites in the bedding, although other farmers seem to be concerned more about coccidiosis in the bedding pack.

Doug cleans out his rabbit barn every other week. He sells screened and bagged rabbit pellets using old feed bags. He doesn't add bedding to the manure pack, but a lot of carbon in the form of hay falls from the rabbit cages and blends with the manure. Doug says that cleaning the manure from the barn is the biggest chore in his operation.

Doug cut the tops off the water bottle drinkers so he can just walk by with a hose and fill them up. This is a big time saver because it transforms a five-step chore into one step. Doug has set some of his pens up with a simple semiautomatic water system made from a 5-gallon bucket plumbed to water lines and nipple drinkers. The bucket is set up higher than the cages to provide gravity pressure and the water lines are flexible, so they won't burst when frozen. This type of setup could also be made fully automatic by utilizing a float valve in the bucket.

Doug raises mostly New Zealand rabbits, but he has learned that many customers want pretty, colorful rabbits for pets, and meat customers don't care either way, so it makes sense for him to breed in a little color. He uses a Chinchilla or a Satin buck for his cross-bred rabbits. Doug plans his spring breeding around the county fair season. He can sell a lot of rabbits to 4H kids if he has the right-aged animals at the right time. He says buyers will drive from far away to purchase fair animals because he is able to sell big groups of right-aged rabbits. He also breeds registered rabbits for 4H kids who want to show a purebred animal. At the end of the summer he sells a lot of rabbits for meat and also for breeding.

Doug plans to breed three to five does a week in order to have enough rabbits for his markets. His does have an average of four litters a year. Doug weans his young rabbits at about 4 to 5 weeks and then rebreeds the doe 2 weeks later.

> Doug's main operation is producing red wiggler worms that he raises in the rabbit manure and compost, hence the name Northwest Red Worms. He recently started raising black soldier flies (BSF) and building black soldier fly containers. He says that BSF is a pretty new thing outside of the southern United States. He feels that there will be a strong market for BSF, but cautions that a lot of these things come in cycles. He has seen the popularity of red worms wax and wane over the years and thinks the same will be true of BSF. There could be a big future in BSF for use as fish food on commercial fish farms and in poultry and swine feed as a more sustainable replacement for fish meal, meat meal, or soybean meal.

Post-Harvest Handling

Rabbit meat is often considered a delicacy. It is a finely grained, mild-flavored white meat. It is high in protein and low in fat, cholesterol, sodium, and calories. Some buyers only want fully white-fleshed breeds. Some breeds have dark points. Find out which breeds they prefer before you raise a single rabbit for them. Carcasses are usually sold whole, often to restaurants.

BISON

Bison are in the same family as cattle (the Bovinae). Their Latin name is *Bison bison*, of which there are two subspecies in the United States: the plains bison and the wood bison. Most domesticated bison are the plains type. Although Americans usually call bison meat "buffalo," the scientifically correct term is bison. Buffalo is sort of a misnomer because the true buffalo are different species: either water buffalo or African buffalo. If you want to educate your consumers about the animals that make your meat, stick to the correct term "bison." Consumer demand for bison is increasing as people become aware of its superior nutritional profile. Prices for bison meat are usually much higher than beef, but costs of production can be higher too.

Bison can be raised in about the same way as cattle, so we will only point out the differences in this section. Here are the main differences and how you manage for them.

1. **Fencing for bison must be taller and much sturdier than it is for cattle.** We recommend that you use at least 6-foot-tall high-tensile fencing, and 7 feet is even better for your perimeter. Shorter, temporary fencing can be used to subdivide interior pastures.
2. **Bison behavior is much more unpredictable than cattle.** They can become aggressive, and extreme care should be taken when inside a paddock with bison. Be especially careful during the rutting (mating) season. If you have more than one bull, they might fight quite a bit during the rut.
3. **Bison are hardier than cattle, and thus don't need shelter.** It is nice to have some woods or shelterbelts of trees for them to get out of the wind or rain, though.
4. **Bison like to make wallows.** This can be hard on the land and make your pastures uneven

Exotics: Rabbits, Bison, Elk, and Deer

(making reseeding or haying difficult). The wallows can also become infected with parasites and even anthrax, a naturally occurring soil organism that can infect bison herds.

5. **Bison are both grazers and browsers.** They like a mix of pastures and shrublands more than cattle do.

6. **Bison can be grain-finished like cattle, but do extremely well on fully grass-based diets.** Unlike many modern-day cattle breeds, bison have not been bred for CAFO production, and they gain weight just fine on well-managed pastures.

7. **Bison are less susceptible to predation than cattle.** This is due to their size and more ferocious behavior.

8. **Bison do not have a continuous growth curve like cattle.** Their growth starts to slow down considerably after 18 months old. They also lose quite a bit of weight during the winter, up to 15 percent of live weight. This means you probably want to time things right to finish your animals before they enter a second winter, but it may take them a full 24 months or more to be finished.

9. **Bison cows only go into heat during a specified period of time (usually August through October).** Beef cows can go into heat every 21 days or so. That means you have fewer opportunities to get your cows bred and a more defined calving period. Bison also don't usually require any assistance during calving, and bison calves can hit the ground running.
10. **Bison have different disease pressures than cattle.** That said, they have less disease pressure than cattle if managed well. Bison are susceptible to some of the same cattle diseases, such as tuberculosis and brucellosis, but also are susceptible to diseases like MCF (malignant catarrhal fever) and mycoplasma-induced arthritis, which is specific to bison. MCF can spread from sheep (which are carriers) to bison (who die from it). It is best not to raise these two animals together or in rotation.
11. **Bison have similar dressing percentages as beef, but slightly more cooler shrink.** This is due to their leaner carcasses, which cause more water to evaporate. Overall, you will get similar amounts of saleable meat off a bison carcass. Bison have slightly larger forequarters than beef animals, but hindquarters are nearly the same.
12. **Tenderness is nearly the same between bison and beef, despite having lower fat marbling.** Meat color tends to be darker and taste panels often show a preference for bison meat over beef (Koch et al., 1995). Bison meat is said to be slightly sweeter than beef.
13. **Bison breeding stock are much more expensive than cattle.** Bison meat commands a much higher price than beef as well. Keep all this in mind when running your numbers.

ELK AND DEER

Elk and deer are species from the Cervidae family. There are other cervid species that people eat, but not commonly in this country. These include moose and reindeer. Elk and deer meat is commonly referred to as "venison"; you will see it on restaurant menus as either elk or deer or just generic "venison." There are many species and subspecies of elk and deer. The most commonly farmed ones in the United States are discussed in the breeds section below.

Some crop and livestock producers are turning to cervid farming as a way to diversify their existing operation in a lower-input and more environmentally sound manner. Cervid farming also is an expandable enterprise that can be transitioned into with moderate land and capital investments, and with species that fit different markets. Because cervids require less forage than cattle, they also require less land. A 60- to 80-acre farm with a mixture of pasture and woods might be a good fit for cervids. Also, since cervids are browsers, they can do fine on more marginal lands and in brushy areas. That said, raising wild game is not for the faint of heart or the slim of pocketbook. Startup costs are higher than most other meat species, and the intense regulatory framework takes a lawyer-like brain to navigate (adapted from Burden, 2012).

There are many different markets for cervid products. These include meat (which will be the focus of this chapter), shed antlers, velvet antler, urine and glands, breeding stock, and what are affectionately referred to as "shooter bulls" for captive hunting preserves. To be profitable, most cervid farmers take advantage of several of these markets.

Additionally, many deer farm businesses combine their farming operation with a tourism component that may incorporate leisure, educational, or private hunting activities for school groups and other people. It should be noted that providing "canned hunts" or selling shooter bulls to hunting preserves may not be allowed in your state. However, many would contend that selling shooter bulls to be hunted within the confines of a fenced private hunting ranch is an unethical practice. Others would say it's better than killing truly wild animals. You will have to weigh those ethical choices yourself.

The main market for venison meat is high-end restaurants and the processed sausage market. However, we have seen elk or venison burgers on menus at family-friendly cafes and diners as well, particularly around the West. Venison also is sold at some specialty grocers and at individual farm outlets like farmstands, farmers markets, or via mail-order. There are several large venison distributors in the United States as well as numerous smaller companies; one distributor on the West Coast, Nicky USA, will be featured later in this chapter. The vast majority of venison meat consumed in the United States is imported from New Zealand, Ireland, or Germany, where cervid farms are more plentiful and regulations less onerous. There is considerably more consumer demand than there is domestic supply.

Texas is the state with the greatest number of cervid farms, followed by Pennsylvania and then Minnesota. Several states don't allow cervid farming at all or prohibit the establishment of new cervid farms (the older ones are grandfathered in). As of 1997 (more recent data could not be found), the states that ban or restrict cervid farming include: Alabama, Arizona, California, Maryland, Massachusetts, Oregon, Virginia, Washington, West Virginia, and Wyoming (North American Elk Breeders Association website: www.naelk.org). Before you do any planning, check with your state to see what the current laws allow.

Special Regulations

Raising elk and deer can be enjoyable and moderately profitable, but an extra layer of permitting, testing, and regulations can hinder both of those things. For the most part, these regulations are needed to protect farmed animals from mixing with wild animals and to prevent the spread of certain diseases. They are also important for protecting the much more lucrative cattle industry. If a cervid farm tests positive for diseases such as tuberculosis, brucellosis, or chronic wasting disease, there are severe repercussions to that state's cattle imports and exports, such as losing their brucellosis-free status for the entire state. With that in mind, it's probably a good idea not to raise cervids and cattle on the same farm because their diseases are transmissible.

The increasing discovery of chronic wasting disease (CWD) in new states and the expansion of the cervid-farming industry have pressured the USDA to design a new national CWD herd certification program in 2012 as a voluntary, cooperative program among the USDA, state agencies, and the cervid-farming industry. Although the program is voluntary, if a state wants to have a CWD herd certification program, it has to comply, and therefore the cervid farmers in that state also have to comply. You should find the program rules on the Internet; print them up if you are thinking of raising cervids.

The national CWD herd certification program requirements include fencing, individual animal IDs, triennial inventories of herds, and testing of all animals older than a year that die for any reason. Also, an animal over 12 months old at slaughter must have a sample sent to an Animal Plant Health Inspection Service-approved lab to verify that it was CWD-free (there are about 21 labs in the country that do this). Testing for CWD can cost between $30 and $50 per animal—add this to your cost of processing. To move animals across state lines, such as breeding stock, you need a certificate of veterinary inspection, unless they are going straight to slaughter.

All animals must be 100 percent traceable and individually ID'ed. This is a good idea for many reasons, including finding your animals if they somehow escape.

With each year of successful surveillance, a participating herd will advance in status until reaching 5 years with no evidence of CWD. At that point, the herd is certified as being low-risk for CWD and permitted for interstate movement and sale of animals.

According to the USDA, twenty-three states have achieved approved status in the CWD herd certification program as of this writing. The rest of the states hold temporary provisional approval, which means they have submitted a complete application for approved status but have not yet received approval.

In addition to CWD, some states will require annual tuberculosis and brucellosis testing, which can be costly and cumbersome. You have to either tranquilize your animals to take the blood sample or you have to corral them and run them through a chute in order to draw the samples. Getting your cervids into a corral system requires lots of repetition training and stress-free handling (as much as possible).

In more than forty states, regulatory authority over captive cervid facilities rests with state agriculture agencies or is shared between agriculture and wildlife agencies. Understand that the culture of those agencies will likely vary: Some may be antagonistic toward captive breeding of wild animals, others will consider it an "alternative livestock" system and be more supportive. There are an estimated ten thousand cervid breeding farms in the United States, yet some states prohibit the captivity of certain game species altogether. This should be obvious, but before you buy a single animal, make sure you find out the laws for your state.

At present twenty-one states prohibit the importation of certain species of deer, whitetail being the most frequent. They don't want farmed whitetail deer around areas where wild whitetails live as well. Other states require only non-native breeds so that if they do escape into the wild they are easily identifiable and in some cases don't readily breed with the native species and subspecies.

In states that restrict the import and export of animals across state lines, cervid farmers can still grow big bucks to satisfy the trophy market's demand. But instead of moving live animals across state lines, they trade in semen and breed and shoot the bucks in-state.

Many states require you to pay an annual licensing fee with whatever department regulates game farming. These licenses or permits can range from $50 to $300 a year. Likewise, states will usually require a certain type and height of fencing. Cervids usually require an 8-foot-tall high-tensile woven-wire fence with more frequent fence posts than a traditional woven-wire fence.

Breeds and Breeding

Elk

Manitoban elk are the largest farmed elk species and are very hardy. They probably would not enjoy life in the hot, humid South. Roosevelt elk are

TABLE 6.1. Characteristics of Different Elk and Deer Breeds

Breed	Average size (live weight)	Other special characteristics
Roosevelt elk	1,000 lb bulls 600 lb cows	• native to Northwest coast range • can handle wet forest conditions • bred more for meat than for antlers
Manitoban elk	1,200 lb bulls 700 lb cows	• hardier • can survive colder climates
Rocky Mountain elk	800 lb bulls 450 lb cows	• hardy • cold-climate elk • largest antlers of farmed elk species
Axis deer	150 lb bucks 100 lb does	• considered best-tasting in blind taste tests • less than 1% fat • tropical species
Fallow deer	150 lb bucks 80 lb does	• 7% fat • hardy • disease-free • efficient grazers
Red deer	450 lb bucks 250 lb does	• 5% fat • larger than other deer • can breed with mule deer so may be banned in some states
Sika deer	200 lb bucks 100 lb does	• from the tropics • not hardy in cold climates
Whitetail deer	Northern: 250 lb bucks 175 lb does Southern: 150 lb bucks 100 lb does	• native to the United States, so don't let them escape into wild • northern animals are much larger than southern animals • some consider this deer meat more "gamey"

also hardy and can handle wet, temperate environments. They do fine in wooded landscapes. They also grow to a fairly large size, just below the Manitoban. Rocky Mountain elk are fairly versatile in habitats and grow to a medium size, so they will require less feed.

Deer

Red deer are the biggest farmed deer species, while Fallow deer are easy to care for and don't require much feed or meds, and Axis deer are tasty, tender, and mild. Axis are relatively disease-resistant and have high fertility. They are considered a serious pest in Hawaii.

Because a lot of states have live animal import bans and because breeding for extreme type genetics is the norm, many cervid farmers use artificial insemination (AI). That way they don't have to import bulls and they can select for specific traits they are looking for. It seems that many cervid farmers are selecting for huge size and enormous antlers. If your market is meat, then body size may matter more than antler size. It appears that a big reason why the United States doesn't supply enough venison to meet the demand is that cervid farmers can make considerably more money selling breeding stock or shooter bulls for hunting preserves. There is so much less money in selling animals for meat that there is little reason to do it. If you are going to utilize AI, you need a squeeze chute to put your cows or does into and you have to figure out when they are in heat, which is generally in the fall.

If you are naturally breeding, the cows and bulls will let you know when the time is right to put them together. For both deer and elk, right after the bull has shed the velvet off his antlers is about when the rut (breeding season) begins. The rest of the year it is best to keep males and females separated to reduce fighting, skin damage from antlers, and non-seasonal pregnancies. You don't have to pull the fawns off the does when you AI them, but you may want to do this when naturally mating just to protect the fawns from the bucks. This separation doesn't happen in nature, but in confinement your animals have fewer places to hide or move away from agitated animals.

Housing

Housing? What housing? That is part of the allure of raising wild game. They are hardy creatures. The only exceptions to that rule are the Axis deer, which come from tropical climates. They may get cold, so give them a bit of shelter for the winter. Better yet, don't raise them in regions of the country with snowfall. The Southwest, Texas, and the Southeast are the best regions for raising Axis deer. On the opposite spectrum, Manitoban and Rocky Mountain elk may not enjoy the hot, humid South all that much. Make sure you provide them with some woods or shade.

Feeding and Watering

For each beef cow-calf pair that a pasture can sustain, it can normally carry six to eight mature deer does and their offspring. Deer are browsers, not grazers, and their feed consumption decreases in winter months. They are hardy, intelligent, and adaptable in most climates. They are highly productive and generally have twins and often triplets, making deer a good investment with an above-average rate of return. This investment can be maintained on less acreage and on poor-quality land that is unsuitable for other types of farming.

Estimate around $500 per year to keep a doe with her offspring alive in feed, veterinary, and medical costs.

If you need to supplement your pastures, buy baled alfalfa but not regular grass hay—they won't eat it and will waste most of it. Plant clover and fescue in their pastures to withstand the beating and provide some food value. Rape mustard is also good because it comes back readily after several grazings.

If you want lean, healthy grassfed meat, then feed your cervids hay, pasture, and browse. There's no need to grain-finish them. By encouraging more browsing higher up on plants, you can reduce parasite infestations, just as with goats. Also, if you do feed hay or alfalfa, try to keep it off the ground as well. Apply hydrated lime every few years to kill bacteria in the soil that may build up from excess manure.

All water, as with every animal, should be fresh and cool. Build a shade structure over your water trough if you have to. Likewise, if you live in colder climates, make sure you have frost-free hydrants.

Cervids prefer natural watercourses, but they can also be hard on them, by overgrazing and eroding the banks. Riparian vegetation that grows along watercourses is some of their favorite browse. If you do allow access to natural watercourses, make sure you rotate and rest those areas frequently or only allow seasonal access. People who ignore this give livestock farming a bad reputation.

Keeping Healthy

Cervids have quite a few disease and parasite issues, particularly in captivity. Because you have to build such heavy-duty fencing, you are less apt to rotationally graze them over extensive pastures as you would with cattle. Therefore, diseases transmitted by bacteria and feces, along with parasites, will build up in the fenced pens used to contain cervids and will need to be aggressively managed. Parasites can easily build up in the soil, and diseases can be transmitted through contact with feces, saliva, or nibbling on bones or a dead carcass. Practice impeccable biosecurity standards and immediately remove any dead animals and have them tested. Also remove animals that appear sick and separate them out to a confined quarantine area until they regain health (if they do). Cull out animals that seem to be constantly loaded with parasites or any that contract one of the following diseases: brucellosis, tuberculosis, or CWD.

CWD in cervids is similar to bovine spongiform encephalopathy (BSE) in cattle. It can spread between cervids but supposedly not to humans. CWD is an always-fatal brain disease of cervids that is spread through direct contact with infected animals. There is no live-animal test, nor is there any vaccine against the disease, though researchers are working on both and showing some progress. We anticipate in the next 5 to 10 years something will be commercialized.

Elk are hardy livestock with natural immunity to most diseases. Although they can contract normal bovine diseases, they are not prone to do so. There have been cases of domestic elk with tuberculosis (*M. bovis*) as with other domestic stock (e.g., cattle). Testing requirements and tests have been developed to better identify and eliminate tuberculosis in elk. Elk now have a federal herd accreditation program for tuberculosis. As a result, tuberculosis has nearly been eradicated in domestic herds.

Internal and external parasites are of concern if you tend to overgraze your pastures or keep your elk too closely confined. Consult your veterinarian for the proper drug to treat the specific parasites. This is easier said than done, though, as most veterinarians have little to no experience with cervids. As mentioned above, applying hydrated lime every couple of years can help reduce some pathogens that may build up in your soil. It can also help kill mites, fleas, and ticks, which all can be problematic external pests for cervids.

Just as with poultry, coccidiosis can be a problem for cervids. Practice good pasture rotation and manure management to keep the animals from coming into contact with this manure-based bacteria. Nematodes (worms) can also be troublesome. You can't just use the same nematicide over and over; you need to rotate wormers or the nematodes will build up resistance to the drug.

Observation of your herd is important. The warning signs of health problems are the same as for any domestic herd animal.

- Animal off by itself
- Head down, ears back, watery eyes, limping

- Off feed and not eating
- Swollen body parts, abscesses, lumps
- Loss of hair or body condition

If you do suspect a problem, call a vet, an experienced cervid rancher, or your state deer or elk farming association to get advice.

Fencing

Fencing is of huge importance for raising wild game like cervids. You don't want your animals escaping and potentially breeding with or passing disease to wild herds. Likewise, you don't want to lose any of your expensive investment in animals. You should check your fences daily and also have at least one to two strands of electric poly wire in addition to high-tensile woven-wire at the top. Use deer fence with smaller squares at the bottom to not only keep predators out but to keep inquisitive fawns inside. There are electric fence detectors that can tell you if your fence is not hot for some reason. One way animals can easily escape is when a tree branch falls on the fence. Trim branches up and away from your fences if possible.

Corner and end posts should be taller and thicker than line posts and probably set in concrete for strength and longevity. Fence installation revolves around the posts that support it. Use high-quality posts (spaced closely enough to support the weight of the fence without sagging) and galvanized fence staples to secure fencing. Work the fence from center to ends, creating a rubber band-like tension. Pay for quality fencing: Don't try to cut corners. This is your most serious investment in cervid farming and should be done well. It will also make your neighbors and regulators more content.

Handling and Slaughtering

You need a squeeze box for weighing, vaccinating, worming, artificial insemination, and for harvesting velvet antler. You may have to modify things to be able to accommodate very large antler racks.

Elk cows can be harvested for up to 8 years without being tough, whereas bulls are more likely better harvested at 2 to 6 years of age. Harvest all animals on the gain: This can be done with 1 to 2 months of some light grain feeding if need be or on your best pasture. Elk cows put on weight more easily from June to December and bulls put on more weight from late winter until the rut (mating time) in September. Young elk rapidly gain weight until around 18 months of age, then they start to slow down. Do not handle bull elks during the rut: They are aggressive and hormonal. Let them do their thing!

Deer are often harvested between the ages of 2 and 4 years old. They are smaller animals and put muscle on much more slowly. Bucks ideally should not be harvested during the rut.

The animals going for processing should be sorted into one group at least a week before the processing date so that they are pre-socialized and not fighting for dominance during loading and transportation. However, time spent in the trailer and holding area should be minimized. There is no benefit to meat quality from an overnight stay in holding pens or standing in a trailer or pens waiting to move to the knocking area: The deer will be stressed. Stressed animals create tougher meat. Animals should never be overcrowded during hauling. If elk are loaded to the point that they are "packed in" and cannot stand comfortably, they mill and jostle around in the trailer, which increases their stress levels. If any bulls or bucks have antlers, they can injure each other and cause bruising or tears, which also damage the meat.

Post-Harvest Handling

Meat quality problems have been identified with the meat industry standard, which is to use blast chilling to reduce carcass temperatures to as close to 34 degrees F (1 degree C) as soon as possible. This can cause cold shortening of the muscle fibers and reduces tenderness, particularly in ruminants like cattle and deer. The optimal cooling method is to hold hanging carcasses at 43 degrees F (6 degrees C) for 24 hours and then lower it to normal cooler temperatures. Aging of lean carcasses such as elk is best done in vacuum-sealed packages (wet aging) because they are simply too lean for regular dry aging.

NICKY USA

Nicky USA is a Portland, Oregon, based meat fabrication and distribution business that specializes in wild game and unique specialty meats. Geoff Latham and his partner started the business over 25 years ago selling locally raised rabbits and quail under an Oregon Department of Agriculture license to a handful of innovative restaurants in the Portland area. Now a couple of decades older and wiser, the company sells twenty-two different meat species to restaurants around the Northwest and markets via mail-order to chefs and specialty retailers around the country—from antelope to venison and nearly everything in between.

Wild boar and antelope from Texas are really the only truly "wild game" domestic meats in the United States that are legal to sell, according to Geoff. All of the rest are being farmed. Texas has special laws that allow for these animals to be live-trapped, corralled, loaded onto trailers, and brought into a USDA-inspected plant. No other state allows this, meaning Texas has a hold on this market. So even though Nicky USA sells "wild game," it is sort of a misnomer, says Geoff.

Wild game in other countries is sometimes raised in a more feral style, but still often comes from specialty breeds that are selected for meat production, and they are managed in various ways by humans, such as perimeter fencing and moving around with herders.

Nicky USA operates a USDA-inspected fabrication plant in an industrial part of Portland. Most people don't even know it's there. Other than poultry and rabbits, no other slaughter takes place in the facility; therefore, it's a pretty small, innocuous building. With a skilled crew of eight staff, they receive carcasses and break them down into fresh cuts and ground meats, and they cure and smoke meats, too. On the days that they are processing exotic non-amenable species, they don't have the USDA inspector on-site since USDA inspection is not required for non-amenables. A typical weekly schedule looks like this:

Mondays: kill and process five hundred rabbits and grind exotic meats under state and federal inspection (federal inspection allows them to sell meat across state lines).

Tuesdays: butcher twenty large hogs for Tails & Trotters and make their specialty sausages under USDA inspection.

Wednesdays: butcher water buffalo or bison and maybe some Fallow deer or Roosevelt elk under state and federal inspection.

Thursdays: butcher lamb and goat under USDA inspection.

Fridays: this is their biggest delivery day, when they run around and deliver all this fresh meat to restaurants as well as run some up to the Nicky Seattle warehouse. They deliver meat the rest of the week as well, except on Sundays.

Nicky USA buys from farmers and ranchers around the Northwest as well as a few more distant locations, like wild boar from Texas, fresh and cured *jamón ibérico de bellota* (acorn-finished pork) from Spain, and Red deer from New Zealand. There are simply not enough producers with enough volume of these game animals in the region to supply the growing demand. Geoff knows these game farmers well and realizes that they are not really making enough money to scale up in any appreciable way. Factor in the high cost of breeding stock (for example, a quality breeding elk buck may cost $10,000), add to that expensive fencing costs (around $10,000 per linear mile), long grow-out periods for relatively small quantities of meat, permits and licenses, and the challenge of finding slaughterhouses that will receive your animals, and you have the makings for a field of agriculture that is not attracting a lot of new participants. Even though demand outstrips supply, farmers can't just keep raising their prices when they often have to compete with cheaper imported sources, such as from New Zealand.

The future of Nicky USA looks exciting. They just finished building what will be the first USDA-inspected mobile slaughter unit (MSU) that kills red meat species under 300 pounds and poultry too. Most MSUs just do one or the other. The MSU is built within a fifth-wheel trailer attached to their Ford Superduty truck.

They also just bought the farm, literally. Nicky USA purchased a 36-acre farm in Aurora, Oregon, about a 30-minute drive south of Portland. There they intend to raise rabbits, goats, and poultry for the business. More importantly, it will become a showcase farm that their clients can visit and even stay at. The 114-year-old farmhouse is being fixed up to include a commercial kitchen and guest rooms. Geoff understands the importance of connecting his buyers to how and where animals are raised. Accordingly, he will organize wild game tastings and chef events, and he will let the public come out and see how Nicky USA's animals are raised. He does not anticipate growing all the animals the company needs: The Aurora farm will focus less on commercial production levels and more on quality, innovation, and humane animal care. They will also build holding pens at the farm for the MSU processing. Other farmers in the region will be able to bring their animals to the Aurora farm and contract for USDA slaughter, instead of the MSU traveling to their farms.

CHAPTER 7
Regulations

There exists a complicated layering of federal, state, and often county-level regulations that govern meat processing and sales. It is enough to scare away many would-be farmer direct-marketers due to the sheer complexity and reams of paperwork and licensing and fees that are often involved. The rules can be so complex that sometimes the agency staff doesn't even understand them. Don't let the regulations scare you off before you at least take the time to understand them. Once you sort them out and have a year or two under your belt dealing with them, the regulatory hurdles will seem less and less insurmountable. This book will cover the federal and state level regulations as precisely as possible—but that is an important caveat, as the rules change frequently.

It's wise to call your county health department and ask them if they have any additional licenses or regulations that govern the production, processing, transport, storage, and sales of meat. We would recommend you don't give your name when you make this kind of exploratory call, so that you don't get onto the regulators' radars just yet. We personally like to be thoroughly knowledgeable of the law before we have to interact with regulators, so we start off with anonymous information-gathering calls. Even if you plan to exploit the gray areas of meat sales (more on this later), you should at least know the law. A good way to approach the regulators by phone is to say something like this: "Hello, my name is _____ (first name only). I am new to _____ county and am considering starting a small poultry farm. I would like to process my birds on-farm under the USDA exemption, if possible, and wondered if there are any specific permits or licenses I need to first obtain from the county health department." Or, "Hello. I am a new sheep producer in your county and have requests by customers who want to buy live animals and personally kill them on our farm. Is that practice legal in your county?" What often transpires next is what we call the "five-person run-around," meaning that your inquiry will have to get tossed around to at least five government employees before you receive a proper answer. This could take weeks and a lot of phone calling and persistence on your part.

It's best to enter into the inquiry with as much legal information as you can so that you can educate regulators on certain federal and state laws that they may be unaware of. For example, many county health departments are unaware that poultry can be killed on-farm without inspection (up to twenty thousand birds annually) under the federal Public

Definitions

Abattoir a building where animals are slaughtered and sometimes butchered too

Amenable species a term used within the context of the USDA's meat and poultry inspection program to signify animal species that require federal inspection at slaughter, which is taxpayer-funded

Custom-exempt plant slaughter and butcher plants that are expected to meet the same sanitation requirements as inspected plants, but are "exempt" from continuous animal-by-animal inspection. Meat processed in custom-exempt plants cannot be sold. However, custom-exempt plants can sell inspected meat that they purchased. For example, they can bring in stamped carcasses and break them down for sale through an on-site butcher counter.

Cut and wrap, butchering, or fabrication the cutting up of animals into primals, subprimals, or retail cuts; includes packaging for food safety and cold storage

FDA the US Food and Drug Administration; they regulate shell eggs and game meats (non-amenable species), but not other meat species (odd, but true)

FSIS Food Safety Inspection Services, a division of the USDA; the public health agency responsible for ensuring that the nation's commercial supply of meat (excluding game meats, such as venison), poultry, and egg products is safe, wholesome, and correctly labeled and packaged. They provide inspectors to "USDA-inspected" abattoirs free of cost to the plant.

Inspection mandatory state or federal inspection of each animal that is slaughtered for sale as meat or meat products; a trained individual checks the living animal, the slaughter process, the carcass, and food safety

---

Law 90-492. More on that law later. Make sure you write down people's names that you talk to, the date of the conversation, and what they said. If they cite specific codes or laws, ask them to either give you a web link to the specific language or to email you those documents for your files. If you give them your email, you may no longer be anonymous. Put together a binder of pertinent regulations, organized into sections, and keep that available in your office (or the dashboard of your truck) just in case you need to reference it later.

From experience, we have found (along with many other farmers we talked to) that you often have to go to a higher-level agency to get your questions answered. County-level staff frequently don't have adequate information, or they may simply have no written laws on the books for what you want to do. Starting off with your state Department of Agriculture is a good way to go. Likewise, the Niche Meat Processor Assistance Network (NMPAN) has state-level coordinators in over forty US states that are available to take your questions and point you in the right direction if they can't answer them. These are mostly extension agents, well versed in animal science or meat processing, that are truly there to help, not to enforce regulations. While doing research for this book, we called several of them to find out the latest information on meat-processing laws in their states. They were always very helpful.

Four-Leggeds (i.e., cattle, pigs, sheep, goats)

To begin, we will provide a flow chart explaining the difference between custom slaughter and federally inspected slaughter. These are the two main routes for meat processing. If your state has a state-inspected meat program, it must meet or exceed the federal inspection rules, so we will lump the state programs in with the federal route unless they differ significantly.

To understand the diagram, you have to ask yourself *who* the meat is intended for. If for any reason you decide you would like to sell the meat in the future, it must be slaughtered and butchered *under inspection* of some sort. There are three types of meat inspection:

1. **State inspection** for intrastate meat sales only, except in the few states that have instituted an interstate cooperative meat shipment program (Ohio, Wisconsin, North Dakota, and Indiana thus far; more may be added later).
2. **Talmadge-Aiken plants (TA plants)** are where state food safety inspectors act as agents for the USDA FSIS; they are active in nine states only. Therefore, processing in these plants will allow producers to sell into any markets where a federal inspection seal is required.
3. **USDA-inspected plants,** which in 2010, the latest year with data, numbered 841 federally inspected slaughter plants in the United States, of which 310 did poultry.

Custom (or sometimes referred to as "custom-exempt") facilities do not have inspection and thus the meat you have processed in those facilities cannot be for sale. This will be explained in more detail further along in this chapter.

If you want to sell across state lines, your meat must be processed under USDA inspection (except for a few states that have interstate agreements mentioned under point 1). If you want flexibility when it comes to meat marketing channels, your best route is paying for inspected processing. You do not pay for the inspector's time; they are assigned for free by the state or the USDA (unless you are processing non-amenable species—more on that later).

Red meat regulations are essentially the same for all states with little variation. All red meat must be processed under either state or federal inspection. Some states don't have state meat inspection programs, so they let the USDA do all of that work for them. If a state says "none" under the State Inspection Program column in table B.1 (page 289), that means all inspection is handled by USDA FSIS. Some states have kept their red meat inspection programs but surrendered their poultry inspection to the USDA (they are labeled "Red Meat only").

As mentioned above, some states have what are known as Talmadge-Aiken plants, which are state-inspected plants that meet USDA requirements and are thus able to process meat that can go across state lines. They are known as "equivalent" in terms of inspection. A few states have joined a similar program, called the Cooperative Interstate Shipment (CIS) program. These are state-inspected plants that are now deemed "equivalent" to USDA

The New Livestock Farmer

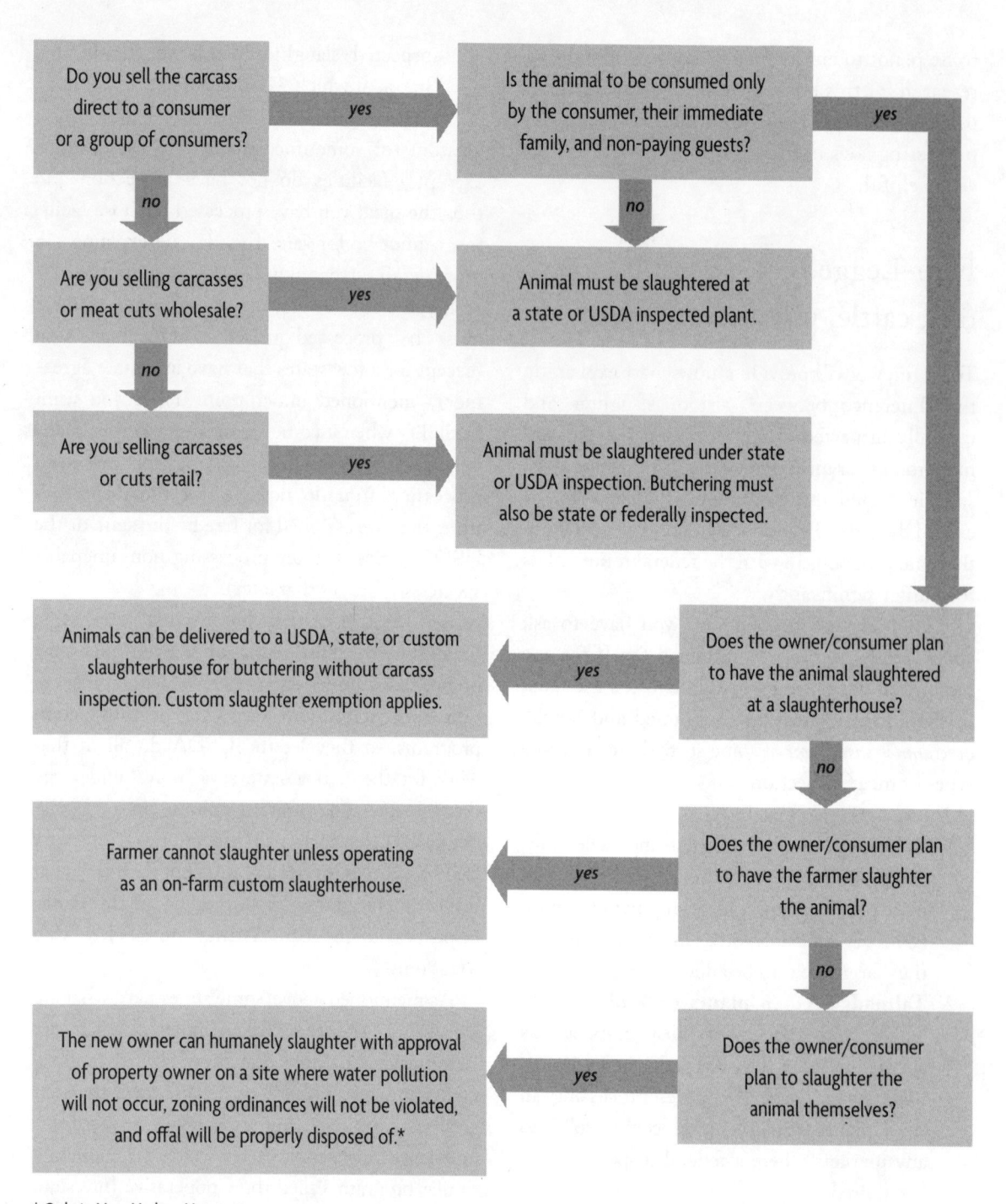

Do you sell the carcass direct to a consumer or a group of consumers?
- yes → **Is the animal to be consumed only by the consumer, their immediate family, and non-paying guests?**
  - yes → (continues to slaughterhouse question below)
  - no → **Animal must be slaughtered at a state or USDA inspected plant.**
- no → **Are you selling carcasses or meat cuts wholesale?**
  - yes → **Animal must be slaughtered at a state or USDA inspected plant.**
  - no → **Are you selling carcasses or cuts retail?**
    - yes → **Animal must be slaughtered under state or USDA inspection. Butchering must also be state or federally inspected.**

Does the owner/consumer plan to have the animal slaughtered at a slaughterhouse?
- yes → **Animals can be delivered to a USDA, state, or custom slaughterhouse for butchering without carcass inspection. Custom slaughter exemption applies.**
- no → **Does the owner/consumer plan to have the farmer slaughter the animal?**
  - yes → **Farmer cannot slaughter unless operating as an on-farm custom slaughterhouse.**
  - no → **Does the owner/consumer plan to slaughter the animal themselves?**
    - yes → **The new owner can humanely slaughter with approval of property owner on a site where water pollution will not occur, zoning ordinances will not be violated, and offal will be properly disposed of.***

* Only in New York or Vermont

Figure 7.1. Meat Inspection Decision Chart. *Courtesy of Michelle McGrath*

inspection and are now allowed to sell their meat across state lines as well. All told, there are thirteen states that have either TA plants or are part of the CIS program, providing a lot more opportunities for producers to get their meat processed in those states and opening doors for where they get to sell that meat. Some believe there is no benefit to having state meat inspection programs because they have to be equivalent with USDA rules anyway. Yet plant owners we talked to have all told us that they prefer to work with their state departments of agriculture over USDA FSIS because the state agencies are more eager and willing to provide technical assistance and be supportive of their efforts.

As we were creating table B.1, we originally had two additional information columns. One was on whether or not states allow on-farm slaughter of red meat species by either the farmer or the buyer. We took that column off the table because, as our research turned up, there are only two states in the country that specifically allow this. New York and Vermont allow farmers to sell a live animal to a buyer who can then humanely slaughter that animal on the farm of origin. The farmers are not supposed to do the killing themselves: It is the buyer who carries that out. In Vermont there is a limit to your annual sales of this type—up to 3,500 pounds of live animal weight can be sold annually in this fashion—so maybe three cattle or thirty sheep or goats or a dozen pigs, or some combination of those species. Obviously this allowance is not a commercial-scale venture, but it does provide some flexibility for farmers and satisfies a specific market niche, especially for those who want to sell animals for cultural holidays. In all other states either a licensed custom mobile slaughter contractor needs to slaughter the animal, or the animal must be purchased live and taken to the buyer's own property for slaughter. There may be a few states that simply turn the cheek to this activity, making it neither illegal nor legal. Yet to build a market niche based on the gray area of law can be quite risky. While one health or agriculture department may turn the cheek, a new inspector could change all that.

The other column we took off this table was related to non-amenable species such as rabbits, bison, elk, deer, llamas, and the like. The general rule is that if your state has a state meat inspection program, they probably allow these species to be processed under state inspection. Here is where things get confusing. Under federal law, non-amenable species are not required to be inspected at slaughter in order to be sold: They are actually regulated by the FDA as "food." However, many states will not allow uninspected meat to be sold within their state, period. So if you want to be able to sell meat to a wide variety of markets, your best option is to look for a state-inspected plant (if your state has a meat inspection program) or to request voluntary inspection at a USDA plant to sell to any state you want. Not all USDA plants are willing to process non-amenable species; you will have to ask around. If your regional USDA inspectors already have a full schedule, you will have difficulty scheduling any additional, voluntary hours from them. Finally, voluntary inspection is not free, unlike regular amenable species inspection. You will have to pay by the hour for that inspection, to the tune of around $60 an hour (this wage will likely increase each year). That could make sense for high-value animals like bison, but probably not for lower-value animals like small deer, rabbits, or quail. You must keep these things in mind as you develop your species mix and your markets.

To determine the best route of processing for your animals, you need to consider who the end consumer is of that meat. This is illustrated in figure 7.1. Ask yourself the following questions:

Is the meat for personal home-farm consumption only? Then on-farm slaughter is OK, as many as you would like (or as many as your county will allow you to raise given your land zoning). You are also allowed to give this meat away to friends, family, and non-paying guests. You cannot charge for it, nor can you technically barter for it since that implies a financial transaction.

Will you sell a live animal to a buyer who will come to your farm to pick it up (and presumably take it home and kill it themselves)? This is legal everywhere. Because you are selling a live animal and don't have any involvement in the slaughter of that animal, you are not liable for it after the animal leaves the farm. So go ahead and sell as many animals as you are legally able to raise in your county, according to your land zoning.

Will you sell a live animal to a buyer who will kill it on your own farm (without your help)? On-farm slaughter by the buyer is technically legal only in the states of New York and Vermont. Even though it happens in many other places, it is not legal. If this is part of your business model, be careful how you present it to the public, or contract with a licensed mobile slaughter contractor to do the slaughter to protect your business. The person buying the animal must perform the slaughter themselves, or they can hire a mobile slaughter truck to do the slaughter for them. The farmers themselves are *not* allowed to kill for a customer. In New York, the farmer can intervene if the slaughter is not going well. On-farm slaughter is more common in places with a high number of ethnic populations that desire ritual slaughter (kosher or halal) or who want to save money on their meat and want it fresh. If you are going to allow customers to kill animals on your farm, you should have a good liability policy to cover any potential accidents. We can't imagine the risk of allowing people to wield knives or guns on a livestock farm. In our previous operation we never allowed this, even though many people asked. People could buy live animals from us, but they had to take them to their own properties to dispatch them. We were not willing to take on that liability, nor did we want to be present for potentially inhumane slaughter. Not everybody knows how to kill an animal properly.

Will you sell a live animal to a buyer and then arrange for the slaughter of it? Custom processing is the typical route for this transaction, but you could use a state or federally inspected facility too if you have one near that you prefer. Often it depends on what facilities are available nearby. We used a USDA-inspected slaughterhouse for custom sales, not because we had to, but because they were the closest facility that processed pigs, could scald and leave the skin on, and did it all at a very affordable price due to the size of the facility. The slaughter can be conducted on-farm with a licensed custom mobile slaughter contractor or the animal can be transported to a custom slaughterhouse. The farmer cannot do the slaughter herself (see the paragraph above). **Gray area:** Who does the carcass transport? For example, if a mobile slaughter outfit comes to your farm to kill an animal for a customer, do they have to take it to the butcher shop, or can the farmer or the buyer throw the carcass into the back of their truck on a tarp or in a cooler and take it themselves? Does the customer have to pay the kill fee directly (probably) or can they pay the farmer who just includes it in their price? More on the legalities and nuances of selling custom animals will be in chapter 8 on sales outlets.

Will you sell a carcass to a retail buyer, such as a store or restaurant, where they will further break it down? The slaughter must be USDA-inspected (or a state equivalent) and the carcass must be stamped. That same processor may be willing to break the carcass down into primals or subprimals for your buyers, but that too must be under USDA inspection. Stores, restaurants, and butcher shops don't have to have USDA inspection. Unless they wholesale to other retailers, they are what are known as "retail exempt" locations that are allowed to do further processing. They still have to have approval by either their county or state health departments and usually have to have a Hazard Analysis and Critical Control Points (HACCP) plan or a "HACCP-like" plan for each process. For example, in New York, these are called 20C licenses for meat fabrication or butchery. In the case of 20C licensees, they have to work with USDA-stamped carcasses. **Gray area:** Who transports the carcass to the store or restaurant? Can it be the farmer or does it have to be the slaughterhouse? Odd story along those lines: We had a USDA-inspected slaughterhouse that offered no cut and wrap services. We found out they delivered carcasses around the San Francisco Bay Area to several restaurants, butcher shops, and meat distributors for a nominal charge. So we arranged for a bunch of restaurants to receive our fresh pork on the same day from the slaughterhouse delivery truck. We decided to meet the truck at one of the restaurants to make sure our pork was going to the right buyers. We expected to find a refrigerated box truck with washable walls and pigs either hanging on rails or lying on the floor wrapped in food-grade plastic. Instead, we found a barely refrigerated truck with pig halves lying on the floor only partially on top of cardboard but much of it lying directly on the truck floor. We couldn't believe transporting animal carcasses in that manner could pass USDA inspection, even though this facility was USDA-inspected. Conversely, our nice refrigerated van with thick, washable walls seemed like a much better way to transport carcasses. Your buyers may not be thrilled with this either. It's probably a good idea to ask how your plant will deliver the carcasses, in order to possibly avoid an unpleasant surprise like we encountered.

Will you sell cuts of meat to a customer? Cuts of meat, whether they be primals, subprimals, or retail cuts, will require inspected slaughter and butchering (either state or federal). Regardless of where the final butchering happens, even if it's completed at a restaurant or grocery, the slaughter of the animal always has to be done at an inspected facility. The inspectors will stamp the animal carcass with a food-grade stamp if the animal passes inspection. If the meat will cross state lines, it must be either USDA-inspected or done at a TA or CIS approved plant. **Gray area:** Can the cut and wrap be in a county-inspected butcher shop or restaurant? Can the farmer pay for that as a service and then go on to sell that meat, or does the farmer need to be an "owner" in that process for it to be legal? This seemed to be a big loophole in California while we were selling meat at farmers markets. While all of our meat had to be USDA-inspected with a USDA seal (California has no state meat inspection program), there was a retail butcher shop (county-inspected only) selling meat at the same market, a caterer who made value-added meat products in a county-inspected commercial kitchen, and an itinerant chef who made sausages at a commercial kitchen he rented who also sold at the market. For years we wondered why we couldn't just do the same thing as them: hire a chef or rent a kitchen and make sausages, salamis, cured jowls, pâtés, head-cheese, confit, and other interesting value-added products that could have improved our bottom line. However, since everything we sold had to be USDA-inspected (or so we were told), we could not compete with the fancy value-added meat vendors since our processor was not able to do any of those things under their approved HACCP plans. Likewise, farmers markets have rules that the farmer vendors have to produce the products or ingredients they sell, but the other value-added meat people were not restricted by the same rule. They often bought CAFO meat from the cheapest source they could find and marked it up to prices at or above ours. Great for their profit margins; not so great for supporting regional, ethical agriculture.

Poultry: Birds of a Different Feather

Selling poultry, including chickens, turkeys, ducks, geese, ratites (emu, ostrich), and guinea fowl, follows a different set of rules than four-legged creatures. You can process them in a USDA-inspected facility, especially if you have one nearby and the price is right. However, independent USDA-inspected poultry facilities are very rare; many states don't have a single one. The rest are privately owned by the large integrated poultry companies, and, no, they won't process your birds. Don't even bother asking Tyson or Pilgrim's Pride to do this. But, unlike with red meat, you have another option for poultry processing in most states.

That option for poultry works like this: A federal rule, named Public Law 90-492 (or P.L. 90-492) sets a standard for the nation exempting poultry from continuous bird-by-bird inspection if the farmer meets a narrow set of guidelines. States can choose to adopt that federal rule or make rules *more* stringent than it, but they cannot make a rule less stringent than the federal law. Public Law 90-492 provides exemptions to the federal Poultry Products Inspection Act (PPIA) for growers processing fewer than twenty thousand birds a year. Most states that adopt the twenty thousand bird exemption will still almost always have additional requirements such as building and sanitation, annual facility inspections, and permitting fees (Johnson et al., 2012).

Domesticated poultry that fall under this law include chicken, turkey, ducks, geese, guinea fowl, ostrich, emu, and squab. Squab is an interesting category because they are actually young pigeons, but the definition of domesticated poultry at the USDA added only the word "squab" in 2002. Where an older pigeon falls under this rule is undefined. We doubt there are many USDA inspectors roaming the land, trying to determine whether your pigeon is a four-week-old squab or an eight-week-old pigeon. We'll leave that interpretation up to you.

There are actually five different categorical exemptions under P.L. 90-492, but for this book we will focus on two main classes that most growers or processors would fall under: a 1,000 bird and fewer category and a 20,000 bird and fewer category. You cannot operate under more than one exemption in a calendar year, so don't think you can raise and slaughter 21,000 birds using the 1,000-bird exemption plus the 20,000-bird exemption. The 1,000-bird rule closely resembles the 20,000-bird rule, the only difference being the number of birds you can process. Many states only accept the 1,000-bird federal exemption, but not the 20,000-bird exemption, or they have much more stringent standards for the 20,000-bird category. However, because raising 1,000 birds or less is not really a commercially viable enterprise on its own, this section will focus on the 1,001 to 20,000 bird range. You should keep records of how many birds you process in a year in case a state or federal poultry inspector wants to make sure you are falling under the bird numerical limits. You can *raise* more than 20,000 birds on your farm; you just cannot *process* more than 20,000 birds on your farm. So if you wanted to bring some to a USDA-inspected facility for slaughter and also do some on your own farm, that is fine. Here are the specific rules, with our commentary in parentheses:

Producer/Grower—20,000 Limit Exemption: A poultry grower may slaughter and process more than 1,000 birds as exempt product for distribution as human food when the following eight criteria are met [9 CFR §381.10(a)(5) and (b)(1) and (2)].

CRITERIA:

1. The producer/grower slaughters and processes, on his or her own premises, no more than 20,000 poultry, raised by him or her, in a calendar year. (*Your birds only, not other people's birds.*)
2. The producer/grower sells, in a calendar year, only poultry or poultry products he or she prepares according to the criteria for the Producer/Grower—20,000 Limit Exemption; he or she may not buy or sell poultry products prepared under another exemption in the same calendar year in which he or she claims the Producer/Grower—20,000 Limit Exemption. (*Again, your birds, not other people's birds; only one exemption per year.*)
3. The poultry products are distributed solely by the producer/grower and only within the District of Columbia or the State or Territory in which the poultry product is produced. (*So no interstate sales of exempt poultry; you can have your birds processed at a small enterprise under the exemption in a neighboring state, but you must sell them in the state where the farm is located. For example: You farm in Washington State and have your birds processed at a small poultry plant in Oregon. You can sell those birds in Washington but not in Oregon, unless that plant has USDA inspection, which is the basis for interstate sales.*)
4. The poultry are healthy when slaughtered. (*No sick, dying, or dead birds, please.*)
5. The slaughter and processing at the producer/grower's premises are conducted using sanitary standards, practices, and procedures that produce poultry products that are sound, clean, and fit for use as human food (not adulterated). (*Clean equipment, clean water, proper chilling, etc.*)
6. The producer only distributes poultry products he or she produced under the Producer/Grower Exemption. (*No distributing for other farmers—this includes having a meat CSA with poultry from other growers, unless that poultry is USDA-inspected.*)
7. The facility used to slaughter or process the poultry is not used to slaughter or process another person's poultry unless the

Administrator of FSIS grants an exemption [9 CFR 381.10(b)(2)]. (*You can't process other people's birds.* **Gray area:** *Could you become legal owner of other people's birds prior to slaughter? If so, at what age can you buy them for them to be considered "yours"?*)

8. The shipping containers or bags, when distributed in intrastate commerce (instead of the required features of a label of inspected product) bear:

   a. Producer's name
   b. Producer's address
   c. The statement, Exempt P.L. 90-492
   d. Safe handling instructions

A common misconception is that an exempt poultry operation is exempt from *all* requirements of the Poultry Products Inspection Act (PPIA). "Exempt" means that certain types of poultry slaughter and processing operations qualify to operate *without*:

- Daily federal inspection
- A grant of federal inspection
- Continuous bird-by-bird inspection
- Presence of a federal inspector

However, as an exempt operator, you are not exempt from running a clean facility and processing the birds in a way that minimizes microbial contamination. Just because you are exempt from federal inspection does not mean you should have your dog hanging out under the evisceration table, use contaminated water, or throw your finished birds into a non-food-grade barrel with tepid water. You still must use good sanitary practices. A good place to start is to follow the chicken-processing steps outlined in our poultry chapter or read through the small book *Mobile Poultry Slaughterhouse* by Ali Berlow (2013). Be cognizant that chicken meat is the most contaminated meat in the marketplace, so you should do everything within your power to keep the carcasses clean. Some small producers are moving to mechanical evisceration to speed up their lines, but this practice has been shown to increase opportunities for microbial contamination because sometimes the organs (especially the intestines) can be ruptured in that process (Posch et al., 2006). One marketing advantage of small-scale poultry producers is that we can produce cleaner birds, something consumers are keenly looking for in this age of *Salmonella* and *E. coli* scares. Remember to keep food safety to the highest standard as you process your birds.

There are a few other key points about operating under the federal exemption. Nutrition labeling is optional for exempt products (this is not true for USDA-inspected red meat and poultry; see Chapter 11 for more information). P.L. 90-492 allows you to sell exempt birds directly to consumers as well as to hotels, restaurants, and similar institutions (called HRI) but only if your state and county allow that. They may want you to stick to direct-to-consumer sales only (see table B.2 on page 291 for the exact details). Also, even if your state allows HRI sales, keep in mind that many retail food establishments are not willing to purchase uninspected poultry, so just because it may be legal does not mean there is a market for it. Many states only allow direct-to-consumer sales either on the farm or off the farm at places like farmers markets or through other direct-to-consumer purchasing arrangements like a CSA or buying club. Additionally, you can utilize a rented mobile poultry-processing unit and still operate under the exemption (some states require that the

MPU itself be inspected). You don't have to own the equipment, but you do have to conduct the processing on your own farm. In one state—California—we found that you can only do exempt processing with "immediate family" labor, not hired help, which can make things challenging if you seek to scale up. While you may be able to process one thousand birds a year on your own, ten thousand birds will likely necessitate some hired help. Perhaps this is a good excuse to have more kids?

Game birds (quail, grouse, chukar, pheasant, etc.) do not fall under the definition of domesticated poultry, nor do rabbits. They are considered exotic species and are classed with the rest of non-amenable species such as bison and deer. As such, federal inspection is not required, and is considered "voluntary." You will have to pay for that voluntary inspection, usually around $60 an hour, as mentioned earlier in the chapter, which may cut into the profitability of that venture considerably.

Keep in mind that laws change and that this is not meant to be a comprehensive list of every last legal detail (that would be a book unto itself). Consult with your own health department or state department of agriculture for further information. Livestock or poultry extension agents may also be helpful, and since they do not work for a regulatory agency, they are a little less risky to talk to. They may help you sort out some of the "legalese," the regulatory language that is tough to understand.

The main differences between the states are those that have a state poultry inspection program (PIP) or those that do not. States with PIPs usually honor the federal exemption but also require annual inspection by their state Department of Agriculture's PIP. To pass a state inspection, your poultry-processing "facility" may need to be an actual four-walled building complete with concrete floors, washable walls, drainage, proper wastewater disposal, and potable water testing of some sort. Are you willing to make that kind of investment just to sell up to twenty thousand birds—an enterprise with relatively slim margins? The annual depreciation costs will add a considerable expense to your enterprise's profitability. While a simple outdoor setup with stainless steel equipment may cost $500 in annual depreciation costs, a building may cost around $5,000 a year. That may add $0.50 to $1.50 per bird in added expenses. Perhaps there are other ways to spread those costs among a wider range of enterprises. Could you turn the poultry plant into a commercial kitchen in the off-season and make value-added products? We know of a few examples where producers have built "USDA-ready" facilities in case they decide to scale up beyond twenty thousand birds annually and start processing for other farmers too.

States that don't have a poultry inspection program will usually honor the federal exemptions but sometimes still require an annual permit, license, or on-site inspection via their Department of Agriculture or their Department of Public Health. Just because they don't have a bonafide state poultry inspection program does not mean that permitting and inspection don't happen. These usually still are part of the animal division of the state's Department of Agriculture. Without experts in poultry, however, you may have to do a little educating of the agency personnel about the nuances of P.L. 90-492. Your state poultry extensionist may be able to help you build those relationships and do that educating. Still other states, like California, honor the federal exemption and don't have a state poultry inspection program, but allow individual county departments of health to determine whether or not they will allow sales of uninspected poultry within their county. You still could process birds in that county and sell to other counties, just maybe not the county your actual farm is in. We had a comparable experience when we were raising broiler chickens under the federal twenty-thousand-bird exemption in California. Our county allowed us to process and sell birds within our county, but our main farmers markets were in the adjacent county where their health department did not allow us to sell uninspected birds. That roadblock contributed toward the decision to halt our broiler production enterprise, along with the fact that profit margins were low per hour of labor invested.

In summarizing table B.2, on page 291, it would appear that most states accept part or all of the P.L. 90-492 exemptions. Some states limit it to one thousand birds a year under exemption (Georgia, New Hampshire, New York, Oklahoma, Oregon, West Virginia—which is currently devising regulations to adopt the twenty-thousand-bird exemption—and Wisconsin); Illinois is five thousand birds; Texas is ten thousand birds; and the rest allow up to twenty thousand birds to be processed annually on-farm under exemption (thirty-eight states). There are only two states that have no exemptions and do not allow on-farm processing of any amount of poultry for sale: Colorado and Nevada. Sadly, both of those states do not have a single USDA poultry processor willing to do processing for other producers. Therefore, farmers in Colorado and Nevada either have to build their own USDA-inspected plant (to the tune of a $100,000 to $200,000 minimum investment) or they are unable to raise chickens for sale or have to resort to "alternative" purchasing schemes to avoid regulatory burden (fancy talk for under-the-table or gray-area sales).

Another thing you will notice from table B.2 under the "Sales restrictions?" column is that some states allow you to sell exempt birds to a wide variety of customers—those allowed under the federal law (consumers, hospitals, restaurants, institutions, retailers). Others allow you to sell direct to the consumer at your farm, farmstand, or farmers market, while others restrict sales to "on the farm only." If your farm is rural or in an out-of-the-way place, requiring customers to come out to your farm for purchases may severely hamper your ability to grow. Where we currently live in our thriving metropolis of five hundred people, we would be hard-pressed to convince just ten people to come out to the farm once a week to pick up processed birds. At those levels, why bother?

The last column of table B.2 is about any permitting, licensing, inspections, or fees you may be responsible for pursuing. In most states, your Department of Agriculture is who to call first to find out if you need to fill out any paperwork or prepare your site for an inspection. Although this information could change from year to year depending on your state legislative process, if you can make a phone call at least knowing some of the right questions to ask, it can make the process less confusing for everybody. Your phone call could go something like this: "Hello. I read that I have to apply for a permit in order to process more than one thousand chickens a year on my farm under the federal public law 90-492 exemptions. Can you email or mail me a copy of that application?"

Inspection

What is meant by the term "inspected slaughter or meat processing"? It generally consists of four main components (true for red meat and poultry):

Regulations

1. **Ante-mortem inspection.** The term ante-mortem (AM) means "before death." Ante-mortem inspection is the inspection of live animals and poultry prior to being slaughtered. All livestock presented for slaughter at the establishment must receive AM inspection: Each animal is inspected. AM inspection of poultry is performed on a lot basis. AM inspection is performed either by an FSIS public health veterinarian (PHV) or a food inspector under veterinary supervision. However, if a food inspector performs AM inspection, the PHV must be notified of any disease conditions that are observed. Observation includes all livestock at rest and in motion. Inspectors observe overall condition of each animal, including the head, eyes, legs, and body of the animal; the degree of alertness, mobility, and breathing; and whether there are any unusual swellings or abnormalities. If anything looks amiss, the inspector will require an animal to be segregated and further inspected. If the animal appears sick, dying, or non-ambulatory (not walking), it may be considered "suspect" or "condemned" for slaughter by the inspector. As a farmer, do yourself and the slaughterhouse a favor: Don't bring sick animals to the slaughterhouse.

2. **Humane handling and slaughter.** The inspector will check all livestock pens, driveways, and ramps. They will make sure the slaughterhouse has measures in place to deal with inclement weather and animal comfort. They will watch to ensure animals are unloaded properly from transport vehicles. They will also make sure that animals have feed and water available while they are waiting for slaughter. The inspector will observe the proper movement of animals into the slaughter area and ensure that animals are properly stunned and unconscious before loading onto the rail.

3. **Post-mortem inspection.** This involves a much more complicated series of steps, differing for each animal. It involves an inspector checking the head, glands, organs, and meat of the animal post-slaughter. The inspector looks for signs of parasites, lesions, foreign bodies, and other diseases. If they suspect something, they may require testing (as a next step).

4. **Residues, pathogens, and microbiological testing.** Not every animal is tested for hormone and antimicrobial residues, nor are they all tested for microbiology or pathogens. The acting PHV for the plant will determine whether or not to do any testing based on visual inspections and any sampling mandated by their front line supervisor. The amount of sampling varies over time, based on the occurrence of any recent outbreaks and mathematical formulas. Small plants are often subject to a higher rate of sampling due to their volume relative to the level of inspection each animal receives. If part of your meat is sent off to a lab, you will not recover that meat (or the lost potential revenue from it). They could take a chunk out of your chicken and perhaps make it unsalable after that point. Protect yourself by bringing in healthy animals in the first place.

CHAPTER 8

Sales Outlets and Market Options

This book is intended for animal producers who want to have some measure of control over their market, whether selling direct to the end consumer or to intermediate markets such as butcher shops, restaurants, or grocery stores or even to brokers and meat distributors. It is for those who want to raise an animal through maturity and know where it goes after slaughter. Our goal is for animal producers to have more power in the meat supply chain and be price-makers (to a much larger extent) rather than price-takers.

However, sometimes cash markets such as auctions or feedlots have strong prices, and you may very well want to sell some of your animals through that channel without having to take on all of the marketing expenses and risks as you would with direct marketing. That is perfectly fine. Just keep in mind that if you spend all the time and money developing a brand and a customer base, but then shortchange them because you sold half your animals at auction, you may do damage to your future market. Instead, the auction yard might be a good place to sell animals that don't meet your quality specifications or animals that you had to medicate for some reason or another and don't want to sell through your branded channels. It's a wise strategy to have a plan B market, such as selling some animals via cash markets when needed or if the sale barn prices are high. For long-term financial success though, don't make it a habit. According to research, 8 years out of 10 cash spot markets will be buying at *below* the costs of production. Building up your herd or flock to satisfy your regional markets should probably be your priority: It is one of the biggest challenges in scaling up and creating a bonafide meat-marketing side of your business.

There is an increasing diversity of outlets and customers that meat producers have to choose from today, even as some more traditional ways of selling meat are falling out of fashion. Every time we pick up a book or read a magazine or blog we find a new way of selling meat that we had not thought of before, or a new combination of channels. Bacon-of-the-month club, a full-diet CSA, chicken pot pies, a mobile BBQ trailer, ground beef sales to hospitals, restaurant-supported agriculture, and more. Successful livestock producers are usually creative businesspeople, too, always thinking of new ways to sell their animals or new ways to add value to their meat or expand their base of customers.

The average American consumer today is not likely to go out to a farm, pick out a live animal, and kill it themselves. However, the increasing ethnic diversity of the United States means that the average consumer no longer eats the same way as in years past. Despite what seems to be a "homogenization" of our food culture around the country, where the regional specialties seem to be disappearing, the diversity of cultural foodways are increasing due to immigration. This is a great development for savvy meat producers and salespeople. As we learned in a recent Wharton Business School marketing class, "customer heterogeneity" (also known as diversity) creates a great niche for the innovative entrepreneur of today. Put simply, different people like different things.

Americans also don't buy sides of animals as commonly as they once did, or store that meat at a local "locker plant" (a small custom butcher shop with a walk-in cooler). Most Americans have no idea what a locker plant even is, except for in a few rural areas where this practice still exists. Americans ordinarily don't own chest freezers anymore (partly because most of us don't farm or hunt as much); indeed, many surveys on why people don't buy meat in bulk show that the main reason is the lack of freezer space. A new project in Ithaca, New York, called "Meat Locker" is trying to change this constraint by creating a not-for-profit community cold-storage space, enabling farmers to once again sell meat by the side and consumers to buy meat in bulk at a more affordable price. More on this promising model is in chapter 11.

On the supply side, the meat supply has become much more consolidated, vertically integrated, and filled with many more intermediaries to move it from one step in the chain to the other. There may be five to eight different steps in that supply chain.

Now, if everybody along that highly specialized supply chain needs to make a profit margin along the way, where does that possibly leave the farmer? For example, eight independent businesses must make their profit margin on ground beef priced at $3.99 per pound at the retail store. What then could a farmer reasonably expect to make on that beef—$0.39 per pound, maybe $0.59 per pound? With margins that low, the only reasonable strategy to cope is to increase animal volume. Instead of finishing one hundred beeves on grass, you now must feedlot one thousand beeves on grain. Instead of raising one hundred feeder pigs on your pasture, you now must raise ten thousand to make a reasonable living. Thus the genesis of high-volume operations (CAFOs), to spread costs among a larger number of animals. Something, however, has got to give. Animal health, water quality, soil tilth, worker health, air quality, owner sanity—you get the picture.

How do you opt out of the typical lengthy supply chain described above and create your own? In the following section, we will describe some of the diverse meat sales channels that farmers have created around the country. If you have the passion and persistence, you can create entirely new channels for selling meat. Just like any farm product, it's wise to develop more than one market to provide some stability and flexibility.

Extending the "Season"

The seasonal nature of finishing animals can make securing markets difficult. Indeed, many of the sales outlets described in this chapter will demand year-round meat, although a few methods work just fine for seasonal production. As we discussed in the production chapters, there are some areas of the United States where you can't raise and finish animals very well year-round. Every region has a less than ideal season of the year, whether it is too hot, or too rainy, or too cold for the good health and gain of livestock and poultry. You can extend the "season" of meat through good freezing and packaging, quality cold storage, and metering out frozen meat during those times of the year when you can't practically finish animals.

All animals should be harvested when they are "on the gain." If your animal is losing body condition due to weather, poor grass production, or lactation, it's not the time to harvest it. The southeastern states have a great advantage in that regard when it comes to ruminants, since they can produce grass nearly year-round. In the Mediterranean or desert climates of the West where the grass-finishing season is about 8 to 12 weeks long (if they are lucky to not be in a drought year) it will mean that finishing during the rest of the year will be difficult, unless irrigated pasture is involved. You shouldn't finish a ruminant on brown grass. Yet buying in lots of quality hay or silage from somewhere else, or finding some irrigated pasture to rent, can be quite expensive. Relying on purchased feeds to finish an animal when there is no grass, whether that's the dry season of California or the snow season of Minnesota, means an increased cost to the farmer, to the point where it may no longer be financially feasible to do grass-finishing. Indeed, the short grass season in many western states is one reason that feedlots became popular in the first place—to ensure good weight gains despite the limited grass production.

Some producers in the western states do manage to rotate their livestock onto irrigated cropland, hayland, high-elevation meadows, and even into orchards. In these cases, they get to take advantage of another irrigated crop, which is not only more sustainable but may also be cheaper. We have heard of quite a few sheep producers running their stock in fruit and nut orchards, as well as vineyards, which are already irrigated for the crop. Pigs could be run into crop fields after the vegetables have been harvested to "hog down" any remaining plant material. The options are limitless as long as you find cooperative neighbors and a way to transport your animals.

When raising monogastrics (such as pigs and poultry), seasonality becomes less of an issue. However, feed efficiency drops dramatically during the winter when animals are burning up energy to stay warm. Most meat birds are not going to perform well when it's wet or cold. Again, if you don't have the right infrastructure to keep them warm, you may be running a lot of feed through those animals and end up losing money because they don't efficiently convert the feed to weight gain. We knew some farmers who kept a batch of young pigs through a Vermont winter. Despite eating a lot of feed, the pigs hardly grew at all. The farmers lost a considerable amount of money on that deal, money they did not have to lose. Make sure you are ready to go to more year-round production before you embark on that effort: Have the right infrastructure in place and the proper cash flow to feed your animals through the winter. Otherwise you may just want to consider seasonal production from the spring through the fall with feeder animals.

Also, just because the demand is there for year-round meat does not mean you have to match it with supply. Sometimes selling out can be good for your business. It creates more of a climate of scarcity around your products, which can potentially elevate them to a cult-like status. It also allows you to raise your prices in order to control the supply for a little longer over the year. Besides, it gives you a seasonal break, something you probably need to remain in this business for the long haul. However, there is a balance to be found here. If you are constantly sold out, your customers may seek a different supplier who they can rely on more. You want to prevent them from switching their allegiances. You'll find more on marketing psychology in chapter 12.

Despite these seasonal challenges, farmers are pushing the seasonality envelope in many locations: We feature several of them throughout this book. Forage can be stockpiled into winter, you can cut your own hay or make your own silage, some pasture can be irrigated, and grains can be brought in for supplementation if need be. You could also strive to finish your largest batch (or your highest quality) of animals when your grass is the most nutrient-dense and then smaller batches during other parts of the year on different forages. Some producers sow summer and winter annuals so they have more seasonally adapted forages. Cull cows or sheep could be slaughtered in the fall for ground meat and roasts to sell over the winter. Also, as mentioned further along in this chapter, you could collaborate with other producers to generate a more year-round supply. They may have infrastructure or microclimates that afford them the ability to raise animals at different times of the year than you.

In the next part of this chapter we first start out describing sales right on the farm and then work outward from there. You may have to market far away from where you actually raise your animals in order to tap into the right markets. Not all places in the United States are ready to pay a premium for local food. You may have to look farther afield than those folks in your neck of the woods to find people who care about the meat they buy and who are willing to pay a fair price for it. How to figure out your target customers will be described in chapter 12. Keep in mind that this chapter will not be discussing conventional sales outlets such as auction yards, feedlots, or meat-packers. Nor will we be delving into specialized animal industries such as selling breeding stock, animal fibers, eggs, milk, collagen, or other animal byproducts. These are all potentially great ways to add value to your flock or herd, but the focus of this chapter (and indeed the entire book) is on selling animals in the form of meat.

At the Farm

On-Farm Slaughter

Also sometimes referred to as "ritual slaughter," this is when an individual comes to your farm seeking to buy a specific animal. They pick out the live animal and kill it on your farm according to their customs. Jewish and Muslim peoples often do this with lamb or poultry. Polynesians frequently do this with roasting pigs. Having ethnic diversity in your region will play in your favor if you want to market animals this way. The legality of this practice varies by state: In most states there are no written laws that *specifically* prohibit this practice, but if you call their Department of Agriculture they will tell you it is not allowed; therefore it resides in the gray area of meat law (see chapter 7). Only the states of New York and Vermont explicitly allow a limited amount of on-farm slaughter for direct-to-customer sales. More often it may be your county zoning or public health laws that effectively prohibit this practice. Many densely populated cities, for example, don't allow you to kill animals on city lots even if it is your own property and your own animal. You may wish to handle the animal slaughter yourself, particularly if you are good at it or have humane handling preferences. However, in both New York and Vermont, the law explicitly states that the buyer needs to do the killing, not the producer, because technically the producer is selling a *live* animal.

Where we used to live in California this was a common method of selling live animals such as goats and chickens to Latino immigrants who wanted whole animals and had experience slaughtering animals. Since the humane handling skills and customs of people vary widely, we are not strong advocates of this method of selling animals. We don't want an animal that we lovingly cared for to be handled or killed in a painful way (although it is true that inhumane handling can happen in a slaughterhouse as well). For example, we had a landlord in California who sold cattle, goats, and sheep to other immigrants. One day he sold a veal calf to a gentleman who proceeded to stick the calf in the heart with a knife. The sound of the loud bawling of that animal as it slowly died haunts us to this day. We believe firmly in rendering an animal senseless first with either a bullet or bolt gun to the head before bleeding it out. Famed animal handling expert Temple Grandin agrees with us, so we know we are on the right track.

The advantages of selling live animals for on-farm slaughter is the ease of logistics, minimal marketing costs (word of mouth usually works), and cash at the time of the sale. You don't have to own a livestock trailer or big truck if you are killing your animals on-farm. You do, however, have to have some way to pen or catch them. The disadvantages are the questionable legality of the practice, dealing with people who may not share your values and often want a bargain price, potential inhumane handling and slaughtering, and taking time out of your busy day for a single sale (capturing the animal, waiting for them to kill it, collecting payment). But it might be a good way to get rid of a certain group of animals, such as piglets accidently born in the fall that you don't want to overwinter, or older laying hens you plan to cull anyway. People of some cultures will also buy older animals from you, such as ewes that you want to cull or a buck that you decide not to keep any longer. To make the logistics easier, you could designate one day a week where you do on-farm sales and be closed the rest of the time. People will still come by on those other days looking for animals to buy; perhaps put a sign out stating that you are "closed" and not to bother you. These kinds of sales are usually generated by word of mouth, so make sure you tell people what hours you are open for business and that message should spread. If you have specific rules for slaughtering, make them clear to people and explain how they must perform it. If your customers don't speak English, try to get those rules written up in their native language and make them readily available.

You also should create a sanitary, semi-enclosed space for the animal slaughter to take place. The whole practice should be done as cleanly as possible and shielded from any neighboring properties or the road. This is helpful for community relations and keeping the peace. While we believe strongly in the transparency of our food system, not everybody wants to see how animals become meat.

Freezer Meat (Custom-Exempt Sales)

This is also known as "locker meat," meaning the animals are for private customers who want to fill their freezer full of meat. In the old days, people would buy a side of beef and store it at their local locker plant, but that practice does not really exist anymore, with hundreds of locker plants disappearing since the 1970s with the extreme consolidation in the meatpacking industry. The advent of hazard analysis and critical control points (HACCP) requirements in the late 1990s also forced a dramatic exit of small plants from federal inspection programs (Muth et al., 2002). Many of those that survived switched to custom processing and don't offer meat storage anymore, beyond just aging carcasses for customers.

Yet there has been a turnaround in consumer interest for buying meat in bulk. Today, an increasing number of farmers are selling animals by the whole, by the half, by the quarter, or sometimes by the eighth (usually with large animals like cattle). Technically, the federal law states that uninspected meat cannot be sold; therefore you are selling a live animal "on the hoof" to a customer or group of customers. Thus, you are supposed to be charging based on the live weight of the animal, not on the carcass or the final cut-up meat weight. This is challenging for a producer because dressing percentages vary by the animal and you may end up losing $50 to $150 per animal if you sell based on the live weight. Most farmers we talk to around the country who sell sides of meat charge based on the hanging weight, not the live weight. A method that farmer Joel Salatin once used to try to circumvent the rule of having to sell based on the live weight (we are not sure if he still sells custom meat) was by charging $1 a head for the live animal and then adding on "shipping and handling" charges based on the carcass weight. In this situation, we think it's best to check with your state or with a livestock extensionist in your state to determine what the exact laws are (if there are any). In Oregon, where we live, you can sell your animal based on the hanging weight price because it is considered the most accurate. Additionally, in most states it is legal for several different people to own the same animal jointly, but they must be buying an individual live animal, not the final packaged meat. Again, remember that only "inspected" meat can be sold, and since custom shops do not have inspectors it is not considered inspected meat. Therefore you are selling the animal, not the meat. Always make that clear on any marketing materials you create (website, brochure, flyers, ads, etc.).

To facilitate selling live animals, either based on their "on the hoof" weights or their hanging carcass weights, Kathleen Harris of the Northeast Livestock Processing Service Co. says, "the most important marketing equipment for an animal farmer is a scale" (NMPAN, 2011). A basic scale for hogs, sheep, and goats can run from as low as $700 up to $3,700 for a "legal for trade" scale. A scale large enough for cattle will cost even more—between $4,000 and $5,000. If you are actually going to sell based on live weights, you need a "legal for trade" scale. If you sell based on hanging weights, the butcher shop will have a "legal for trade" scale at their shop to weigh your carcasses, so your scale does not have to be a legal scale. Although legal for trade scales are much more accurate, they are beyond what is typically necessary for a small operator. However, scales are important not only for knowing the live weight of the animal, but for figuring out if your feeding program is working well or not and for making sure your animal is finishing properly in the 60 to 90 days before slaughter when most of the intramuscular marbling will occur. Scales that come with a squeeze box are also handy when doing things like artificial insemination, vaccinating, and pregnancy checking, or to facilitate a host of other animal management practices if you don't have a regular squeeze chute.

Whether you sell by the live weight or the hanging weight, the process typically goes something like this: A customer or group of customers pays a deposit on a specific animal to reserve it. You either take that animal to a slaughterhouse or you have a mobile slaughterman come to your farm to dispatch the animal. Your customers pay the slaughter charge or you tack it onto your overall price. Technically, your customers are arranging and paying for the slaughter of *their* animal. Then the carcass gets delivered to a custom butcher shop. You can either choose a processor ahead of time or suggest a few to your customers and have them select the butcher they prefer. You (as the farmer) should not deliver the carcass yourself. Either the mobile slaughterman should do this as part of their service or your customer should do this. Remember, it is no longer your animal. Your customers communicate with the butcher shop as to how they want the animal cut up. They can do this by phone, by fax, or in person, and some shops offer online cutting forms. As the farmer, you should not be arranging the cut and wrap. You can provide sample cutting instructions, which might be important for people new to buying meat in bulk, but you should not be dictating how the animal is cut up. See appendix C for sample cutting instructions. Again, doing so blurs the line about who is the owner of the animal. Then your customer picks up the finished meat from the butcher and pays the butcher directly for their services. They do not pay you for the butchering work and you should not be including that in your price (if you want to safely follow the regulations).

A rather seamless way that we conducted this process when we were farming pigs in California (we sold about fifty live animals a year as custom pigs) was this: We advertised to our customer list when a group of pigs was going to be ready for slaughter (usually spring and fall). We provided a contract form to interested buyers for them to fill out. On the form buyers indicated whether or not they wanted a large pig or a smaller-sized pig, a half or whole, and which butcher shop they wanted the carcass to go to. They also indicated if they wanted the head, the offal, and the blood, which were all options to them for an additional price (we had to pay the slaughterhouse to give us these things back from our pigs). The customers paid a deposit that was equal to about 50 percent of the total price to reserve their pig (or pig half). They would keep half of the form as their receipt and the other half they mailed in. We eventually went to an online PayPal system so folks could pay online and get an email receipt right away.

When we had all the pigs in that group pre-sold, we arranged transport to slaughter at a local USDA-inspected slaughterhouse. (USDA inspection is not necessary for custom sales, but they did the best job, scalded and scraped, were cheaper than on-farm and even better for more than three animals, and had the most affordable kill fee in our region.) The slaughterhouse then dropped off carcasses at one of two butcher shops that customers pre-chose. We met the slaughterhouse truck when they got to the shop to make sure they were dropping off the right number of carcasses to the right shop. We then got the hanging weights while at the shop and told the butcher which pig belonged to whom so they could tag their legs with the customer's name and phone number. We then called or emailed the customer with the remaining balance due based on the final hanging weight and told them to contact the butcher shop immediately to give cutting instructions. Although we no longer owned the animals, the butcher shops agreed to not let the meat of an animal go out of the shop without the customer paying us in full—we called the butcher when that happened. Customers then paid the butcher shop separately for their services. The hanging weight price we charged included a $50 fee for slaughter and transport of the carcass to the butcher shop. In our four years of selling freezer meat we only had one customer who we struggled to collect full payment from. The rest of the time this system worked very well. We typically offered a batch of pigs in June before summer BBQ season and a batch in fall, around twenty to twenty-five pigs each time, which was the number we could fit in our livestock trailer.

When comparing custom sales with farmers market retail meat sales, custom sales were considerably more profitable due to vastly reduced processing, cold-storage, labor, and marketing costs. Unfortunately we didn't do an intense financial analysis until our last year of production. If we had known this information sooner in our farming enterprise, we would have devoted significantly more attention to building our custom sales consumer base and reduced or eliminated altogether our farmers market meat sales. Our net profit margins were about 30 percent higher for freezer pork over portion-cut packaged meats. To elaborate on this important subject, we will discuss financial management and pricing in chapter 13.

One challenge that farmers face when trying to build their base of custom sales is that many people don't have large chest freezers anymore. A regular-sized freezer as part of a freezer–refrigerator combo can hold a whole lamb, whole goat, half a pig, and probably one quarter of beef. It's important to let people know this. We took a picture of an entire half pig cut and wrapped and loaded into our freezer and then wrote a blog post about it as a way to educate our customers. See the picture of our freezer on page 13 of the color insert. Another innovative pig farmer, Walter Jeffries of Sugar Mountain Farm in Vermont, posted a picture of a whole pig cut and wrapped and laid out on his kitchen table just to provide spatial context for potential customers. Make it easy for people to see how much meat they should expect out of this arrangement.

Would-be customers also may hesitate to purchase a side of meat if they don't feel experienced enough giving cutting instructions and working with the butcher. Again, try to make it as seamless as possible while still falling within the confines of the law, such as providing sample-cutting instructions. Explain the process to them in person, over the phone, or on your website. Especially with new customers or "novice" bulk buyers, explain all the nuances that we talk about in chapter 9, such as aging time, tenderization, grinding, portioning, packaging options, and so on, so they will be well versed to deal with the butcher themselves. You don't want a customer to go away disappointed because they didn't know how to tell the butcher how they like their meat. If the butcher uses nitrites in the bacon or MSG in the sausage or cuts the steaks too thin or takes off too much of the fat cap, all of those details will influence the satisfaction level of your customer. This might turn them off to future purchases. In all our time selling custom whole and half pigs, we only had one disappointed customer who asked for the skin to be left on some of their roasts but then found out that there were a couple of coarse pig hairs still left on the skin. Sadly, they wanted a refund for their meat, even though (as many of you probably understand) a few stray pig hairs are not big deal and they cook right off during roasting. We ate that half (happily, of course). You may offer a money-back guarantee to make potential customers feel more confident.

Another challenge is that people don't want to commit to that much meat in bulk before they have had a chance to try your meat. Make sure you provide opportunities for people to buy or sample different cuts of your meat before you embark on selling meat in bulk. We did this by having years of experience selling meat in various farmers markets, where we built up a significant customer base that had tried our meat. If we had a customer who spent considerable money purchasing retail meat, we would mention to them the opportunity to buy our pork in bulk as a way to save money. Customers were very appreciative that we let them in on that secret. Many of our devoted customers went on to buy whole and half pigs, and once they tried it, they would often reorder the following season or the next year. They would tell their friends about the experience and then many of those friends became freezer pork customers. We personally would never buy more than twenty pounds of meat from a producer if we have not previously sampled their meat. This is where selling to restaurants or selling cuts at farmers markets may come in handy, because it gives people a chance to try your meat before committing to bulk purchases. You could also sample your meat at county fairs, church potlucks, CrossFit clubs, or other local events where you can introduce your product and your story to potential customers.

Yet another challenge is that people don't always know other people who will split an animal with them. Either you need to find those other people to split the animal with (maybe down to quarters or eighths) or facilitate them in connecting with other like-minded folks. Philly Meatshare and Long Island Meatshare are a couple of groups that help identify people to go in on an animal together. Some call this "cowpooling." We only split our animals into halves so we could easily handle the logistics ourselves. When splitting into quarters or eighths, it can be challenging to figure out who gets which cuts. Some farmers call them "split halves" for mixed quarters that include parts of the frontquarters, middle meats, and hindquarters. Others offer the option of just getting the frontquarters since some religions don't consume the hindquarters of ruminant animals.

Mike Lorentz of Lorentz Meats commented that if a farmer has ten or fewer beeves a year to sell, they should be selling them all as freezer beef (NMPAN, 2013). At this size, don't bother with getting your animals fabricated into retail cuts because you will lose money paying for all that processing, labeling, and marketing of those cuts. Our experience with pigs validates that statement, and indeed we think we could have built our whole meat business mostly around custom sales. That certainly would have given us our weekends back: For 6 years we hardly had a single weekend to play, spend quality time with our family, or just simply sleep in due to back-to-back farmers markets. Selling freezer meat is a great way to establish your business while you are building your herd, and the profit margins are often better than selling retail cuts via other market channels like farmstands or farmers markets. This is mainly due to avoiding the processing, transportation, cold-storage, and marketing costs, all of which can add up to a substantial portion of your expenses per pound (25 to 50 percent).

Start advertising with friends and family. Once they have a couple of good experiences and you get a better handle on the logistics, you may expand custom sales from there. Encourage your friends and family to spread the word for you:

Word of mouth marketing is always the most powerful. For more on marketing techniques see chapter 12.

Farmstand or On-Farm Shop

Most states allow farmers to have a retail component on their farms. Check with your county first to see if this is an allowed use given your zoning. However, government officials may not understand that selling meat from an on-farm stand is really no different from selling hay out of a barn. Their initial reaction may be one of distrust or ignorance: Your job will be to educate them. Make sure you are selling inspected meat that is kept properly cold, just as you would at a farmers market or other location. Likewise, many of the farmstand laws dictate that you have to produce everything you sell at the stand or that you can sell or consign product for other producers in your region only. Where we live in Oregon, farmstands are usually allowed on farmland or rural residential parcels, but the products either have to be homegrown or sold on consignment from other local producers. They can't be purchased and then resold, otherwise the farmstand is no longer a farmstand. It gets lumped into the retail grocer category, which requires a bunch of other legalities like handicapped access, restrooms, potable water, and so on. So you may want to stick to selling your own stuff or consigning other local products in order to keep the legalities fairly simple.

A farmstand can sell frozen or fresh meat, provided you have the right infrastructure that meets county code, but it all must be USDA-inspected meat or processed in a state-inspected equivalent. Your state may require you to get some sort of "meat seller's" license or other type of business license to sell meat, typically a relatively minor expense of $100 to $300 a year. Often this is not required if you are just selling fruits and vegetables, but meat gets grouped into a different risk category, requiring cold storage.

Meat at a farmstand or farmers market booth needs to be maintained at certain temperatures, which can sometimes be tricky. Fresh meat needs to stay between 26 and 41 degrees F (between -3 and 5 degrees C) to stay safe. Frozen meat needs to be held at 20 degrees F (-7 degrees C) or less (according to USDA FSIS rules). Make sure you have a working thermometer near your meat so you can monitor the temperature and in case a health inspector decides to pay you a visit. We once had a health inspector check the temperature of our coolers at the farmers market on a hot day. Luckily we had remembered to put the thermometer into the cooler and our temperatures were below 20 degrees F.

For a farmstand to work, you need to be located on a busy enough road, but one that is safe enough for a driver to pull off onto your driveway. We have seen farmstands fail because although thousands of people drove by every day, those people could not safely exit the road to pull into the farmstand parking lot. Good signage is important, as is keeping regular hours so people can depend on you. Just like any retail sales, you should have friendly, knowledgeable staff on hand to make sales. Or you could conduct it on the honor system, which works best in areas with low crime rates (we never could have done this where we lived in California; everything on our farm that was not bolted down disappeared in the middle of the night). One farm shop in Maine, Misty Brook Farm, stocks refrigerators with eggs, milk, cheese, and yogurt and a couple freezers with meat, in addition to crates of fresh vegetables. The meat is labeled with the weights and prices so it is very easy for customers to add up their purchases. Calculators and scratch pads are provided for folks to tally up their numbers. The customers pay their money into a locking cash box and Misty Brook does not pay a dime in labor for watching the store. The farmers and their employees are often present on the farm doing chores like milking, harvesting vegetables, and restocking the farm shop, so they can—to some extent—keep an eye on things. The benefits of an honor system store outweigh any potential theft that might take place, and, to their knowledge, the cash box income has always matched the inventory sold in the shop. People can drop by any time of the day to pick up necessities, not having to worry about remembering a schedule. Having a good selection throughout the year is key to the farmstand's success; that way customers will make repeat visits because they can trust that there will be something they want or need for going out of their way to stop in.

Stocking food items that tourists enjoy might also boost your sales. They may be more interested in shelf-stable products, dehydrated foods, or things they can pop right into their mouths. Packages of raw meat are not exactly that, but beef jerky, salamis, or canned pasta sauce with Italian sausage in it just might do the trick. You could also carry things that complement meat, such as flavorful hard cheeses, crusty breads, fresh-pressed olive oils, or homemade spice rubs. Some people have added commercial kitchens onto their farmstands to further process their meats or cook them into take-out meals. A mobile food truck is another, albeit risky and expensive, option—one that you could park on your farm as a food stand and take to special events now and again. Vanguard Ranch, the Virginia goat farm described in chapter 3, is a successful example of making the mobile food concession trailer work in a variety of locations, including on their own farm during concert events they have organized.

An on-farm butcher shop is a totally different regulatory beast altogether, but it can and probably should include a retail component. One model that we visited in Vermont, called Vermont Salumi, has an on-farm USDA-inspected commercial kitchen for processing fresh sausages, built inside of an existing barn. Instead of paying an employee to stand there and sell meat to farm visitors, a large chest freezer with clear signage directs people how to pick out their meat and pay into a locked cash box with a slit on top. Exact change is required, but this honor system meat market saved considerable dollars on staffing expenses. According to the owner, he hasn't experienced any loss with this system. We know that not all parts of this country are appropriate for honor system farm shops. In those cases, if you had a butcher in the shop cutting meat or making value-added products, he or she could attend to a customer now and again. If butcher shop sales are steady, you might have to dedicate an employee to making those sales. You obviously have to figure out if that labor cost is worth the added sales.

Although much larger in scale, the Georgia-based White Oak Pastures on-farm abattoir has a retail butcher counter and a restaurant serving daily specials 6 days a week. Adding a restaurant to the butcher shop serves a few purposes, according to Will Harris, the owner. First, he can feed his staff good, healthy food while they are on shift; second, it provides a place for people to try his various meats in a ready-to-eat form; and third, it allows his butcher shop to reduce their shrink by cooking day-old meats, meat scrap, bones, and the like into delicious, seasonal meals. Location is key for any butcher shop: Being located on an out-of-the-way farm may not make any sense if foot traffic is a key component of your sales strategy.

Off-Farm Sales

Community-Supported Agriculture (CSA)

Meat CSAs involve some way of collecting payment from customers ahead of time and raising up enough animals to meet the volume people have pre-purchased. Originally devised for vegetable farmers to share risk with their consumers, meat CSAs are generally less about risk sharing but do involve some aspects of seasonality, education, and connection to the farm and animals that provide the sustenance. CSAs are typically a direct connection between a farmer or group of farmers and their consumers. Some farms establish CSA prices based on predetermined poundage of meat; others do it by a value of meat over a certain time frame (such as $500 worth of meat over 6 months). Many meat CSAs offer a variety of red and poultry meats and different cuts too. Some allow buyers to customize to a certain extent, such as which meats *not* to include—there is sometimes a "no pork" or "no goat" option since there are people who have strong aversions to those meats.

Below is a description of some of the meat CSA options we have surveyed around the country, so you can see how other farmers organize them. Unlike a buying club, which will be described below, a meat CSA gives some basic options for customers but usually does not have complete customer choice of meats, portion sizes, sausage flavors, and other variables. This gives the farmer the flexibility to choose the meats based on seasonality, what they have left in their inventory, the capability of their processor, and so on.

Meat-of-the-Month Club: 12 pounds of mixed-meat cuts once a month:

- 1 broiler chicken (approximately 4 pounds)
- 1 roast, 3–4 pounds (beef, lamb, or pork)
- 2 steaks or chops, ~2 pounds total (beef, pork, lamb)
- 2 pounds of ground meat (beef, pork, lamb)

Must sign up for a minimum of 3 months and up to 12 months at a time. A 12-month prepayment gets a discount of some sort, usually 5 to 10 percent.

Family Share: 20 pounds of mixed meats monthly

Small Share: 10 pounds of mixed meats monthly

Poultry-Only Share: 24 whole broiler chickens, 2 ducks, 1 goose, 1 turkey, 2 guinea fowl over the course of the year

Griller's Share: monthly delivery for 3 months over the summer grilling season, with each month including a mix of:

- 4 steaks or chops
- 2–3 pounds of ribs (baby back ribs, spareribs, beef ribs)
- 4 pounds of ground meat (pork, lamb, beef)
- 2 pounds of sausages

Economy Share: monthly delivery, maybe just over the fall to winter season:

2 roasts
2–3 pounds of ribs
2–3 pounds of smoked shanks or hocks
4 pounds of ground meat
2 pounds of stew meat
5 pounds of soup bones

Dog Owner's Share: monthly delivery of a mix of the following:

5 pounds of trim meat
5 pounds of meaty bones
2 smoked pigs' ears
1 smoked large bone

Four-Season Share: a meat CSA that honors the seasons of animals:

Spring: milk-fed veal, spring lamb
Summer: beef
Fall: pork
Winter: frozen meat selection

One cattle rancher in California, Joe Morris (profiled in chapter 5), has a beef CSA arrangement in which his customers get an installment of beef twice a year, with around 40 to 45 pounds in each allotment. This is called a split half, which is a mixed quarter of a steer so a customer gets both hindquarter and forequarter cuts. Joe also offers traditional whole and half carcasses available in one pickup for customers that have large chest freezers for a slightly cheaper price. The installment arrangement, however, saves the customer from having to purchase a chest freezer while still having access to beef in bulk at an affordable price. The beeves are processed and the frozen meat is stored at a rented cold storage facility with −20 degrees F (−29 degrees C) freezing capability. Then, twice a year, Rancher Joe goes to the cold-storage facility with a few helpers to sort and then load up mixed boxes of beef into a refrigerated truck. They meet groups of customers at dedicated pickup spots around the San Francisco Bay Area where labeled boxes go to their intended families who have prepaid for the meat. Every 6 months customers get the rest of their pre-purchased beef until they finish getting all of the cuts from their beef share. The meat is slaughtered and butchered under USDA inspection so it is legal for sale. The customers don't have to be involved in any of the processing logistics or cutting instructions, as they would have to do under custom-exempt sales. This model also demonstrates that you can do a meat CSA with just one type of meat, in this case beef.

Keep in mind you don't have to set up a meat CSA all by yourself. You could partner with a couple of other meat producers who are producing high-quality, ethical meats, so you can offer a wider selection of meat more year-round without stressing your own land or management capacity. Or you could partner with an existing vegetable-based CSA to offer meat as an add-on option. This will help that CSA farm attract and retain customers and reduce your marketing costs since it won't be all on you to drum up business. The CSA farm will likely mark up the meat share to cover their time in marketing and logistics—something in the range of 10 to 30 percent would be reasonable. This means that if you sold twenty meat CSA shares worth $500 each, you would take home around $7,000 of the $10,000 income the CSA farm receives. If this seems like a big chunk of lost revenue to you, try to estimate what it would cost you to do all the marketing and logistics yourself, not to mention how long it would take to find twenty new customers. This may be a good way for you to get started, and then later you could branch off on your own, once you have the marketing skills and the customer awareness solidified. People who buy vegetable CSAs may be more apt to purchase a meat CSA because they understand the variety and seasonality of cooking. That said, they may not have the funds to buy both a vegetable CSA and a meat share at the same time. So there could be a finite number of meat shares you can add to an existing vegetable CSA.

When we produced eggs in California, we partnered with a well-established CSA who would mark our eggs up about 25 percent over the price they paid us. Once our egg demand grew, we no longer needed to sell through that channel, but it was a great way to get ourselves established as a brand and to provide needed cash flow in the wintertime when sales were at their lowest (the CSA collected payment in winter for summer delivery; we received a portion of that up front each January to buy new chicks). Keep in mind, just like all CSA arrangements, that if you get all the cash up front, it may challenge your cash flow later on in the season. This was certainly the case for us.

Meat CSAs can be set up as a regular CSA arrangement in which customers pay all the money up front and you deliver a roughly set value share of seasonal meats throughout the season. You could also charge people a "credit" up front, say $400, that gets reduced over the season as they select meats to purchase from you at your farm, farmstand, or dedicated pickup locations. Usually customers receive a discount on their meat when they pay a credit up front—somewhere on the order of 5 to 10 percent is fair. You could offer add-ons that people pay for by the cut or 25-pound mixed boxes; you could collaborate with other farmers for a multi-farm CSA; or any other number of financial arrangements. You have to think about the level of risk you want to take, how long you want to lock in prices, and what the perceived value is to the customer for joining your CSA.

On our own farm, we tried for a time to sell mixed-meat boxes through a CSA-like local food delivery service. Their customers could elect to purchase a box of mixed meat in 10-pound increments up to once a month. We never sold very much volume through that channel, but we also never tried that hard to maximize it either, since our farmers market sales were consuming most of our meat. The mixed-meat boxes were a good way to get rid of some of the harder-to-move cuts, because we put several different meat cuts in the box: some high-end cuts, some medium-end, and a few low-end cuts for variety. We would try to put a similar mix in each bag so that every customer was getting close to the same thing. We didn't want some customers to be really stoked by having 10 pounds of steaks and chops, while other customers had a bunch of smoked hocks and shanks in their bags. One of the drawbacks of the mixed frozen meat box was having to go to our cold-storage facility where we rented pallet space, pull off a bunch of boxes, and then sort through them. The cold-storage facility didn't particularly like that we were pulling out product and rearranging it, and using a bunch of their tables to make the boxes. We then had to reconsolidate boxes and put them back on the pallet in the freezer. It was time-consuming, freezing cold work, and all of that handling on delicate vacuum-sealed bags is not ideal for maintaining their seal. Consider how you will sort and pack your meats if you develop a CSA or mixed-meat box program.

The brooder setup at Afton Field Farm. *Courtesy of Alicia Jones*

Dry bedding and warm lights in the brooder at Afton Field Farm. *Courtesy of Alicia Jones*

Three-week-old chicks in a brooder with diverse natural feed. *Courtesy of Harvey Ussery, themodern homestead.us*

A lightweight yet sturdy chicken tractor design. *Courtesy of John Suscovich, FarmMarketingSolutions.com*

Efficient design and layout make quick daily chicken chores at Afton Field Farm. *Courtesy of Alicia Jones*

Chickens, turkeys, and geese all in the same pasture at Afton Field Farm. *Courtesy of Alicia Jones*

Ducks and geese share a pasture while enjoying a comfrey treat. *Courtesy of Bonnie Long*

Great Pyrenees dog guarding his flock of Ancona ducks. *Courtesy of Evan Gregoire, Boondockers Farm*

A simple waterfowl bath keeps birds clean and reduces mud. *Courtesy of Harvey Ussery, themodernhomestead.us*

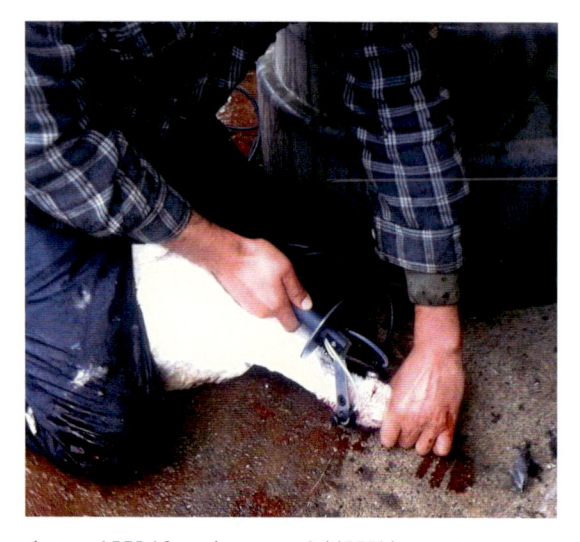

Animal Welfare Approved (AWA) requires more humane slaughter methods, such as this electrical stunning of a turkey carried out at Deck Family Farm in Oregon. *Courtesy of John Deck*

Post-harvest cleanup in a low-cost, on-farm poultry slaughter structure. *Courtesy of Harvey Ussery, themodernhomestead.us*

Setting up a single strand of hotwire for cattle at Afton Field Farm. *Courtesy of Alicia Jones*

A mobile small livestock shelter on skids is ideal for sheep or goats. *Courtesy of Westphalia Trading Co., LLC*

Sheep grazing in an old apple orchard at Afton Field Farm. *Courtesy of Alicia Jones*

Weaner pigs in their deeply bedded winter hoophouse at Afton Field Farm. *Courtesy of Alicia Jones*

A three-sided mobile group pig shelter.

Pastured Berkshire pigs contained with a single strand of electrified wire. *Courtesy of Alicia Jones*

Sow farrowing hut prototypes made from scrap materials.

An electrified offset wire attached to a woven-wire perimeter fence contains most livestock and protects your fence from animals rubbing, digging, and/or chewing it.

Feeder pigs grazing next to an electric fence, a simple way to practice rotational grazing with pigs.
Courtesy of John Deck

A grassfed Beefmaster cow showing excellent body condition on Lasater Ranch, Colorado. *Courtesy of Kendra Kimbirauskas*

Grassfed cows with their calves at Afton Field Farm. *Courtesy of Alicia Jones*

Cattle grazing stockpiled forage in an Idaho winter. *Courtesy of Jim Gerrish*

Leader-follower system with cattle and poultry at Afton Field Farm. *Photo courtesy of Alicia Jones*

Cattle grazing in Sudan grass, a productive summer forage. *Courtesy of John Deck*

Moving cattle to the next pasture in a rotational grazing setup at Afton Field Farm. *Courtesy of Alicia Jones*

Pastured rabbits in a mobile rabbit tractor at Polyface Farm. *Courtesy of Jessica Reeder/Wikimedia Commons*

Spring bison calves. *Courtesy of Northstar Bison*

Bison in a well-fenced secure pasture. *Courtesy of Northstar Bison*

Aging beef subprimals in The Maple Grille's cooler. *Courtesy of Mike Stricklin*

Cooling lamb carcasses and offal in a walk-in cooler. *Courtesy of Chris Fuller*

Half a pig cut and wrapped fits nicely into a normal freezer space.

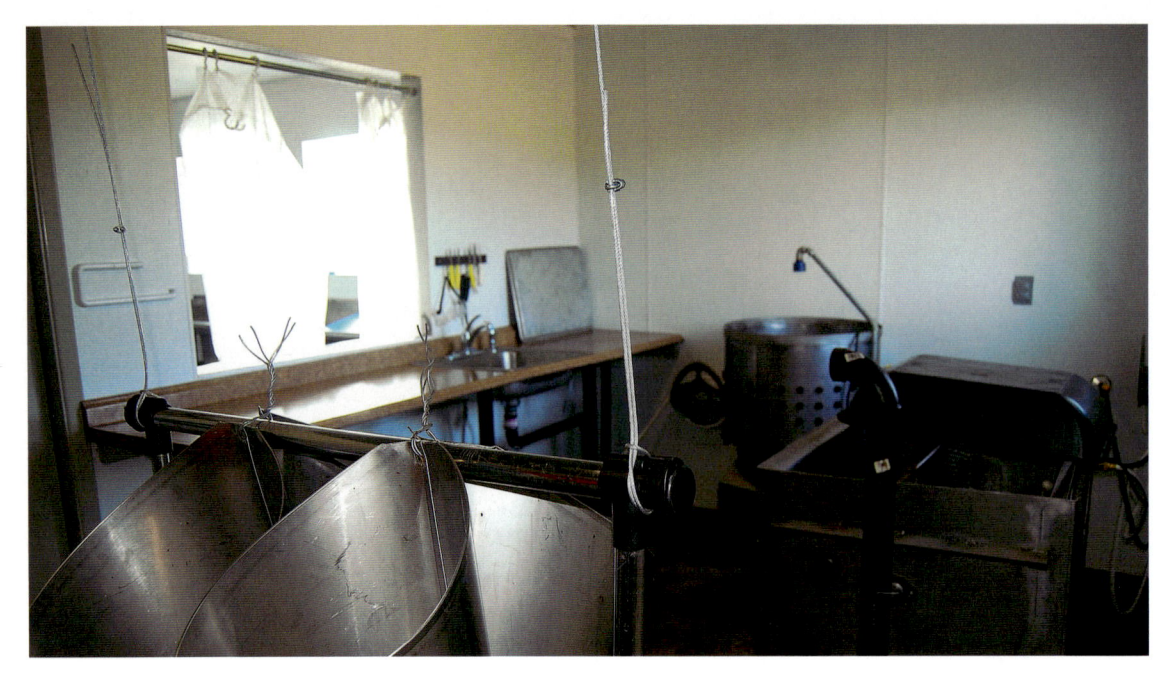

Little Farms LLC's poultry-processing building in rural Washington can process up to 20,000 birds per year and cost around $100,000 to build and equip.

Northstar Bison's heavy-duty handling and loading area for bison keeps animals and humans safe.
Courtesy of Northstar Bison

A mobile processing unit (MPU) for poultry. *Courtesy of Mike Badger, Badger's Hillside Farm*

An intensely marbled acorn-fed American Guinea Hog pork chop.

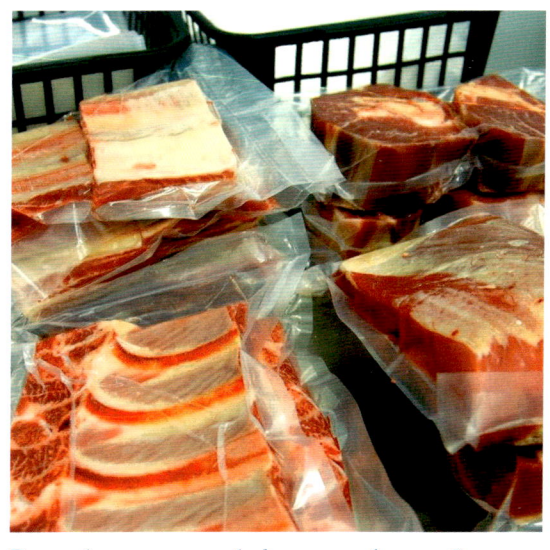

Properly vacuum-sealed meat packages. *Courtesy of Chris Fuller*

Clean, well-packaged, pasture-raised chicken from Afton Field Farm. *Courtesy of Alicia Jones*

Building trust with customers through farm tours. *Courtesy of Linda Ozaki*

Jim, Rebecca, and daughter, Fiona, on their old farm in California. *Courtesy of Jenn Ireland*

Hettie Belle Farm

Hettie Belle Farm is a newish family farm on 48 acres of leased land in Warwick, Massachusetts, and the farm is designed around the concept of CSA. After 4 years of development and growth, they now support their goal number of 120 families with a variety of meats. Seven different animal species graze in rotation around their hilly pastures and woodlands. Most of their animals are slaughtered in the fall, and are packaged and frozen for CSA disbursements over the winter, spring, and following summer. Members have the option of a 60-pound share, a 45-pound share, and a small summer share of around 15 pounds. The small share is particularly attractive for the summertime vacation-home owners and for anybody curious about joining a meat CSA without a big investment. A variety of meats are included in each share, and certain meats can be left out if a member prefers that. Members also get special benefits, such as being invited to farm events like a Fall Farm Day and priority in reserving holiday poultry, and they receive an informative monthly newsletter with good recipes. The farm has five pickup locations in nearby New Hampshire, Vermont, and Massachusetts. Meat is delivered in a reusable freezer bag.

The owners of Hettie Belle, Jennifer Core and Olivier Flagollet, worked as agricultural educators for years. This gave them the time and resources to carefully plan out their debt-free farming model. Jennifer still works off the farm for the benefits it provides and to keep stress to a manageable level. An off-farm job also evens out cash flow, something CSA farms have particular challenges with. They get all their sign-ups and payments during one part of the year, yet they have year-round farm expenses they need to pay for. Having a new summer CSA share also helps with Hettie Belle Farm's summertime cash flow, and they attend a local farmers market too during the summer.

The higher-end cuts are reserved for their CSA customers (especially steaks and bacon), while things like whole chickens, ground beef, and pork chops are also sold at their farmers market booth. These are all popular items for the summertime grilling season anyway. Hettie Belle offers a mix of cuts in their CSA share, and it varies throughout the season. If members have a really strong aversion to certain things, the owners will leave them out. Conversely, if members really want something, the owners will try to accommodate that as well. They bring a giant chart of all their members' preferences and what they received in previous disbursements into their cold storage area while they are sorting and packing shares, so they can keep track of who gets what. Being a small, member-focused CSA allows them to provide this level of customization.

CSA farms across the country sometimes struggle with member retention. Hettie Belle is no different, although their numbers are pretty solid—around a 60 percent retention rate of members coming back year-to-year. Occasionally members don't come back because they don't understand that an animal is made up of many different parts. Others who are used to home cooking, experimenting with recipes, and menu planning, do just fine with the variety of meats offered in the share. People who have been members of a vegetable CSA and are used to the CSA model tend to be a good fit for their meat CSA too. Jennifer explained that communicating the whole animal with a diversity of cuts concept to existing and potential members is really key for customer satisfaction and retention. Often the people who leave are those who are just too used to going into their local grocer once or twice a week to buy fresh cuts of their favorite high-end steak or bacon or those who only enjoy cooking a certain cut of meat in a specific way. Others leave because they realize they don't actually eat that much meat. Offering a smaller-sized share and a summer share can help retain some of those members. Hettie Belle does what they can to accommodate members and always gets compliments on their outstanding meat quality. Sometimes, however, a meat CSA is just a mismatch for a customer.

The reason we never started our own meat CSA was that we didn't feel we had the animal diversity or the year-round selection to support one. In hindsight, we think we could have done a seasonal CSA as described above; even smarter would have been partnering with other farmers who were raising animals that we didn't have. We'd already worked with a grassfed beef farmer and a lamb farmer, and so we had pork, lamb, and beef as options at our farmers market booth. But we stopped raising poultry for meat, and there weren't that many other producers around doing poultry on any appreciable level that they would have sold to us for our CSA. Would pork, lamb, and beef (and chicken eggs) have been enough to attract CSA customers for us?

A meat CSA can be a great marketing outlet, creating loyal customers and helping you get rid of all the parts of your animals. And, as we've outlined above, you don't have to raise all the diversity of animals yourself either; you could partner with other like-minded producers to put together the variety of meats to make your meat CSA more attractive.

Buying Club

Buying clubs can take all sorts of iterations these days, from those that resemble more of a traditional CSA to ones where folks can pick and choose what meat cuts they want and when. Many are simply aggregator businesses that don't actually produce any of the food. But since this book is about what farmers can do, we will describe farmer-based models.

Black Earth Meats in Wisconsin is a full-service USDA-inspected abattoir with a small retail counter and another retail shop in the state capital of Madison (as of this writing they are changing to doing USDA butchery only in a different location). They have recently set up a buying club for consumers to get the benefits of buying in bulk at more affordable prices than if they were to buy retail. Black Earth benefits by having an outlet for a wide variety of cuts, including ones that might not traditionally move as quickly in a retail setting. They offer mixed cuts of steaks and chops, roasts, grind, sausage, and other odds and ends of a wide variety of meats that come from the growers they work with.

Farmers still get paid adequately for selling meat to Black Earth while not having to take on any of the investment, management, or risk of developing their own CSA. Consumers get to experience different heritage breeds of animals and learn how to appreciate the whole animal. They can add-on different items to their box at a 10 percent retail discount and they can shop in the store for 10 percent off as well. There is a 3-month commitment to start that auto-renews after that unless canceled. Two different monthly box sizes are offered, priced at $75 or $125 a month.

Another buying club located in California, The Foragers, was set up a slightly different way for a couple years. The owners have since changed to a more traditional CSA-style arrangement, where they select the meats that go into the boxes. However, we think their original idea is an interesting model that is worth sharing. Customers paid money into an account as a credit, with a minimum amount of $100. Each month, the owners of The Foragers sent out an email with the delivery dates, times, and products available. Customers chose their delivery site (around the Bay Area) and picked what meats and other foods they wanted (choices of beef, pork, lamb, goat, chickens, turkeys, eggs, cheese, olive oil, and a few other locally sourced foods), but they wouldn't know the exact weight of each cut until they picked up the meat. The Foragers then loaded up a custom-built cold box trailer with the ordered items and delivered to their Bay Area drop sites in a back-to-back 2-day route. Customers arrived at their designated delivery location (usually a private house) and picked out the cuts or whole birds that they wanted within the available size ranges that day. The owners of The Foragers marked down what meats and weights customers took, and when they got back to the farm they deducted the appropriate amount of money off of the accounts. Customers could then access their accounts through a secure server to see how much credit they had left. They had the option to reload their account at any time. This system allowed people to have more choice. Customers also got to pick portion sizes, a further customization that they enjoyed. The Foragers also benefited by not having to pack CSA boxes, but rather being able to just load the truck with boxes of various cuts and whole birds. Customers would do the picking and packing into their own bags or coolers.

The Foragers sell their own ranch-raised meats and "curate" the foods of other nearby producers. They seek out other good producers using organic and nearly organic practices and producing a tasty, superior product that people can rarely find in any grocery store. The Foragers website profiles these other producers, so everything is completely transparent. Because all the foods are pre-ordered using an easy online store, there is no loss or waste in the selling, unlike a farmers market. It took awhile to get the number of customers up to a critical mass, but once that was achieved, their once-a-month sales trip was grossing between $4,000 and $5,000. Not bad for 2 days of sales. Conversely, you may have to attend four to ten farmers markets to gross the same amount, depending on where you live.

Farmers Markets

Farmers markets are the best of times and the worst of times all rolled into one. They can be a great place to establish your brand when you are just starting out and finding new customers. You can communicate directly with your customers and establish mutual trust. You can show off your meat. You can make friends. However, to commit to standing in one spot for 3 to 5 hours every week, where the amount you sell is a gamble, may not be the best use of your time. It may rain. There may be a marathon going through town that snarls traffic. There may be competing farmers markets that draw in more customers. You may drive away with only $300 in your pocket after 5 hours of standing there with a faux smile on your face. A lot of farmers markets have started up in the last 5 years, to great hoopla and fanfare, but many turn out to be inadequate for farmers. The good ones may have a long waiting list (we know; we sat on one waiting list for 5 years, to no avail). Likewise, there is usually a dose of politics, favoritism, and rule bending occurring that could take you for a spin and spit you out hopeless on the other side. It used to be there was only one meat vendor at a market; now there are sometimes three to five of them. That competition may impact your sales. It's not so easy anymore to "corner the market."

Farmers markets require good people to staff them. If the thought of hanging out in the hot sun or the cold rain does not appeal to you, or if you are not the extroverted social butterfly that it takes to smile all day at everybody who passes by, then you might consider hiring a staff person to work the market for you. Many farmers we meet think they don't have time to do markets; they are too busy doing chores every day. They also incorrectly believe it has to be their rosy face that customers see each week. Yes, it helps for the farmer to make an appearance now and then, but it does not have to be you, the business owner, staffing the stand. If you are the farm owner-operator, is it really worth your time making $8 to $15 an hour when you could hire a charismatic salesperson to work for those wages while you hopefully make more for your knowledge and expertise "back at the farm"? However, realize that your market staff are an extension of your "brand." If your brand is about certain values of honesty, transparency, expertise, and quality, then you should hire people who possess those same values. If your brand is about fun, spontaneity, and creativity, then there are folks out there who embody those values too. Likewise, the purpose of your sales staff is to *sell* your products. You need to reinforce that through good training, making sales targets, and creating incentive structures. There are ways to pay your market staff that will incentivize them to sell more: How about a base wage equal to or above minimum wage, with a commission based on volume moved or gross sales returned? That is what we started doing our last couple of years with our market staff. Depending on the market, we either paid a bonus based on clearing a minimum cash amount in the till or a bonus based on bringing home less than 5 percent of the meat they started with at the beginning of the day. Paying a commission based on the amount of cash in the till at the end of the day is probably the wisest decision; that way you don't encourage staff to pack the cooler with only easy-to-sell cuts or to discount or give away meat. Asking your salesperson to "clear the coolers" may just encourage them to give away meat to friends or trade for other farm products to take home.

Likewise, another thing we learned with marketing staff is that it pays to hire folks who know how to cook meat. Our last couple of years we had some excellent employees: One was a butcher and chef who could sell any cut of meat and explain precisely how to cook it; the other was a multilingual home cook and health nut who could also sell the whole animal, especially to

Sales Outlets and Market Options

Chinese immigrants who would buy up the harder-to-move cuts like pigs' feet and fresh shanks. Don't make our mistake of hiring a vegetarian to sell meat. We tried that for a couple of weeks—bad idea. We thought he might appeal to the crunchy vegetarian-cum-bacon-eater crowd, but alas that was not the case.

Since farmers markets were our main sales outlet, generating around 75 percent of our income, we will relate to you some of our perspectives and techniques for selling at them successfully. At first we started selling just fresh chickens and eggs at markets, as the only vendors specializing in pastured, organic poultry. We would butcher the chickens on Fridays and pack the whole broilers on ice in coolers. Then on Saturdays and Sundays we would attend a market together: This is when we were just starting out and did not understand the value of our time, but we enjoyed spending time together. Our Saturday market was in a very ethnically diverse part of San Jose, California. We would always sell out of chickens within the first hour, often opening up to a line of ten to twenty people already waiting. We quickly discovered the law of supply-and-demand pricing. To prevent selling out so quickly, resulting in no chicken to take to our Sunday market, we inched the chicken price ever upward to control the demand. Some of our Chinese, Korean, and Vietnamese customers balked at the price increase, but most were thrilled to have chicken that fresh. If we could have brought live birds to the market and killed them on-site, that would have been even better for many customers. We also learned that our customers wanted the chicken heads and feet left on—even better for us, because it raised the weight of the birds we were selling by the pound. What had been primarily dog food up until then became additional revenue. The price of our eggs was high too, but our chicken customers would then try our eggs and get hooked on both. Soon we had a chicken *and* egg problem! Way more demand than supply! That is always a good place to be in this business. A couple of years later we stopped raising broiler chickens because of regulatory hoops and low margins and proceeded to add pork, beef, and lamb to our mix of meats at markets. This meat diversity enlarged our pool of customers and added value to our time at market. For example, customers who for religious reasons did not eat pork would have either beef or lamb to choose from. Overall daily revenues went up.

If you are selling out of product (or of popular cuts) well before you pack up for the day, you are not maximizing revenue for your time standing there. To illustrate this point, let's say you sell out in the first hour at market, grossing $300. Then you must sit there for another 2 or 3 hours, telling people "sorry, all sold out." If you had a greater variety of products to sell for those additional hours, your gross revenue might look more like $800 to $1,200. Think of ways to maximize your time vending and set yourself apart from the competition—not only by having more volume of your mainstay products, but also by having more diverse products that are complementary and will sell too. The key is to increase the market basket for each and every customer. Imagine your average customer purchase is $15 worth of goods. What can you do to get each sale closer to $30, effectively doubling your returns for the same amount of work? A rice farmer friend of ours did just that. Massa Organics first started out by bringing bags of rice to farmers markets and coming home with a few hundred dollars in the cash box since rice is not a very high-value crop.

Then they added almonds and almond butter, bringing home closer to double or more for each market (organic almonds are a very high-value crop lately). Then they added animals to their organic farming system: ducks for weed control in the rice paddies, sheep for weed control in the almonds, and pigs to control weeds on the rice berms and to eat cracked rice. Now Massa Organics can bring meat to many of those same markets they were already in (not all of the markets have allowed the Massas to bring meat), and they have again effectively doubled their sales per market! Make the time spent at farmers markets worth it for you.

Other successful farmers market vendors have paired up their CSAs or chef sales with their farmers markets. CSA customers come to the market booth to pick up their prepaid box. Chefs come to the booth to pick up their preordered boxes of produce. Meat farmers could do the same. Since restaurant sales can require so much transport for small volumes, why not ask your chefs to come down to market where they may already be coming to pick up fresh produce and get their meat order at the same time? You could offer a small discount for avoiding the transportation hassles, which may serve as an incentive for the chef, say 5 percent off their invoice.

Make sure your farmers market booth is eye-catching and attractive. It is especially hard for meat sellers to make an eye-catching booth: Your products are mostly hiding in coolers or plug-in freezers. Use good signage and an attention-grabbing banner that you can see from far away, and bring pictures of your farm. The feedback we often received was that our photos were the most important selling point for the folks who were looking for ethically produced meats. For those who didn't want to know their meat came from real animals, the photos could be disturbing. However, those probably weren't the customers we were looking for anyway.

Display the names and prices of all of your cuts for the day so that people walking by can read them. Make it painfully obvious that what you are selling is meat through whatever signage you can develop to get that point across, be it banners, flags, a giant photo collage, or displaying some of your packages on ice (those might be the ones you take home to eat that day or offer up at a discount to price-sensitive customers). We suggest bringing a handout or poster of animal cut charts to help educate people about the different cuts of meat, what they are called, and where they come from on the animal. You can also suggest different meals that can be made from your meat and whatever produce is in season at the market. This will help educate customers about seasonality and strengthen your relationships with other farmers if you promote their products too. You can order some nice posters from the pork, beef, or lamb checkoff programs, laminate them, and hang them up or have them covering your sales table. For a time, we also offered a great meat cookbook, *The Grassfed Gourmet*, at cost to our customers who needed a little help or inspiration in cooking with grassfed and pasture-raised meats. The key is to make selecting and cooking your meat as easy as possible. You and your staff should know a good recipe for every cut of the animal, so you can help your customers cook it right. When they have a great eating experience, they will come back for more. Also, don't make your booth too cluttered with stuff that doesn't have anything to do with what you are selling. We have seen people bring a bunch of antiques and various doohickeys to put on their table, but these sometimes confuse customers about what you are actually selling. You are not selling antiques and the farmers market is not a flea market: Look professional, clean, and meat-focused.

Your staff, whether it is you yourself or hired labor, should be friendly, engaging, and informative for good customer interactions. If a line develops at your booth, you should let everyone know that you will be right with them by acknowledging their presence with a friendly "Hello, we'll be right with you." Don't dress like a slouch; do be clean. Meat needs to convey a strong sense of sanitation and that relates to how your staff looks and how your stand is set up. If you keep your meat in coolers, make sure they are clean. We would give our coolers a soapy rinse each week, both inside and out, to keep them sparkly clean. Wash your pop-up tent canvas at least once a year too and try to keep it protected in its carrying case for as long as possible—the constant pulling in and out of vehicles gets it really dirty. Consider making or buying a carrying case that is more protective and easier to fit the canopy in. At a cost of $100 to $200 for a good pop-up, you want them to last a few years if possible. You can often order a new canvas piece direct from the company if you just want to replace the fabric.

How can you best display your meat at a farmers market? We have seen many different styles from the hundreds of meat producers we have met and interviewed: You will have to decide which one works best for you. You could even change up your look periodically to see how customers respond and if sales increase as a result. John Deck of Deck Family Farm in Oregon recently told us that their meat sales went up by around 20 percent when they switched from keeping all their meat hidden in coolers to presenting some cuts of meat in bins of ice on the table. He told us some new customers said that they never knew he was selling meat before he began displaying packages of meat!

Most farmers keep their frozen meat either packed into large coolers or in portable chest freezers. Either the customers open up the cooler or freezer and pick out their meat or the market staff does it for them. We personally don't think you want people picking through your sealed meat packages: It can be very hard on the packages and cause the seals to go bad or punctures to occur, resulting in potential loss of revenue. Customers also may not be as vigilant about closing the coolers after they look through them. Make sure they are shut so your meat does not begin to defrost. If your meat is hidden inside coolers or freezers, make sure you have large, clear signage about what you have hiding inside those coolers. Customers want to feel like they have some control over the selection. When we had a customer ask for a specific cut, we would pull a few packages of that cut out for them to inspect on the table and select which one they wanted. That is one reason why transparent plastic packaging is important for retail sales: People want to see what the meat looks like—color, fat content, size, number of cuts, and other visual characteristics.

Other farmers purchase or make small glass meat display shelves and put the frozen meat inside with ice packs or bins of ice on the bottom shelf or directly under the meat cuts. This is a better way to let the customers see the meat, but the condensation and fog inside the display sometimes makes it hard to see the meat very well. Plus, you don't want your meat thawing out at all, especially if the customer is going to freeze it when they get home. Thawing out and refreezing deteriorates the meat quality, especially with beef. Thawed meat will ruin the vacuum seal and then ice up, making the meat unsellable at a later market. Even when kept as frozen as possible the meat will not be able to keep its quality after a couple of trips to the market. However, if a customer wants to cook with that meat that night or within the next couple of days, having it partially thawed out can be an advantage for them. More discussion on cold storage can be found in chapter 11.

One of the coolest displays we have seen was in Italy, but we've never seen it replicated in the United States. There we observed various mobile butcher shop trucks attending farmers markets with sides that opened up and an awning to protect customers from rain and shade at the meat counter. One whole side of the truck was a glass meat case and the refrigerator unit ran off a quiet diesel motor. Customers could see the meat, the meat stayed cool, and it all looked very professional and clean. It was even licensed to do some on-site custom cutting for customers, such as deboning or trimming. We think a group of farmers who wanted to market into a large metropolitan area like Chicago, Philadelphia, Atlanta, Denver, Austin, or Seattle could make a go of an expensive marketing apparatus such as this. If farmers markets are your mainstay and you are grossing $20,000 a month from them, maybe purchasing and outfitting a meat truck for $60,000 could be worth it. Some cities, like Portland, Oregon, have a plethora of licensed food trucks for sale that could be modified into a meat truck. Once you had the truck, who's to say you couldn't just park around town like a food truck and sell direct, not just at farmers markets but at other locations like in front of hospitals, corporate headquarters, college campuses, and so on?

Although most farmers market meat sales are with frozen meat, we know one meat producer in California who sells fresh meat at farmers markets and does quite well with it, probably because he is distinguishing his meat from all the rest. This gentleman goes to the slaughterhouse once a week with a small load of animals and picks up the fresh meat he had brought in for slaughter the week before. He displays the fresh meat in vacuum-sealed packages on tubs of ice. If you have a slaughterhouse close by, this could be a viable option for you and would set your meat apart from the other farmers selling frozen meat. However, not all USDA-inspected cut and wrap facilities will even give you your meat fresh (unfrozen). Also, you have the potential issue of meat loss: If a farmer does not sell out of meat by the end of the weekend, what will he do with the fresh meat? Just like a butcher shop, you should not freeze meat after it has been sitting around fresh for a week or two. Perhaps the farmer could always bring a little less than he needed so he could be assured of selling out. At least he would not have to mess around with worrying about freezer storage, which means one less marketing cost for him. But the once-a-week trip to the slaughterhouse sounds incredibly time-consuming and burdensome for a small owner operation. Also, frozen meat works well for a customer who isn't going straight home and plans on defrosting the meat and eating it within a week. Fresh meat can't be left unrefrigerated for long.

Unless a farmer can sell fresh meat for a higher price than frozen, and unless this is the only way to penetrate a particular market, we don't see the value of the increased costs, vehicle wear and tear, and owner stress of dealing with fresh meat sales—not to mention the difficulties of sorting and loading finished animals each week. This same issue applies with selling fresh meat to other outlets such as restaurants and retailers, not just farmers markets. We will discuss those challenges below, as well as how some producers have made it work.

Some other ways to add value to your market stand include having a greater selection of products—adding homegrown produce, herbs, BBQ sauce, or other value-added products that go well with meat (as long as your market manager approves them, of course). We think selling custom-made spice rubs or marinades would be a nice addition, as would fresh herbs that go well with meat such as thyme, rosemary, dried chiles, and garlic. Encourage customers to try new cuts and tell them how to cook them to perfection. Offer a money-back guarantee if somebody is unsure; if your meat is of high quality you should never have to give people's money back. Making that assurance can help push through a sale, particularly with brand-new customers. Occasionally offer freebies or extras to your most loyal customers. We would also offer a free half dozen eggs to new customers to get them to try our eggs. Nine times out of ten they came back for more! Make sure you get people's contact information: Put out a signup sheet and let them know you send out a monthly newsletter with recipes, news from the farm, and coupons. Give them a 5 percent discount right then and there for signing up on the sheet (make sure their email is legible!). That customer list will be invaluable in the future for many other marketing efforts that you make if you ever want to branch out into a buying club, CSA, on-farm sales, or even agritourism. In fact, it could be the most important tool for when you turn from the farmers markets to focus on other sales methods. Take advantage of collecting people's contact information from day one.

A word of caution about farmers markets: Don't waste your time with bad markets. We would suggest giving a market a full trial month of attendance before you decide to stick with it or not. If you are not making a minimum of $100 in gross sales for each hour you are standing there, it is probably not worth your time. We would drop a market if it grossed less than $800 per day, which was the amount that made it profitable for us. You will have to decide what that baseline amount is for your operation, but please don't shortchange yourself. Just because the market is in the town you grew up in or your best friend is the market manager or all your friends will be there, if it doesn't make you enough money to be there, don't feel you have to remain loyal to it. That said, do your part to help promote the market, too. If you have a customer list or website, let them know that you attend that market and hope to see them there often.

Likewise, factor your drive time and mileage into your overall farmers market financial equation. If you have to drive a 4-hour round trip and then stand there for an additional 4 hours to sell $300 worth of product, that means you made only $37.50 an hour (gross) for your time. We also see many farming couples doing markets together, which might be fun and great for the relationship (or not), but that represents two people's labor and even less returns per hour of labor. Plus that means the owners are not at the farm doing other high-value work. Our point is that you need to think of farmers markets as a business venture and not just a fun way to pass the time. If you get into that mindset that it's just a Saturday pastime or social hour, it will never be a very profitable way to sell much meat. Market managers might try to get you to do a marginal market by allowing you into a better market they manage. Also, negotiate with the manager if you have something you know he or she wants at their market. Negotiate for the ability to bring all your products, a preferred stall location, or better markets within the organization without being too pushy.

Internet and Mail Order

Many meat producers are now using online and mail-order sales as one channel in their diversified sales strategy. Online ordering is now being done for mail-ordered meat as well as for pre-sales with local pickup. Some farms use online platforms to take orders for local pickup only and don't want to do any shipping. Taking payment online is getting easier and easier, so why not get people's money in advance of a sale? Of course, make sure you can follow through with the product.

The owner of White Oak Pastures, Will Harris, revealed to us that the fastest growing segment of their sales is online. It is not usually the only sales channel for a meat producer, but it can be an important one nonetheless. Northstar Bison, profiled in chapter 10, also sells an increasing amount of their meat via online sales. The majority of their meat sales are now generated from the easy-to-navigate web sales portal. They sell everything from individual cuts to sides of animals to meat sample packs. Misty Brook Farm of Maine takes orders online for their Maine and Massachusetts drop sites. People use the online ordering system to be sure they get what they need when they go to the pickup site. The farmer benefits because people are more likely to show up to markets and drop sites if they have pre-ordered. Some even prepay, although most pay when they pick up. The key to all these examples is giving people a variety of options to buy your products.

A challenge with online ordering systems is keeping constant tabs on your inventory so you are not advertising something that you are sold out of: You must keep your website and web store up to date. Having a good selection throughout the year may also prove challenging. Perhaps don't allow online ordering for quick-to-sell-out items like tenderloin or bacon, but only for more ubiquitous cuts and ground meat. If you want people to keep returning and ordering from your site, they have to be able to get most of what they want. Too many times reading "Sold Out" for a bunch of your cuts will probably drive them away, just as would happen at a farmers market. Keep the prices, inventory, and other bits of information fresh, which will require constantly updating the website. Some website templates are easier than others to update on your own: The free Wordpress blogging site is one such way, although their templates lack good sales functions, you have to buy separate plugins such as WooCommerce for that. There are other ones that are tied to an inventory management system on the back end of the site and process online transactions, such as CSAware or Small Farm Central's Member Assembler, so that it is always up to date with inventory (as long as somebody is inputting that information on the back end). Both PayPal and Square offer simple web store formats that are easy to update and that receive payments too: You can add them as a tab, link, or icon on your website. To make things easier on yourself, perhaps you can find a web designer to keep your site fresh in exchange for a meat share.

Shipping is the most important consideration if you are going to process online sales. Although you could take orders online for local pickup, as Misty Brook Farm of Maine does, the majority of online sales involve some sort of shipping. You have to factor in the labor to select items and box them, the costs of shipping, the minimum order size, how far you are willing to ship, if you will ship during the hot summer months, and other considerations. The physical logistics of online sales and packaging for shipment are daunting. First you have to take the order and collect prepayment. Will it be by mail, phone, email, or via an online ordering platform? Who takes that order? Second you have to gather the meat order and box it up. Is the meat at your farm or is it in a cold-storage facility somewhere else? Is the meat at your farm or is it in a cold-storage facility somewhere else? How do you keep the meat frozen during the pick-up and boxing process? Where will the boxes come from? Will you use dry ice or frozen gel packs to keep the meat frozen? Third you have to send it via a shipping company. Some carriers will pick up boxes from your farm (FedEx and UPS); with other carriers you have to bring the boxes into a store (US Postal Service). Your region may have a regional shipping company that can get products delivered even faster and cheaper than a national carrier. You may want to investigate that possibility. How much packaging will you have to use and how do you feel about generating all of that waste? Finally, you have to consider whether you can sell enough volume to make all this effort worthwhile, particularly the commitment of a good, up-to-date website with e-commerce capabilities.

In Northstar Bison's experience selling online, they found that the initial shipping quote would drive away many would-be customers. The shipping quote was based on the weight of the box and the distance the box had to travel, so it varied by order. So the owners of Northstar instead have gone to a flat-fee shipping rate while increasing their meat prices to absorb some of the actual shipping costs. They also require a $75 minimum order to cover their shipping costs. Now, because Northstar has increased in sales volume, they have been able to negotiate reduced-rate shipping charges with their preferred shipping providers. They use a regional shipping company in the Midwest and FedEx for the rest of the country. Customers can choose fresh (regional shipping only) or frozen, and all the meat is packaged in an insulated cooler with liner and dry ice or gel packs.

Perhaps do a couple of test runs before you advertise that you are doing mail-order sales. Send some meat to family and friends around the country. Keep track of all your labor, packaging, and shipment costs to do so, and find out if the delivered meat stayed sufficiently frozen and in good quality. You will probably have to build some loss into your cost, in case any meat arrives not frozen and you have to refund a customer's money. You are relying on another company to deliver that meat, such as FedEx or UPS, and things sometimes happen like snowstorms or flight delays that are out of everybody's control. Also, consider that there may be several months out of the year when you can't ship frozen meat or can only do it regionally. Just as some wineries won't ship to southern states in the middle of summer because they don't want their wine to deteriorate and customers to be disappointed, so too can meat quality deteriorate and food safety be compromised if it's simply too hot to ship. Perhaps those few months of summer when shipping is not feasible you could concentrate on your more regional, direct-sales channels and leave the online sales for the fall through spring months.

Another innovative model we have seen for direct-to-consumer online sales is cowpooling, a term used by Mike Callicrate of Callicrate Beef and Ranch Foods Direct in Colorado. His website www.cowpool.org allows groups of people who don't necessarily know each other to split an animal up into sections and pre-own a portion of a live animal (down to eighths). While not the same as shipping meat, this online system provides the cost savings of custom bulk meat sales and online ordering while also offering more flexibility and streamlining the process for the consumer because they don't have to provide the cut-and-wrap instructions to a butcher. All the meat is processed under USDA inspection and the cost per pound includes all of the processing costs. Customers pick up their meat at a few locations in Colorado.

Retail Butcher Shops

More and more retail butchers want to take control of their meat supply and the quality of meat they work with according to an interview we conducted with meat consultant and butcher Chris Fuller. New butcher shops opening up in the last few years sport names like "The Conscious Carnivore," "The Local Butchershop," and "The Piggery," conveying some of their values in their names. As retail butcher shop owners and staff learn more about animal production, slaughtering, and the renaissance of their own craft of butchery, many are wanting to return to a system of buying whole animals straight from the farmer. However, there are just as many butcher shops that have gotten used to buying inexpensive boxed meat from large meat distribution companies. Unless you thrive on challenge, those shops will probably be less likely to want to work directly with a farmer like you. That said, they may be interested in offering some "niche" meats or diversifying a bit to keep up with the competition. Offering your meat just might be a way for them to do that. An example of this is the Millers of Ohio (mentioned in the cattle chapter) who have just begun working in a nearby town with a traditional butcher shop that is trying to offer local, grassfed beef to their customers. Business is starting off slowly, and the Millers make a point of occasionally doing in-store sampling where they can talk to customers about their beef. They realize the importance of education in order to help drive sales and help the shop keep grassfed beef in the case. Obviously, if customers choose grainfed over the local, grassfed beef, the shop won't be able to continue the relationship. If you want to sell meat to butcher shops, you are probably going to have to spend some time on those relationships yourself and do some educating of not only the butcher shop staff but also their customers in order to help promote your meats. Invite the staff of the shop out to your farm for a BBQ and tour, or—even better—a whole animal slaughter and butchery class. Sometimes butchers that "get it" will see the carcasses you produce and recognize the quality. One butcher that fabricated our cut-and-wrapped meat for us was so impressed by our pig carcasses that he wanted to buy some of them for his own retail meat case.

Butcher shops, just like anything else, are businesses run by individuals. You have to appreciate the culture of those people and you have to understand how the business is run. If the people have a culture of transparency and a desire to showcase local or pastured meats, they may be good to work with. Talk to them about the possibility of selling meat to them—either whole animals or primals. Usually shops want fresh meat only, so find out what kind of volume they are interested in and how often they would want delivery. Will that even work with your slaughtering schedule? How do they want to use your meat—in other words, what kind of cuts or value-added products are they considering? You may have suggestions for them or you may learn a few things from them that you had not thought about before. Many shops have a tradition of making products out of the offal, fat, and bones, so find out if they want those things too. Butcher shops are often willing to take parts that restaurants might not want. Maybe you can set up a delivery route that takes the middle meats to the restaurants while the roasts, bones, fat, and offal go to the butcher shops to be made into ground meats, sausages, stocks, rendered fat, confits, pâtés, and other delicacies.

Unlike restaurants that want to put something on the menu that all diners can order on the same evening, a butcher shop might be more willing to take small quantities of certain items. Once they sell out, they just pull them out of the cold case and refill the space with something else. Customers expect there to be special and seasonal items in a butcher shop.

Just like any other sales account, find out what they will pay you, what the terms of payment are, how they will promote you, and so on. Also, will you be responsible for meat that does not sell? Will they deduct that from the payment they send you? Do they have the know-how and skills to take 2- or 3-day-old fresh case meats and turn them into value-added products so there is little to no loss? You need to feel confident that they will move your meat in a timely manner or turn it into another good product so they are economically viable and so they have the funds to pay their bills, like yours. Lastly, will the shop be willing to put your name on the product and help promote your business? When they run out of your product, do they take your name off the sign next to where your product sat or do they do the "bait and switch"? We knew a small shop in a progressive California town that would occasionally buy local meat but mostly stocked their meat case with meat they obtained from their wholesaler. However, they never changed their signage, so customers continued to buy from them thinking they were supporting local grassfed producers. You should ask about their suppliers and, if possible, get a reference from another farmer they buy from so you can investigate the integrity of the shop. This is important because you don't want your name on CAFO meat. It's your reputation on the line.

Small Retailers and Food Co-ops

Grocery stores typically want meat year-round, of consistent quality, priced cheaply, and with zero risk (meaning if it goes bad or doesn't sell, they might not pay you). These parameters are very challenging to accommodate as an individual producer who is not vertically integrated and does not control their meat processing. Nor is it easy to accomplish with grass-finished meats due to their seasonal nature (as described in the production chapters earlier in the book). Extending your harvest season for animals usually requires at least some irrigation, lots of stored forages (hay, silages), or a huge amount of freezer space. Frozen, seasonal meat might not even be a choice for retailers—just like restaurants, grocery stores often want year-round fresh meat only.

If you approach stores to buy your meat, you need to come in with a lot of knowledge and position yourself as a full-service meat provider and consultant to the store. Dress and act professionally: Leave the muddy work pants at home this time. Bring a portfolio of pictures, testimonials, and any media exposure you have received. Plan to bring some samples for them to try, and don't forget to explain to them how to cook it right. They may not know how to cook the meat correctly, especially if they are not used to working with grassfed or heritage-breed meats. Their first impressions of color, flavor, and texture are very important, so put your best meat forward. Explain to them how you will help them make more money in their meat department and attract new customers, and give them ideas to pursue. If you don't know how to solve their meat marketing problems, you are probably not ready to be selling to stores. Stores with in-house butcher counters may want to work with boxed primals, or subprimals, or they may be willing to give you shelf space for retail-ready cuts. Retail cuts will have to be vacuum-packaged with a nice label to catch the eyes of customers. This may be beyond your processor's capacity or require some research on your part to get it right. Like other sales channels, it's hard to predict the kind of volume you can expect to sell in a retail location: Try to gather as much information as you can with market research before you make expensive investments in packaging and labeling. The artwork alone could cost between $600 and $1,000 for the labels, and there is usually a significant minimum order size for custom labels, too.

Also, you will be expected to carry a much higher level of product liability insurance, likely around $2 million minimum coverage, and you may be responsible for any loss or spoiled product (which insurance will not cover). We've heard of grocery stores ordering meat directly from producers and then refusing to pay because they said the meat went bad in storage. Risks like these make this sales outlet look much less attractive.

Single-store grocers, small chains, or natural-foods cooperatives might be better retailers to deal with because they are more community-driven and may be better able to deal with small farmers since they deal in smaller volumes. We have met several farmers who are happy with their relationships with independent stores and food coops. One example, La Montanita Co-op, is featured in the accompanying profile.

Restaurants and Caterers

According to our interview with longtime chef and University of Kentucky extensionist Bob Perry, if you think you want to approach a certain restaurant to sell to, make sure you eat there first and get a feel for the place. Do they seem to be running a professional ship? Do they highlight who their farmers are already? Do they seem dedicated to good, quality ingredients? Does their staff seem knowledgeable? Ask them a couple of easy questions to find out, such as, "Do you buy from any local farmers?" and "Where does your meat come from?" If the answers to these questions disappoint you, it may not be worth your time barking up that tree.

Identify restaurants and chefs that are more likely to work with local ingredients: Find ones that are members of the Chefs Collaborative, Slow Food, or some other regional food- and agriculture-related nonprofits. Their membership in those groups probably indicates that they have a higher level of commitment to good ingredients and local agriculture.

Find out what form of meat they are willing to work with and how much they need. Are they used to only boxed meat, already pre-portioned? Or will they work with whole carcasses, primals, or subprimals? What kind of volume are they really going through and is it worth your time? The logistics will be steep, so they need to be a trusted partner to go through all the hoops. Because of the low volume per restaurant, you will probably have to line up several deliveries on the same day. One pig once a week probably makes no

LA MONTANITA CO-OP

Despite what we've written in this chapter, not all grocery stores are created the same. Food cooperatives are a completely different breed of grocery store altogether; they have a commitment to the principles of cooperation and are required by law to reinvest profits back into the business or distribute them to their members. They are run by boards of directors who are elected by their members and usually have a very rigorous set of standards by which they purchase foods to resell to their members. La Montanita Co-op in New Mexico is one such food cooperative that operates by a high standard of ethics, especially with regard to their local meat program.

To give you a sense of their scale, La Montanita has around fifteen thousand members plus thousands of other shoppers patronizing their stores. They run four full-service retail stores in Albuquerque and Santa Fe, a small store in Gallup, and an even smaller natural convenience-foods store on the campus of the University of New Mexico. They also own and operate an 18,000-square-foot food distribution center not only to supply their stores but also to help connect local producers with other grocers, food service companies, and restaurants in the region. In 2013, La Montanita purchased over $4.5 million worth of food and other grocery items from nearly twelve hundred regional producers—a big dose of economic energy for this region.

Around 5 years ago, La Montanita organized some grassfed beef producers in northern New Mexico and southwest Colorado to start supplying their store. They even helped that group form their own cooperative, known as the Sweetgrass Beef Cooperative, now composed of eleven ranchers. La Montanita schedules the cattle slaughter at one of the two plants they work with and pays the ranchers based on the hanging weight of each animal. La Montanita usually has the beef dry-aged for a week and then they go pick up the carcasses as primals every 2 weeks with their own refrigerated truck. The four full-service stores have their own butcher counters where they do all their own cutting, grinding, as well as some value-added processing. They supply pre-packaged meat to the small La Montanita stores that don't have their own butcher counters. La Montanita even supplies local meat to "competitor" food stores, as well as to restaurants in the region. Having multiple outlets allows them to move whole animals and increase the returns to the ranchers. Indeed, the La Montanita distribution center is not what one would call a "profit center" for the cooperative, but it is a crucial component of revitalizing New Mexican and Southwestern agriculture. Although profitability is important to this cooperative, they have other values that are equally important.

In addition to fresh beef, pork, lamb, and poultry, La Montanita also sells a selection of specialty meats like frozen bison, goat, and even yak. All of their meats exceed a baseline standard of no antibiotics and no added hormones. Their beef is 100 percent grassfed; their pork is raised outdoors or on deep bedding and is vegetarian-fed; most of the lamb is grassfed and New Mexico–raised; and the chicken is antibiotic-free as well.

La Montanita buys both premium cattle and what they call "value" animals from the

Sweetgrass Co-op. They use ultrasound testing to score the live animals before they are slaughtered so the rancher can manage them appropriately. The premium beef is used for fresh cuts like steaks and the value beef is used more for ground products. This not only provides an outlet for the farmers of both types of animals, but it also provides affordable options for consumers too. Because Sweetgrass is now large enough and has producers raising animals in a wide variety of microclimates, they can finish animals year-round for La Montanita and the other outlets they work with. La Montanita is buying around three hundred beeves a year from Sweetgrass for its stores and the other accounts they service.

Since they own multiple stores and have several other outlets to sell meat to, La Montanita is able to match supply and demand for different parts of the animals and maximize the use of the whole animal. They even do a brisk trade in offal at the store—oxtail and beef tongue are often in short supply. With delis and hot bars in all the stores, they can keep meat waste to a bare minimum. Also, if a cut is not moving well at one store but another one is out, they can shift product around quickly.

Unlike the lopsided power dynamics in most retail negotiations, La Montanita has a great relationship with their meat producers. Producers normally tell the co-op what price they are looking for and the La Montanita buyers give them a reasonable expectation of the volume that will move at that price. La Montanita earns a lower margin than traditional for-profit grocers, aiming for a gross margin of around 34 percent on meat (compared to 50 to 60 percent for mainstream grocers). With the record-high meat prices of late, margins have been squeezed a little. However, meat sales have not declined and now grassfed meat is looking quite a bit more competitive against grainfed meat. The price spread has been compressed. This will hopefully take away some of the price motivation for buying industrial meat and drive more people to the good stuff.

La Montanita has an in-house marketing staff that develops snazzy point-of-purchase marketing materials such as farmer profiles and pictures. They write articles for their newsletter and blog and use other social media platforms to educate consumers about local meat. They even helped Kyzer Farm (pork) and Sweetgrass Beef Co-op develop their own branding materials. In addition to all the other things La Montanita does, they have a loan program to help producers scale up, even ones that don't necessarily sell to the co-op. This is a model worth replicating.

sense unless the abattoir is on your farm. But would eight pigs once a week be worth it for you? Or sixty ducks once a week? What will it take to line up that many restaurants that will take *all* parts of the animal? This requires a large selection of higher-end restaurants to choose from, something not likely to happen in a small town of under fifty thousand people. It also may take some time to get up to the critical mass that makes your transportation costs worthwhile. What will you do in the meantime when you are under your minimum order? For example, it may take 6 to 12 months before you have more than two restaurants ordering on the same day. Will you make that drive for the one or two restaurants that sign on first? Sometimes you have to accommodate the early adopters and make less profits while you are building your market. Those early adopters might pave the way for you: Make sure you gather some testimonials from them and encourage them to talk your meats up if they are successful.

When developing your list of accounts, whether they are restaurants, caterers, food service buyers, or others, it is important to select your target customers based on all the parts of the animal. For example, if 20 percent of the cow is ground beef, then 20 percent of your customers (by volume) must be ground beef customers. Every time you kill a cow, your job becomes to find the highest margin customer for the resulting cuts. It may not be so much about whether it's a farm-to-table restaurant that shares your values as about whether they are willing to buy all your offal cuts in a single delivery. You may have to sell some cuts, such as your top round, at a lower margin, but the point is, you *have* to sell your top round. Also, help your customers figure out what to do with any cuts that don't sell over the week. For example, one restaurant that buys from Oregon grassfed beef producer Carman Ranch will turn unsold beef cuts into higher-value pepperoni instead of just grinding them up for burger.

What if you could convince some restaurants to buy your meat in bulk and consider investing in freezer storage? This will probably never become popular among restaurants, especially because of their preference for fresh meat. Yet restaurants could save considerable money doing it this way. It would require those restaurants to have knowledge of whole-carcass cooking, menuing different cuts of the animal, and properly thawing out frozen meat. The restaurants that have already figured out how to successfully do whole-animal cooking could easily transition into frozen meat in bulk. How about a chest freezer at every farm-to-table restaurant? Although chefs usually have a prejudice against frozen meat, blind taste tests show no difference between fresh and frozen. Carman Ranch demonstrated this to some of their chef customers and was able to convince them to start buying frozen beef for the part of the year when fresh meat isn't available.

Many restaurants want steaks and chops only, so you must match them with other outlets willing to take the roasts and ground. Grocers or institutional food service customers might be good partners for the grind. If you land accounts with stores and restaurants, you will have to produce meat more or less year-round. They might not want to fit you on the shelf or menu if you are just seasonal. However, this is not always the case: Many restaurants are willing to do seasonal and nightly specials with seasonal meats.

Restaurants may need help working with whole animals or primals. Los Angeles–based meat consultant Chris Fuller helps restaurants figure out how to do this in order to incorporate local meats and improve their profitability. The more profitable the restaurant is at selling grassfed meat, the more likely they will continue purchasing it and being committed to it. Chris helps them with the actual meat cutting, menuing, cooking techniques, and making more money from the cuts. Chefs are now learning how to use the whole animal, which is a great improvement on the old system of the last 40 years or so of buying boxed, anonymous meat. You, too, may need to help them along in this process.

Restaurants want fresh meat, frequent delivery (often one to three times a week), and less size

The Maple Grille

The Maple Grille, located in rural Saginaw County, Michigan, did not get its start as a typical restaurant, nor is it run that way. Owner and chef Josh Schaeding started The Maple Grille as a small hobby maple syrup evaporator. He then added chickens and rabbits to his family farm and obtained a Michigan Department of Agriculture approval to process them on-site. People would stop by, smelling the wood-fired syrup evaporator, thinking it was a rib pit. After enough people inquired about food, Josh began cooking wood-fired food for them. Following a couple of summers of running this weekend grill (in 2011 and 2012), Josh added on to the facility to have a full restaurant, kitchen, walk-in cooler, and pantry that can run year-round. Both lunch and dinner are served 5 days a week now, with nearly everything cooked on the enormous wood-fired grill.

What makes this restaurant different, apart from not having a microwave, deep-fryer, or oven and cooking everything over a wood-fired grill, is its commitment to local. Josh estimates that 90 percent of the meat served is from a 10-mile radius around the restaurant. Vegetables are grown on Josh's 4-acre farm behind the restaurant. All of the poultry and rabbits are butchered in a processing room at the restaurant, and the rest of the meat is slaughtered nearby and brought in as primal cuts to be broken down in the kitchen. Josh handles all of the slaughter logistics for his red meat and even picks up the primals directly from the slaughterhouse. The restaurant is closed on Sundays and Mondays, not so Josh can relax and rest, but so he can slaughter poultry and rabbits on Sundays and pick up red meat from the slaughterhouse on Mondays. We have never heard of a restaurant so committed to and intimately involved with its meat supply as The Maple Grille. The restaurant purchases around twenty whole beeves a year, thirty hogs, and around a dozen lambs in addition to the chickens, turkeys, and rabbits raised by Josh and his family. By purchasing locally raised whole animals, The Maple Grille can avoid some of the middleman costs that typical restaurants pay, and thus pass the savings onto their customers.

Although The Maple Grille is not a huge restaurant, it does turn a brisk business, especially in the summer. They have eighty seats and turn each of them about three times a night during the height of the season. In the slow season they serve more like eighty patrons a night. Once they get their license to serve beer and wine, they expect business to double. People come from throughout the county, even as far away as Lansing (1 hour away) for their unique food. There are no other restaurants in this largely blue-collar county serving farm-fresh food like this. Even though fast-food restaurants abound and obesity in Saginaw County is the highest in all of Michigan, loyal patrons flock to The Maple Grille to taste well-loved comfort foods like steaks, pizzas, and potato salad. Every day the menu changes depending on which cuts of meat are left from the animals purchased that week, the fresh local catch of fish, and whatever produce happens to be growing next door.

While some restaurants make excuses about why they don't work with whole animals, The Maple Grille has cleverly figured out how to do

> it. It is not for the faint of heart though; the work is very labor-intensive. The Maple Grille first cuts as many steaks and chops out of the animal as they can, including a lot of the "one-off" cuts like flank, hanger, and tenderloin. They sell the standard steaks and then offer the more limited steak as a "chef's-choice steak" that is available until it is sold out and crossed off the chalkboard menu. Much of the rest of the animal is used for burgers, a very popular menu item. The Maple Grille does lamb, beef, pork sausage, and pork and beef burgers to provide variety for their customers.
>
> Since The Maple Grille does not own a deep fryer, they render their own beef tallow or pork lard and fry up potatoes in a cast iron skillet on the grill with the rendered fats. This also means they don't waste any animal fat. Whatever rendered fat might be left over they give to a soap maker and then sell the private-label all-natural soap at the restaurant. The Maple Grill prepares the livers, hearts, and gizzards of all the animals for the menu as well. The pig heads get turned into head cheese, bones get used for stock, and when they are done with the bones they give them to customers for dog bones or burn the rest until they are ashes for the compost pile. Josh's laying hen flock will eat any cooked meat that doesn't sell by the end of the week, and their rich eggs go right back onto the menu. This is full-circle dining.

variability between primal or subprimal cuts. For example, they don't want a loin that breaks down into small pork chops for one customer, while the patron next to them receives a large chop from a different loin. Restaurants require significantly more logistics to work. For that reason you may mull over working through a restaurant distributor instead, described later in this chapter. Take into account if all the logistical costs and time will be worth it for your business, especially considering restaurants expect to pay wholesale prices and don't pay in a timely manner.

Despite all that has been said about selling to restaurants, some restaurateurs are doing things differently and may make great partners, such as The Maple Grille (profiled above).

Most restaurants need more than one source of meat and have trouble working with a bunch of suppliers. If you don't produce every kind of meat year-round that they desire, can you develop relationships with other ethical producers near you and coordinate distribution for all of you? That not only solidifies your own business, but also can become an added revenue stream. If at all possible, sit down and carry out production planning with each chef you sell to: You are not a meat store, and they can't call you two days ahead of time for an order. Ask the restaurants to co-market with you. You help promote them as a place to buy your meat cooked to perfection and they help promote you by listing your farm on their menu, chalkboard, or somewhere else that diners can see.

Jamison Farm in western Pennsylvania has built the foundation of their lamb business with restaurant sales. They sell lamb to some of the highest-rated, most esteemed restaurants on the East Coast. Because they own the abattoir, they can custom-cut for the chefs, which makes working

Carman Ranch

Cory Carman and her husband Dave Flynn sell grassfed beef into the markets that not many other producers want to deal with. Yet they have made it work. Their family cattle ranch is in Wallowa, Oregon, near the northeast corner of the state. Their consumers are some 300 miles away in Portland. How does a far-flung cattle ranch build a thriving market of mainly restaurants and food service clients that far away? If the Carmans can do this, so can many other farms and ranches that are not located near any major population centers.

Owner Cory Carman shared with us some of the ups and downs of working in these markets. They tried selling to retail grocers, but were not particularly successful in negotiating their peculiarities. Just getting into a store is really hard; it often takes several meetings with the meat buyers, giving away free samples, and lots of follow-up just to get them to place that initial order. They drive a hard bargain on price and like to remind you of how cheaply they can get other meat. They can sometimes struggle to give you any sense of volume and don't want to take on any risk for shrinkage or loss. In one particularly disappointing case, a small grocery store chain finally ordered some Carman Ranch beef trim to process into their own grind. They purchased the trim and put it in freezer storage to supplement their supply in the winter months when grassfed beef is in short supply. When they pulled it out several months later, they said the trim was sour, yet they didn't get around to using the meat in a timely fashion and refused to pay Carman Ranch when the meat went bad in their cooler. Cory knew the trim she sold them was in good condition and did not feel she could be held responsible for product loss months after the sale. That

with them easier. Some chefs buy whole animals, but most get primals or whole racks and cut them down from there. They butcher lambs between 4 and 8 months old, when they have only 40- to 50-pound hanging weights, resulting in very small cuts that chefs love. According to the Jamisons, chefs are fickle, have strong egos, and you should never call them at lunchtime. Approach them as a professional with a cooperative spirit, not an "I only have lamb in May, take it or leave it" approach. Yet, let them know about your production schedule and when they can get in on it. If you have fresh beef for 6 months a year and frozen for the other 6 months, let them know that. If you kill hogs once a month and make deliveries of whole sides the following week, let them know that. If you can only process chickens 30 weeks out of the year because chickens don't like to grow on pasture when there is snow covering it, let them know that. Ask yourself, when a chef can just call Sysco and get whatever they want with ease, why would they want to work with you? How are you making things easier for them? Do you help them figure out how to put your meat on their menu? Can you describe your meat's flavor, texture, tenderness, and appeal? Some chefs want unfamiliar cuts that American butchers don't know how to fabricate. Jamison Farm is willing to figure out what those cuts are and do them for the chefs—things like "palleta of lamb," which is a experience made her much more wary about working with grocery stores.

Carman Ranch sells middle meats and ground beef to a variety of restaurants around Portland, something that took her many years of persistence to cultivate. One small Paleo-inspired burger chain called Dick's Kitchen uses Carman Ranch grassfed beef in large amounts. In general though, she has found that restaurant buyers can be a mixed bag. Devoted chefs are a pleasure to work with, and Cory learns a lot from them as well. On the flip side, some chefs can be very fickle, are hard to work with, often don't place value on sourcing (authenticity) over pricing, or don't do enough to educate their customers about grassfed meat. Grassfed meat requires a different way of cooking, a different presentation, and a unique sales strategy and consumer education. If a chef is not committed to doing those things, their customers won't understand why the eating experience is different from what they are accustomed to, sales of the grassfed meat will be slow, and the chef will ultimately stop ordering.

Carman Ranch also sells grind, roasts, and stew meat to a variety of food service companies and institutions serving a large clientele. Cory remarked that institutional customers were amazing for their business, buying up a lot of volume, especially ground beef. Although they pay less, they buy in large volumes. They also have a good understanding of the quantity of meat they go through, so it's easier to do production planning with them than with restaurants. Colleges, for example, know how many students they have, how many meals they normally serve, and the poundage of meat they go through. If you can get your foot in the door with one product, like ground beef, it can provide the anchor for all your other marketing channels.

shoulder cut around the blade bone with the shank still attached to it.

In order to move whole animals, work with several restaurants that take different parts. While some may just want the loin, others may be willing to buy whole legs to grind into sausage or shoulders to roast, smoke, or shred. One chef that works with the Jamisons says that he sometimes will serve a lamb "trilogy"—three cuts of lamb served different ways. That way the customers will get enough lamb and he can serve lamb to more people on a given night. There are only nine loin chops on a side of lamb, so it is that kind of outside-the-box thinking that allows some chefs to excel at working with whole animals. In the case of lamb, the trilogy is shoulder confit, a couple of slices of roasted sirloin, and one loin chop; the plate costs closer to $5 to $10 instead of $15 to $20 if you just served chops. Restaurants that aren't white-tablecloth establishments may want to work with lamb sausage or ground lamb, making things like meatballs or shepherd's pie. They might be willing to buy whole legs or shoulders and grind them themselves or to buy ground meat in bulk to spice themselves. This is a way for them to save money while still getting good meat on their patrons' plates. As a farmer-seller, sometimes you have to encourage them to get creative or discuss some of the ways to plate your products.

Institutions (Hospitals, Universities, Corporate Campuses, Private Schools)

According to University of Kentucky Extension's Bob Perry—who at one point in his career oversaw twenty-one different restaurants for the Kentucky State Parks department and who worked hard to get them to source more local ingredients—the big three food service companies command the market (Aramark, Sodexo, Compass Group). They can offer rebates to institutions for buying certain products, meat included. Smaller-volume producers like those reading this book can't offer rebates. Chefs are given a product list to work off of and their bonuses reflect their adherence to that list and keeping costs down. The product list is geared toward maximizing rebates, chefs will sometimes get a poor review if they go off the approved product list.

A better idea is to look for institutions that are working with an independent provider or that have a really activist student body that is fighting for change and Real Food. Many institutions are already under contract with large food distributors that provide their meats, outfits such as Sysco or US Foods (currently in the process of merging to become one large monopoly). Therefore they are unable to work directly with a meat provider outside of that contract. It does not hurt to ask, but ask in a professional way.

Becoming an approved vendor for institutions often requires very high insurance coverage (at least $2 million; usually $5 million minimum), third-party audits of both the farm and the slaughterhouse, and a refrigerated truck. To top that off, institutions frequently demand lower prices because they think they are doing you a favor by moving lots of volume and then paying you in 45 to 90 days, like you are some sort of bank. Added to that challenge, they want meat year-round or whenever their school is in session, which may not coincide with your actual animal harvest season.

According to Lauren Gwin of NMPAN, large buyers (like institutions) are challenging to get to commit, to pay what it costs to produce good meat, and to stick with the same farmer suppliers. So are wholesale and institutional markets viable for small farms? In some cases the answer is yes because there is grant money to initiate the project or because the institution is a private hospital that has some money to put into it. They are willing to invest in good food. Lauren says, "There is a real need for mid-scale farms to serve these kind of markets. It takes more than consumer demand, it takes consumer movements. We need to consider actions like shifting price incentives and changing legislation. It also takes producers like Cory Carman who are willing to grow delicious food and work tirelessly to recruit good customers."

To serve these larger markets, you either will need to scale up as an individual producer or coordinate with a group of producers. Like restaurants, institutions want to deal with one person ideally, not a bunch of different farmers. Want to form a cooperative? This might be a good reason to do so.

Look for institutions that already have some sort of commitment to buying local or have sustainability principles—many universities are doing this, with over twenty so far committing to the Real Food Challenge, an effort to change the way that universities procure and deliver food to their students.

Wholesalers and Distributors

A meat distributor is a wholesale middleman, especially for product lines where selection or exclusive distribution is common at the wholesale level and the meat producer expects strong promotional support. They normally buy from branded meat programs to deliver and resell to retailers, restaurants, and food service. Nicky USA, who we interviewed for chapter 6, is one example of a meat distributor that is willing to work directly with select regional producers as well as buying from large vertically integrated brands, such as Cervena elk from New Zealand.

In the US food industry, the terms "wholesaler" and "distributor" tend to be used interchangeably. Both purchase products from the producer or manufacturer and sell to a retailer or another distributor. Some people distinguish between the two, noting that distributors tend to specialize in a specific market category (e.g., deli, bakery, grocery, or convenience) and provide more retail services, such as stocking the retailers' shelves and running in-store promotions. Distributors usually take higher margins than wholesalers in return for these added services. Just like some of the larger-volume markets mentioned above (retailers and institutions), meat distributors typically deal in large volumes.

Meat distributors purchase, warehouse, and deliver meats. Many of them do their own butchering and have USDA-inspected fabrication plants as well. Thus they are willing to buy whole carcasses or primals. They have weekly schedules and delivery days and usually want the same selection of meat year-round. If they service a lot of restaurants, they may be willing to work with seasonal meats as specials that go on restaurant menus.

They also handle the billing and receivables of customer accounts. They are experts in the logistics of transporting and selling products to the retail or food service customers. Distributors charge anywhere from 20 to 35 percent, which raises your product's shelf price considerably when you consider that a retail store will add an additional 30 to 50 percent on top of that.

Farm-raised, branded meats could be considered a gourmet product. Distributors that handle specialty food items may be a good fit for this specialty product. The easiest way to find such a distributor is to call white-tablecloth restaurants in the region you want to sell to and ask them who delivers their meat. These types of distributors are always on the lookout for new and different products, and price is less of an issue. Quality and consistency are more important factors. You'll need to evaluate your individual situation to determine if using a distributor is an effective marketing tool for your operation.

To do business with meat wholesalers you need to present yourself professionally, communicate clearly, and have enough volume for them to be interested. They may handle the slaughter logistics, or you may have to deliver the animals to slaughter yourself, but make sure you deliver the size and quantity of animals that you agreed to. You usually won't get a second chance with a wholesaler. On the positive side, selling your meat through a wholesaler reduces your labor and marketing expenses as well as your risks. They usually handle all of the processing logistics, transport, and cold storage, as well as the marketing and accounts receivable (be sure they have the financial strength to pay you on time as well). Realize you will be paid less than retail prices for all of these services: Make sure they explain all of their fees and charges up front as well as their payment terms. Your products may get lost on the list of foods that a distributor sells, so you may have to help them with some marketing support, such as visiting key clients. If you want the sales staff to keep your products in mind and be able to answer customer questions, conduct periodic training and education with the sales staff. Even better, invite them out to your farm for a personalized farm tour followed by a scrumptious BBQ. This is true for retail sales staff and restaurant staff as well. If they have a good experience visiting your farm and eating your meat, they will have a stronger emotional connection to your business and you will move to the front of their minds.

Aggregated Models

According to University of Kentucky's meat maven Bob Perry, we need to support value chains in order to support mid-scale producers and buyers. The small-scale, direct-to-consumer farmers are great and necessary, but so are mid-scale producers who will be able to tap into more mainstream markets and get good meat to Middle America, rather than just the progressive coasts where much of the good food currently flows. Value chains are just a fancy phrase for meat processors, aggregators, and distributors that care about supporting ethical producers while maintaining transparency and high quality each step of the way. These scaled-up systems often aggregate live animals or meat from a number of producers through a variety of business models: cooperatives, farmer-owned LLCs, and private corporations. To do it right, it typically involves considerable investment in computer technology, equipment, cold storage, and transportation, and building strong relationships with meat processors (or taking on the processing step itself). Successful aggregated models also demand good people, and for that the scale needs to be large enough to support decently paid managers.

There are hundreds of working models out there aggregating organic and pastured meats within nearly every region of the United States. We will highlight a handful of them in the following sections. Maybe working with other producers in your region will help you access new markets that you can't do alone.

BLACK EARTH MEATS, WISCONSIN

Black Earth Meats wears many hats in a Midwestern regional meat supply chain (in late 2014 their operations were shifting as their plant was forced to close by a not-in-my-backyard [NIMBY] town board). They are a USDA-inspected slaughterhouse and processor. They wholesale meat and they retail meat. They have a meat-buying club. And Black Earth also has a new artisan butcher shop in Madison, Wisconsin, called The Conscious Carnivore.

Black Earth Meats was started by several forward-thinking partners in 2008. Bartlett Durand is the managing partner, and the other partners contributed financially to the operation but don't manage day-to-day operations. They purchased a closing-down processing plant in the small village of Black Earth, Wisconsin, that was by no means a turnkey operation. The company had to invest a considerable amount of money into humane handling equipment and outfit the interior with all the latest equipment for high-quality meat fabrication. The whole concept was to become a custom processor for organic producers in the region, of which there are many. Farmers from as far away as Iowa, Illinois, and even North Dakota have used the plant, but their main audience of farmers are those in the Driftless Region of southwest Wisconsin. This is one of the best places in the country to produce grass and therefore is the location of an abundance of grassfed producers.

The fundamental problem nationwide has always been a lack of small-scale processors that want to do fee-for-service processing for small and mid-scale producers. Even though Wisconsin has hundreds of slaughterhouses and meat-processing plants, they are often split into two different categories. One side does custom processing for hunters and backyard producers; thus the meat is not legal for sale. The other side does private processing for brands, but does not do fee-for-service work. Or perhaps they do fee-for-service work but won't do the organic processing and retail-ready processing that a lot of direct-market producers are looking for. Black Earth Meats' partners took over the plant with the purpose of offering "farmer-focused" processing in their region and providing jobs.

Farmers in the region were struggling to direct-market meats and create viable brands. Buyers were looking for local meats but needed a more year-round supply and in an aggregated form so they didn't have to call a bunch of people. Both sides of demand and supply are trying to connect with each other, but are not able to very well because it seems that nobody is the right size (this is true nationwide). That is where Black Earth Meats comes in. They have developed their own branded meat line, now buying from over two hundred farmers in the region. Actually they offer two different programs for buyers: either the Black Earth Meats brand or the individual farmer's own brand (a customer asks for a specific farmer or production method and Black Earth Meats finds that farmer for them). A lot of the food co-ops opt for the farmer brands because they want that level of transparency and want it to be farmer-direct as much as possible.

Black Earth Meats sells three classes of meat. These include certified organic meats, certified grassfed (which also has to be antibiotic and hormone-free), and what they coined "Grandpa's Way"—animals that have been raised without antibiotics, hormones, and confinement, but may be fed a little grain. The Grandpa's Way producers are usually small farms that have a limited number of animals that they raise and are not going to go through the trouble of getting certified because they aren't really farming for a livelihood. Interestingly, despite not being certified for any claims, the Grandpa's Way line of meat is becoming more popular than even organic or grassfed, because it seems more trustworthy for some buyers. It may be that the name itself conjures up some nostalgia and images of supporting neighbors who raise a few animals. Sometimes people want to know more about *who* raises an animal rather than *how* they raise it. Black Earth Meats is able to control for quality in that they see every carcass being processed. If something grades low (such as below Select for beef), they can always choose to grind it for ground beef sales.

The longer that Bartlett is at this, the less likely he is to encourage farmers to do their own wholesale meat marketing. He believes it is too much work, too risky, and requires a certain scale that many farmers struggle to reach. He would rather farmers in his region come to him to sell their animals because Black Earth has the benefit of controlling all of the processing and has wholesale and retail outlets already developed. The farmers can still direct-market whole animals directly to consumers as they wish. If only every region had a Black Earth Meats—there might be no need for this book!

Black Earth Meats had a small retail counter at their actual fabrication plant in the small village of Black Earth (population 1,338) but obviously that was not a large enough population center to move much meat. So in 2013 they opened up The Conscious Carnivore in the capital of Wisconsin,

Madison, which is also a big university and foodie town. This butcher shop not only sells local, organic, and grassfed meats but also focuses heavily on heritage breeds too. The shop is very clean and open, with natural light. Consumers can see the butchery happening in front of them, but Black Earth keeps the blood and gore to a minimum. They are trying to be inviting to all potential good meat buyers and not be intimidating or macho. They want the mom with children to feel good about coming there, along with the vegetarian college student thinking about eating meat again (and everybody else too!). They also host classes to increase people's "meat intelligence" and strengthen their connections to local agriculture. Their motto is "Respect for Every Animal—On Four Feet or Two."

The Black Earth Meats Buyers Club provides the benefits of buying meat in bulk, similar to custom-exempt sales, but with a diversity of cuts and species. Being a butchers' CSA allows them to balance out their inventory and get people to try all of the other cuts they normally wouldn't try. For example, there are around twenty-nine standard cuts on a beef and most Americans know just six. Because there are no other layers of middlemen involved, Black Earth is able to charge just above wholesale pricing for their CSA. Consumers get affordable meat, a diverse selection, and they learn a completely new repertoire of cooking. The buying club is a great educational platform for ethical meats that also appeals to consumers' budget concerns.

NORTHEAST LIVESTOCK PROCESSING SERVICE COMPANY, NEW YORK

The idea for NELPSC started in early 2000 with a committee of farmers and the Hudson Mohawk Resource Conservation and Development Council. Livestock processing appeared to be a real bottleneck in that region of eastern New York, so the RC&D Council organized meetings to discuss the challenges. They successfully obtained funding for a feasibility study to look at the meat-processing logjam and, out of that, determined that there was actually adequate processing capacity in the region. Issues of seasonal highs and lows, scheduling, and communication issues were really the problem. The study suggested forming a servicing company to help with scheduling and processing logistics, a sort of go-between for farmers and processors. Some local livestock farmers invested their own money and formed a board of directors for the company, and NELPSC was launched on June 1, 2005. Although it is a for-profit LLC, NELPSC was funded with grants until 2010 due to its compelling social mission. They are now self-sufficient thanks to a lot of hard work getting to scale and adding more paid services.

In fall of 2006 NELPSC hired Kathleen Harris to run the company due to her extensive background in livestock production and direct marketing. She knows the strengths and weaknesses of all the regional meat processors because she dealt with them while she herself was farming. She also has a background in meat judging, having taken meat science classes while in college. She knows what a good side of beef looks like.

NELPSC mainly serves farmers in the Hudson, Capital, and Mohawk regions of eastern New York. Around 150 farmers are members or use NELPSC services in some way. The initial focus of NELPSC was livestock-processing facilitation. Kathleen will help anybody who calls her, member or not, to find the best plant to meet their needs. In addition to matching farmers with the right processor, NELPSC has an à la carte menu of services, which includes scheduling slaughter, conveying cutting instructions, and even being present on the cutting room floor to oversee the process. She goes over cutting instructions with the farmer and the plant, and keeps a copy herself of the cut sheet should any questions (or disputes) arise. If farmers who are new to direct marketing need advice on developing cutting instructions in the first place, Kathleen can help them figure out the best carcass breakdown for their specific markets. She can even arrange for transport or co-mingled loads if farmers need transport for their animals or the resulting meat. Additionally, Kathleen can give advice on packaging, labeling, and finding cold storage.

In 2008, the NELPSC board of directors decided to add aggregated marketing as a service for their producers. Rather than compete with their farmer members, NELPSC focuses on larger markets with onerous and expensive market entry requirements that farmers won't or can't do themselves, such as institutions. Schools and colleges, for example, want high volumes and low prices—not something most farmers are eager to sign up for. A large part of what NELPSC sells is cull cows in the form of ground beef, patties, stewing beef, and braising roasts. This gives farmers an outlet for those cows, and the institutions an affordable option for local, pasture-raised meats. With every single order comes a signed affidavit from the farmer and a copy of the farm's story. The schools like to use this to educate their guests. NELPSC will also market regular beef cattle, lambs, pigs, and poultry as whole animals, cut and packed to the buyer's specifications. Customers also have the option of selecting what kind of production methods are used, such as 100 percent grassfed or grain-supplemented. All animals that NELPSC sells are free of added hormones and subtherapeutic antibiotics, and farmers sign a legal affidavit stating as much.

Kathleen's advice for successfully obtaining institutional accounts is to first find the places that are "self-ops," those that are not under a food service contract. If the institution already has a contract with one of the big three food service companies, they will be less likely to work with you. If the school uses local suppliers, caterers, or does their own food, you will find more opportunities to work with them. Also, find allies within the organization itself. Network with the school's green clubs, local food clubs, and others to help create some grassroots energy around using local, humanely raised meats. Try to identify and build a relationship with food service directors. Understand that they are under pressure price-wise, just as everybody is. Work to try and find them consistent, affordable options such as ground meats, stewing or braising cuts, and roasts.

ISLAND GROWN FARMERS COOPERATIVE, WASHINGTON

Island Grown Farmers Cooperative (IGFC) does not do aggregated marketing, but it does help its sixty-plus members get their meat processed in a region with few other options. The San Juan Islands and northwest corner of Washington did not have a single USDA-inspected slaughterhouse within 300 miles until IGFC used the collective power of their members and the assistance of a land trust to build the first USDA-approved mobile slaughter trailer in the country.

At the request of a group of local farmers and the San Juan County WSU Extension, IGFC was initiated as a project of the Lopez Island Community Land Trust. The land trust reached the conclusion that meat production offered one of the best options to protect farmland on the islands. After numerous surveys and extended research, and with the help of grants from the USDA, the decision was reached to form the cooperative in 2000, and by 2002 the first animals were being slaughtered under USDA inspection. Shortly after starting, IGFC leased a closed cut-and-wrap facility on the mainland in Bow, Washington.

IGFC provides USDA-inspected mobile animal slaughter services to members and non-members in San Juan, Whatcom, Skagit, and Island counties in Washington. Now that IGFC is operating near capacity, it is harder for non-members to get on the schedule. Meat fabrication, cold storage, and some limited retail sales services are provided at IGFC's facility in Bow. The business is owned by the farmer-members, and is directed by an elected board and its officers. Each member contributed a modest equity investment, and, just like most cooperatives, they each have one vote in decision-making. Profits, when there are any, are reinvested back into the business. Members are not receiving dividends but they could conceivably have equity returned to them at some point. The co-op employs a general manager, who oversees the rest of the staff. Having both slaughtering and butchering services helps keep the staff busy year-round and allows IGFC to pay competitive wages to keep on skilled employees.

Many other mobile units have struggled in other parts of the United States, but not IGFC. First off, producers don't really have other options in that part of Washington other than custom processing, so demand for IGFC services is high. Also, IGFC operates both slaughter and cut and wrap as one integrated business that helps to spread fixed costs over the full processing of an animal, rather than just one part. Many MSUs just slaughter livestock so all their overhead (insurance, licensing, taxes, accounting, etc.) has to be earned on the slaughtering, often making the cost of slaughter non-competitive. Also, with just slaughter, MSUs need to find a relatively nearby cut-and-wrap facility to transport the carcasses to for cooling, fabrication, and aging. If there are no accommodating cut-and-wrap facilities nearby, then the MSU will struggle. Because of IGFC's integrated model and the fact that the farmers are invested in its success, they have a clear advantage. Now into their fourteenth year, things are running smoothly with a good crew processing around five hundred thousand pounds of meat a year.

Farmer-members of IGFC do their own marketing under their own individual brands around northwest Washington and in the Seattle metropolitan region. Because everybody has their own production methods and markets developed, there was no interest in group branding. So far, that suits members just fine. Most of the IGFC farmers are seeing demand readily outstrip their supply—a nice "problem" to have.

FIRSTHAND FOODS, NORTH CAROLINA

The genesis of Firsthand Foods was a growing awareness in what's known as the "Triangle Region" of North Carolina that there was a need for a business to connect pastured producers to local markets. Area farmers want to sell their meats locally, but they often are not adept at or interested in doing all the marketing. There is an equally strong interest on the part of chefs and retailers for local, non-CAFO meat. Buyers want to be connected to local food but don't want to have to call a bunch of farmers on a regular basis to place their orders. Solving problems for both farmers and buyers was the entry space for Firsthand Foods. First incubated by the nonprofit Center for Environmental Farming Systems and its NC Choices initiative and funded with grants, Firsthand Foods got its legs under it in 2010 and is a for-profit business co-owned by Tina Prevatte and Jennifer Curtis. Their services focus on aggregation, marketing, and distribution of pastured, local meats. These women are passionate about creating meat supply chains to help local farmers thrive. They are the embodiment of a "value chain" business, built around transparency, equity, and stewardship.

Around fifty farmers in North Carolina work with Firsthand Foods to market their beef and pork. Chicken was added to the offerings in 2014. Firsthand has production standards that all their producers must abide by. The beef has to be pasture-raised, although modest amounts of supplemental feed are allowed. The pigs also have to be on pasture. All animals have to be free of added hormones and antibiotics. While Firsthand does not dictate specific breeds to their producers, they encourage them to raise healthy, quality animals with a medium level of marbling and backfat. Overly fat animals won't appeal to their consumer base; neither will overly lean ones. In a typical year, 55 percent of all revenues go back to the farmers, 20 to 25 percent to their processing partners, and another 20 percent to Firsthand to cover operating expenses. They hope to break even this year, a noteworthy achievement for a company going into its fourth year in a tight margins industry.

Firsthand Foods' niche is marketing fresh meat weekly to restaurants, specialty grocers, food service, and a "(M)eat Local" box program for consumers. Again, due to their aggregating ability, they are able to serve markets that individual farmers simply could not. They schedule slaughter dates, call their farmers, and have them deliver animals to slaughterhouses. Firsthand provides the cut sheets to the processor, and backhauls the fresh meat to their facility to inventory it and pack orders. They contract with partnering distribution companies for order delivery to their accounts. To optimize food safety, ground beef and sausages are stored and sold frozen. Ironically, Firsthand has had more trouble selling the high-end cuts than the lower-end cuts and ground meat due to the higher price points of their products; sometimes they don't have the volume for a restaurant to easily make a menu item with the middle meats. Home consumers are more apt to pay more for local, pastured beef than they are for pastured pork in North Carolina. This could be because North Carolina is the second-biggest hog production state and pork is readily available and dirt cheap there. Good beef is more the novelty.

Many of the farmers selling through Firsthand have been able to increase their production because they have an assured outlet for much of what they produce. Firsthand provides them feedback on carcass quality and ways to improve it through genetics, pasture management, and so on. The network of farmers is even learning from each other now through this new community of practice. Firsthand highlights all of their farmers through profiles on their website. Buyers receive information each week about the source of their meats: This level of traceability enables them to tell the farmer's story. Firsthand Foods is learning to adapt to the changing demands of the marketplace while improving the economic viability for a growing number of farmers.

A TALE OF CAUTION

Not all aggregated models succeed. Sometimes they are in the wrong place at the wrong time or without the right management. This is a tale of caution that you can learn from, adapted from a Kansas Rural Papers article in 2001.

Tallgrass Prairie Producers Cooperative started in North Dakota as a way for grassfed beef producers in that very rural state to find markets for their beef by aggregating product from a number of producers. The cooperative limped along for 5 years from 1995 to 2000. The farmer-owners helped with the marketing: This is not only advantageous in building trust on the part of buyers but also time-consuming for the farmers who need to also be on the farm managing their cattle. Local markets in the region were too low in volume and high in servicing costs; thus they were not profitable. Yet the more distant markets such as Chicago or the West Coast demanded much larger volumes and lower prices. Tallgrass did not have the volume to get stocked on those butcher counters in distant stores nor the volume to keep their processing costs low. In that regard, they could not compete with the extremely cheap beef processing coming out of places like Iowa.

Scaling up was difficult because they did not yet have the revenue to hire a professional manager. Although the farmer board was full of competent and business-savvy individuals, they were not the same as a professional business manager. It is hard to have a bunch of overworked farmers also volunteer their time to manage a growing business. In retrospect, they probably should have borrowed more money to not only hire a professional manager but to purchase some key equipment and maybe take on the processing end of things, but they were too afraid of debt.

Likewise, they caution other farmers and cooperatives about thinking that local niche marketing is going to be the salvation for all farmers. In very rural places or places that don't have access to nearby meat processing, the logistical costs may be too high and price you out of the market. Other Tallgrass Prairie Producers Coop advice:

- Do honest, full-cost accounting or you may think you're profitable when you are not.
- Focus on investors, not grants. Most grants are not that useful for for-profit businesses and can be more time-consuming than they are worth.
- Marketing efforts that ignore price and convenience issues are doomed and reflect a lack of understanding about the marketplace.
- Restaurants are short-lived and slow to pay; be careful about being stuck with a high accounts receivable on a restaurant that went out of business.
- Growing animals out of season is more expensive, but markets don't want seasonal-only supply.

Ultimately, the building of the cooperative took away from the quality of life that the farmers were originally seeking when they choose to be grassfed cattle producers. The stress and financial risk of the cooperative were detracting from that quality of life. This is often the key conundrum that direct-market meat producers face: Do you live the "quiet, simple" life out on the farm and just auction off or wholesale your animals, or do you focus your energy on meat marketing and sales, which take you away from the farm life?

We think if Tallgrass Prairie were starting their cooperative today, they would see much better market acceptance of their grassfed beef and better sales. Grassfed beef is now 3 to 6 percent of market share in major US cities and will continue to grow as consumers become disillusioned with feedlot beef.

CHAPTER 9
Slaughtering and Butchering Logistics

Preparing for Slaughter

Separate the animals destined for slaughter a day or so in advance of loading them onto the truck or transport trailer, or into crates for poultry and rabbits. This will allow you to withhold feed from those particular animals, help them reduce their stress level, and allow you to watch for any health concerns of those animals prior to slaughter. It will also make your load-up process that much easier, not having to pick out or kick out animals that aren't ready yet. That said, if you don't have sorting infrastructure such as chutes and holding pens, you can sort on the trailer itself. With our pigs, we would often let a bunch of them come into the trailer looking for food and then push out the ones that weren't ready yet based on visuals. We would then lock in the slaughter pigs overnight with bedding and water, but no feed. You want to have as much of the food and waste out of your animals as possible on slaughter day. It makes for a less messy ride and a cleaner carcass at the slaughterhouse, minimizing the opportunity for fecal contamination.

If you can, wash your animals prior to slaughter; this makes processing much easier and reduces the opportunity for pathogen contamination. This will go a long way toward building goodwill with your processor; otherwise they will have to deal with the dirty mess you bring in. They may also charge you a cleaning fee if they have to wash down your animals. Similarly, certain animals like sheep will need to be shorn around the bleed-out regions: Ask your processor what areas they want shorn. Lots of dirty wool around the neck region can both be challenging and provide opportunities for contamination. Keep your animals bedded down on clean straw, which will help keep them clean, dry, and warm prior to slaughter, as well as rub off some of the dirt that may be on their skin or hair.

What kind of vehicle do you own or will you purchase or lease in order to take your animals to the slaughterhouse? When first starting out, you might be able to get by with a reliable heavy-duty truck (minimal $10,000 investment, probably more like $20,000). Some people get by with tall stake sides to keep the animals in the back of

Definitions

Aging loss the percentage of shrink that happens due to aging meat. This is from water loss (through evaporation and dripping) and exterior meat and fat that has to be removed after aging because it has discolored, dried up, or may have molds present on it. This aging loss means less salable weight; however, in many instances dry aging allows a premium price to be charged that can make up for shrink.

Carcass yield also known as dressing percentage, this is the percentage of the live weight left after slaughtering the animal and removing its blood, viscera, hide, and usually the head. On pigs you may leave the head since there is more marketable meat on a pig head than other red meat species. Some ethnic markets prefer lambs and goats to include the head as well. There are regulations that may dictate, dependent on the stunning method used, whether the head may enter commerce or must be disposed of as "inedible."

Cutting yield the percentage of hanging weight that is left after the animal has been cut up, removing a certain amount of bone, cartilage, silver-skin, and other trimmings not generally used for human food

Hanging weight also known as carcass weight, this is how much the animal weighs after slaughter, minus the blood, guts, hide, and usually the head and before it goes into the cooler (sometimes referred to as "hot carcass weight"). This weight is used to determine the carcass yield, as defined above.

Live weight how much the animal weighs when alive, pre-slaughter

USDA grade the USDA has set standards for beef (and to a lesser extent veal and lamb) quality based primarily on marbling in the loin. This is supposed to indicate the level of juiciness, tenderness, and flavor. The most common grades are Select, Choice, and Prime (least fat to most fat). When a carcass is cut and the rib eye is exposed, one can measure the amount of fat, visually or with a machine, to determine where along this spectrum the carcass falls. Choice is the most popular grade of meat in the United States across all consumers. Both the yield grades and quality grades are voluntary add-on services done at the slaughterhouse, thus they will cost a fee.

Yield grade ranges from a 1 to 5 and indicates the amount of usable meat from a carcass. Yield grade 1 is the highest grade and denotes the greatest ratio of lean to fat, meaning highest cutability; yield grade 5 is the lowest yield ratio, meaning lowest cutability (most fat trimmed off). Most producers seek a happy medium based on market preferences. For beef, the yield grade is used in conjunction with the grades of Select, Choice, and Prime to help processors and producers understand the quality of the carcasses.

the truck (or crates for poultry), but the better, safer option is a fully enclosed stock trailer. Your stock trailer should ideally be sized so that it fits the number of animals that you want to take each time to the slaughterhouse. You don't want to undersize your trailer and take too few animals; the transport costs per animal will be too high in most cases (unless your abattoir is nearby). Neither do you want the trailer to be too big—beyond what you plan to scale up to—because it will require a bigger truck and considerably more fuel to run it, even if you don't have that many animals inside. You could start out with a small trailer, like a used horse trailer, and then scale up as your operation does. Just keep what pulls the trailer (the vehicle) in mind. It probably doesn't make sense to have to replace your truck in 2 years because you decided to upgrade your livestock trailer and need a larger truck to pull it, unless you have a brisk local market for used trucks.

Another option is to hire a transport service to pull larger loads of animals for you. This is more common with large animals such as beeves and bison. One of the abattoirs we used for a time in California provided a custom transport option for farmers. We loved the ease of loading up pigs on the farm into their large trailer and saying goodbye to them, rather than driving in the middle of the night for an early morning delivery 4 hours away. Despite our fondness for this transportation option, that processor relationship did not last due to serious humane handling and meat cutting issues at that plant. Ask your slaughterhouse if they can custom-haul your animals and for what price. It may be worth it in the end to not own all of the expensive hauling equipment.

Poultry and rabbits obviously don't need a full livestock trailer and can be crated in small plastic or wooden boxes when hauling them to a processor. We used to load up around one hundred birds into the back of our Ford F250 pickup truck on a nice bed of straw for the night before processing. We fashioned a lid out of chicken wire attached to a wooden frame that fit perfectly on the top of our truck bed. That was the entire infrastructure needed to contain the one hundred birds for on-farm processing, which we did each Friday for fresh sales on Saturday and Sunday farmers markets. We avoided the cost of purchasing poultry crates, which can be quite expensive for high-quality plastic ones. However, if you are hauling birds off the farm to a processor, make the investment in good-quality crates. They will last longer and won't injure the birds during transport.

When you are just starting out, smaller animals such as sheep and goats can be hauled in a van or minivan. Just make sure you have a screen or barrier between the back and the driver's seat; driving around with a bunch of goats sitting on your lap or nibbling on your ear is not a safe practice. You can buy a well-used minivan for around $2,000, pull all the backseats out, throw in some fresh bedding, and end up with an affordable small-animal hauler that will probably fit three or four market-weight goats, sheep, or pigs a trip. Not a bad way to start off when you are still small-scale.

Slaughter Day

Some farmers like their animals killed as soon as they arrive at the slaughterhouse; others believe the animals should be held for a day or two in the holding pens in order to calm down and reduce their stress hormones. For certain animals that are not going to calm down regardless of how long they spend in the holding pens, such as exotic species, wild boar, and range cattle, you might as well have them killed as soon as you deliver them. However, many times the choice is not yours: It is dependent on the number of animals waiting at the slaughterhouse before yours. If you can schedule a specific time for your animals, make sure you show up on time and be ready for your animals to be dispatched immediately. When you have animals that are frequently handled, they will more than likely be calm at the slaughterhouse and exhibit less stress. This will make for better-quality meat in the end. Interestingly, studies have shown in the case of swine that pigs raised with more benefits like access to the outdoors and regular human contact exhibit less stress at the slaughterhouse than confinement-raised pigs (Geverink et al., 1998). It is probably because they are more used to moving around, changing of scenery, and being in different social groups. This could probably apply to other animal species as well.

There have also been instances where animals held overnight have been "spooked" and caused panic, resulting in harm or even death to the animals. If you don't know how your animal will react at the slaughterhouse, don't assume that letting them "calm down" will be a good option. Sometimes it's better to drive them straight to their appointment rather than having them sleep over in unfamiliar holding pens.

If you are selling certified organic meat, you may want to bring your own feed for your animals. Otherwise, the slaughterhouse may feed non-organic hay or grains to your animals while they are waiting in the holding pens for a day or two, something that could jeopardize your organic certification.

Processor Relations

Farmers wanting to turn their animals into meat and meat processors who do the work often don't have the fondest relationships or the best communication strategies to work with each other. They seem to be talking over each other instead of talking to each other. Respect levels are sometimes low. It's really unfortunate that this is frequently the situation even though the two "sides" absolutely need each other in order to sustain their businesses. NMPAN gathered information from both farmers and processors and this is what they say about each other in table 9.1 below.

This disconnect between producers and processors points out a need for better communication and coordination. Some of the ways to achieve this identified in the NMPAN report "From Convenience to Commitment" (Gwin et al., 2013) include:

- Bring processors more business, more livestock, more often to keep them profitable and in business. Your farm business may start

TABLE 9.1. Farmer/Processor Misperceptions and Communication Challenges

Farmers say about processors	Processors say about farmers
There are not enough processing facilities.	There aren't enough farmers bringing in enough livestock.
Processors don't have the right services or inspection status.	Farmers ask me to do new things but they don't have enough volume to cover my costs.
I have to schedule a date too far in advance; I can't get a kill date.	Farmers don't come on time or date or bring fewer animals. Can't count on them.
I can't get a slot in summer or fall.	I have no business in winter or spring.

out small, but if you scale up your processor will be more open to working with you. If you can't bring in more animals, perhaps you can coordinate larger groups of animals by collaborating with neighboring producers.
- Help farmers coordinate among themselves to bring loads of animals in together, reducing transport costs and increasing throughput in the slaughterhouse. This is a good role for nonprofit organizations, food hubs, and for-profit meat distributors to engage in.
- Schedule out further in advance and work to bring more animals in the off-season if possible.
- Provide technical and financial assistance to processors. This is one thing NMPAN, NC Choices, and the Vermont Department of Agriculture are striving to do, but not many other states are tackling the issue of processor technical assistance. As a farmer dependent on good processors, you should be advocating for them to receive technical and financial assistance as well.
- Co-marketing: Help promote good slaughterhouses just as we promote good farmers. As a farmer, if you are happy with one of your processors, let your customers know that. Say something like, "We are proud to partner with Heritage Meats of Rochester, Washington, because of their commitment to working with local farmers and dedication to high-quality meat cutting."
- Consider investing in an aspect of the processing business, such as buying a key piece of processing equipment that will solve one of your needs and help out the abattoir. You could also become a minority partner in a processing LLC. Or if you want a custom processor to upgrade to state or federal inspection, you will need to bring them a lot more animals to justify the cost, and you might consider making an investment in the plant so that they can do this.

As NMPAN wrote in their 2013 report, we need to change the relationship between farmer and processor from one of "convenience" to one of "commitment." This is not always easy. Many meat processors come from a more old-school, commodity-based meat supply chain where they are used to boxed meats and not used to working directly with farmers. As one butcher told me, many processors are cut from the same cloth and to some extent hail from a "good-ol-boy" culture. Many don't even have a customer service component or a person to answer the phone. As a niche meat producer, you may not fit into this culture, which can make relationship-building a challenge. Because you aren't selling your meat into the commodity chain, you are by nature doing something out of the norm. Can the niche meat culture of creativity, innovation, and quality relate to a culture of quantity, tight margins, and working with relatively standardized meat? Read on to see how to better that relationship.

Many meat processors don't do fee-for-service slaughtering and butchering at all. Many are privately held by the likes of Tyson, Foster Farms, Cargill, and Smithfield. Don't waste your time barking up those trees. Similarly, if they do fee-for-service work but don't act like they want your work, they may not be a good choice for you. However, many farmers don't have the luxury of seeking out another more "customer-oriented" facility. When we hear about states like North Carolina, Iowa, or Wisconsin, with over a hundred USDA-inspected abattoirs in each state, we think how wonderful it must be to have all those choices and to have that level of competition between processors. Meat processors can't be complacent, continuously put out bad product, or charge inflated prices when they have stiff competition. Maybe find a meat processor who is active on the NMPAN or the American Association of Meat Processors because at least you will know that they are seeking continuing education. A willingness to learn goes a long way.

With that said, many processors are unable to adapt to new regulations, rising costs, and increasingly hostile neighbors. Sometimes they get their federal inspection pulled on a moment's notice and don't know when they will be up and running again. Some have even been raided by USDA officials or local police without a clear explanation of what happened. This all occurred to one plant in California as we wrote this book. When we farmed in California, we had to withstand the experience of having our pig slaughterhouse shut down on and off several times because they did not get along well with their USDA inspector. It is a relationship and there is give and take. Processors must be willing to make the necessary changes to satisfy the inspector, but also know when and how to state their case. It can be incredibly tough and risky to build your business on the essential services of others, so it's probably a good idea to have one or two other processors in your Plan B file to fall back on in those times yours is unavailable. Your Plan B may be another 4 hours away, but it can keep your meat flow and market going through the other plant's shutdowns (maybe).

Finding a Good Meat Processor

To make this whole farmer-owned meat business thing work, you have to find a good processor to work with. This is, by far, the biggest crux in the direct-market meat world. It can make or break your hardest work and best intentions. First start off by finding the state or federally inspected facilities nearest you. If you are near a state border and the best place is on the other side, as long as it is federally inspected your meat can cross state lines. A handful of states have joined the Cooperative Interstate Shipment program (see chapter 7) allowing interstate sales of state-inspected meat (Indiana, Ohio, North Dakota, and Wisconsin as of this writing). There are also a few states that process under the Talmadge-Aiken law, essentially using state employees to inspect under federal regulations (also mentioned in chapter 7). All of these kinds of plants will also allow your inspected meat to cross state lines. One way to find USDA FSIS-approved facilities is to ask other meat producers in your region who they prefer to do business with. Ask them why they like that facility in particular: Is it price, customer service, following cutting instructions, speed, packaging quality, friendliness? You can also go to the FSIS website and find the approved facilities in your state on an Excel spreadsheet. It's not the most user-friendly document, but it will list all the poultry and red meat facilities currently licensed in your state. Some are approved just for slaughter; others are approved just for butchering; while some are approved for both. Once you find the ones that do what you want (preferably both slaughter and butchering in the same plant), start calling them. Keep a notebook nearby so you can record the answers they give you to the various questions you ask. In a quick introductory phone call you should ascertain the following basics:

1. Where are they located? Calculate how far of a drive that is for you.
2. Do they process for farmers, fee-for-service?
3. What animals (red meat, poultry, both, beef only)?

4. Do they offer full service, just slaughter, or just butchering?
5. What are the processes used (such as scalding, scraping for pigs)?
6. If you require certified organic or animal-welfare-approved processing, do they offer that?
7. What are the kill fees, cut-and-wrap fees, and other costs?
8. Where *exactly* are they located, what are their hours of operation and receiving hours for animals, and can you come for a visit?

The In-Person Visit

Schedule Your Visit

Once you have your list narrowed down to the potentials (it may be very narrow, just one or two facilities that do what you need), schedule a visit. If they won't allow a person to visit, don't do business with them. Remember, they are a service business. If they don't provide customer service, then they are not in the service business. They don't want your money. Likewise, if they don't want you to come see what they are doing, they probably have something to hide. If this is the case, find another processor.

The Drive

When you visit, note how long the drive is, how rough, and how much traffic you encounter. Think about what it would be like for your animals to travel this distance, to bump around on rough roads, or to glide along the superhighway at top speed.

Our main abattoir was an 8-hour round trip, sometimes 10 hours if there was traffic. We would try to time it so that we could pick up our meat from the previous trip and not have to make a second trip to pick up the boxed meat. A few times a grassfed beef producer we collaborated with picked up our meat when he was going to the plant to get his. He lived 10 miles away. Work to make the logistics as painless as possible. Your animals should not have to travel more than 4 hours at a time and you should not be making that round trip with your transport vehicle too many times either. If you factor in your labor to drive and the wear and tear on the vehicle, it may be too much. Try to spread those costs out among the volume of animals that you take each time. Or contract out the hauling to a specialist who maintains their own vehicles. Just make sure they handle the animals gently themselves. Watch how they load animals and ask them how they get them off the trailer, too.

Plant Exterior and Holding Pens

When you pull in, take a look at the parking lot, back-up ramps, holding pens, lighting, and so on. Do the holding pens look safe for the animals and the workers and are they well lit? If any animals are there when you visit, how do they look? Stressed? Given feed and water (good)? Standing knee-deep in mud or manure (bad)?

The Management

Once inside, see how the owner or managers greet you. You might be able to tell right off if they want your business or not. Ask to see the kill floor, to watch an animal get killed, see how the workers move around the animals, and so on. Look at the cleanliness of the facility, the lighting, the smells, where they deposit the blood, offal, and hides. See what kind of rail system they use. If the slaughterhouse won't let you view the kill, this may not be a deal-breaker. Ask them why you can't. Some places have a policy of no visitors on the kill floor for reasons other than trying to hide something—like liability.

Fabrication

Ask to see the workers break down an animal and do the packaging. Ask to see the walk-in cooler and see what the temperature range, humidity, and smells are like in the cooler. A smelly aging room might give your meat an off smell, but keep in mind that aging meat does have a musty odor you may not be used to. What chilling method do they use, such as rapid blast chilling, two-stage chilling, or electrical stimulation to prevent cold shortening (toughening) of the meat? Although the science is not yet conclusive on the best way to chill meat to prevent toughening of the meat (Savell, 2012), it is important for meat quality that your carcasses have undergone full rigor mortis, their pH has dropped to around 5.6, and the meat has reached an internal temperature of around 40 degrees F (4 degrees C). Also, find out how long are they willing to dry-age your beef, or do they only do "wet aging"? For other animals, how long do they age the carcass (lambs and pigs)? You can age lamb carcasses for 5 to 10 days and pigs can normally age for only a few days unless they have their skin on, in which case they can age for longer.

Plant Violations

Ask if they have had any FSIS violations and what they did to redress those situations. Also ask specifically about any meat recalls: If they have had a history of food safety violations you may want to look elsewhere. If your meat gets recalled because of something that may have happened at the plant, it could permanently damage your business and you could lose thousands of dollars in inventory. A plant in California (Rancho Veal Corp) recently had an entire year (2013) of beef recalled not because of contamination but because supposedly some of it did not receive ante-mortem inspection. Despite who was at fault, those nearly nine million pounds of beef came from a farmer or group of farmers somewhere. They will now lose all that money, unless they had some amazing insurance policy that protected them against such post-harvest issues, but that is unlikely.

Regardless of the answers that the plant gives you about previous FSIS violations, you should check on the FSIS website to see which plants have had violations in the past few years. What you discover might shock you, but these are real issues that some slaughterhouses face and you should understand them. We found out on the FSIS website that a plant we worked with for a short time in California had a violation for not properly stunning some goats and sheep prior to slaughter. A problem like this could throw a kink in your production process. If the plant that has your labels gets shut down for a period of time, what will you do with your animals during that time? Do you have a backup plant? Can you get more labels made quickly or will that other plant print up generic labels for you? Can you even get slaughter slots in that other plant?

A slaughterhouse we came to rely on for our pig slaughter (not cut and wrap) was shut down for nearly 3 months by the USDA for some sort of humane-handling violation. The plant manager said it was a personality conflict with his inspector (this happens). This forced us to process at another plant that was 5 hours away and more expensive, which dug into our profits a bit. Unfortunately, these things crop up now and again despite your best intentions to get along with your processor. While we are stressing the importance of building stronger relationships with your processor, some things are still completely out of your control.

Processes and Price Sheets

Ask to see the meat processor's price sheets and ask what they charge for any value-added products such as grinding, smoking, sausages, and jerky. Also, ask what they use for spices and curing agents. If they only do nitrate curing and your customers are asking for a "no-nitrate" cure, then you may not be able to cure there. Likewise, if their sausage spices include DHT or MSG, that may not work for your customer base, either. We still remember when we had some pigs processed at one plant and asked for full organic processing, yet we got several boxes back of pork sausage that had MSG in it. Aagh! We couldn't give that stuff away, nor did we really want to eat it ourselves.

Understand processing terminology, processes, and charges. For example, some plants will pay the farmer a credit for the beef hide or for some of the animal parts that they take (heads, feet, blood, offal, etc.). You may see this on your invoice as a credit. Small plants will often include that credit in the price they charge for the kill. So if it costs them $50 to kill, gut, scald, and scrape your hog but they take the head, trotters, and offal that they can sell for $15, maybe they'll charge you only $35 for the kill fee.

Packaging

You also need to ask if you can get your meat back fresh or if they freeze everything. Also ask what their method of freezing is and what kind of packaging they use. Will they do all the label making and application for you? Can they provide a generic label for the meat with your farm name on it or will you have to pay to create original artwork and labels for the meat? Can they help expedite the FSIS approval of your label or do you have to do that yourself? More on labeling is in chapter 11.

Scheduling

Another huge consideration is scheduling. When can you actually get a slaughter date at the plant, and how far in advance do you need to schedule? Are there months of the year during which they are fully booked? (This happens often in the fall during deer season in some regions of the country.) Are there other times of the year when they could use more volume to process? What are the fees for cancellations, and how far in advance can you cancel? Be kind to your processor and try to always give them a decent amount of advance notice if you have to cancel.

Cutting Instructions

What parts of your animal will you get back? What parts do they keep? Can you buy back parts of your animals? How do they keep your meat separate from the meat of other animals?

Will you want the hides? If so, it helps to deliver relatively clean cattle to improve the hide quality and reduce the amount of spray-down that the slaughterhouse workers have to do. Also realize that hides usually have to be picked up within 24 hours after slaughter and you normally have to pay for them (even if they came from your animal). Likewise, if hide quality is important to you, you may want to make sure they do hand pulling and skinning and not mechanical hide pulling.

How do they prefer to get their cutting instructions? How precisely do they follow them? What are some challenging cuts or processes that they can't do or won't do? For example, if their cutters are used to cutting boxed beef and pork for retailers who want most of the fat cap taken off, will they be able to leave ¾ to 1 inch of fat on your chops or steaks? If you have specific cutting instructions, ask if you can be present the first time they cut up one of your animals so you can watch and make sure they understand them. If they balk at this, they probably don't want your business. (See appendix C: Sample Cutting Instructions, page 296.)

Is there a minimum batch size? And for the value-added products like sausage: What is the minimum pound size to make a batch of sausage?

We had to start using a new pork cut-and-wrap facility because our normal place was booked up for 4 to 5 months. (Even though we had been good customers, they wouldn't give us a space. Customer service was *not* their forte.) The new facility was a little closer, and they were eager to work with us. They told us their specialty was primals and subprimals and that their employees were not familiar with retail cuts. So we showed up at our first appointment when the slaughterhouse delivered a bunch of our pig carcasses to the facility. Jim gave them the cutting instructions and sort of drew out where the cuts would go with his finger on a side of pork. They nodded their heads in agreement. He then stood there all day and watched them cut. It wasn't the prettiest butchering he had ever seen, but they were trying. They wanted to branch into doing retail cutting and knew they needed the practice. Whether our pigs were a good place for them to get that practice is debatable. Needless to say, they sort of followed our cutting instructions, but it was not as good as our preferred abattoir. One nice thing about this place doing our pigs, however, was that since they were not doing retail cutting of any other pigs or further processing, we knew we were getting our own pork back.

Test Run

Once you have had your facility tour and had all your questions answered, if you are satisfied, set up your first processing date. We suggest you bring a small batch of animals to test out their facilities just in case they don't live up to their promises. Stay and watch the slaughter and the fabrication. When you get your first invoice, check it with your inventory and ask them to explain all of the charges so you are clear about the full processing costs. Find out how long they will freeze the meat for you and when you should pick it up. If you want to make the trip once a month and pick up the meat from the previous month but they will only store your meat for one week, then working with them may not be possible, especially if the distance to travel is long.

You will sometimes have to work with several facilities to get all your processing needs met. Often slaughterhouses are fully booked in the fall when many livestock are finishing at the same time as hunting season, so, as we've mentioned before, you will need to find some backup (Plan B) slaughterhouses. We prefer to have three facilities that we have inspected and have a relationship with at all times. Also, not every facility does all of the processing you need. Some just do slaughter, some just do cut and wrap, some just do grinding, and some are called "breaking plants" and just break down carcasses into primals, as illustrated in the following profile.

Processing at Carman Ranch

Cory Carman of Carman Ranch (mentioned in chapter 8), has got her beef processing down to a science. She works with one custom facility in Wallowa to cut and wrap animals that were killed on her family's ranch with a mobile slaughter truck—that is, for custom sales only. She also works with four other processors regularly that do different aspects of the processing. One does kill-only of her beeves and they can break down into primals and subprimals. Another facility does retail cuts for her. Another does grinding, where she sends 1,000-pound totes of trim originating from the retail cutting facility. And yet another does the cold storage for her frozen meat. She manages all these logistics with a full-time salesperson who lives in the city of Portland, where the majority of Carman Ranch beef sales take place. They also employ a part-time delivery driver as well. This has allowed Carman Ranch to grow to over three hundred head of beef a year sold primarily into the food service and restaurant trade of the Portland metropolitan region. They work closely with their processors to schedule slaughter dates and cut the carcass according to their buyers' specifications. Cory suggests spending some time on the kill and cutting floor to make sure you understand your abattoir's processes and they understand your specifications.

Carman Ranch chooses not to age most of their meat, except for a few restaurant clients that want the middle meats aged for around 14 days. For those restaurants that want longer aging, Carman Ranch charges more based on the shrink loss that occurs during aging.

Although Cory is not able to maximize her beef carcass values as much as she would like too (not selling the variety meats nor hides), she does manage to sell quite a bit of bones to her food service customers for them to make stock.

Carman Ranch manages to sell fresh meat for up to 6 months of the year (primarily June through November) and frozen meat for the other 6 months. Some of their restaurant clients choose not to buy the frozen meat, but Carman Ranch has slowly been able to convince the majority of farm-to-table-type restaurants that this is how it needs to be with local beef. Carman Ranch will only harvest animals when they are on the gain and won't harvest them when the pastures are hardly growing or covered under a deep layer of snow, which is why their fresh beef is seasonal.

There is a stigma against frozen meat, especially among chefs. Many chefs have a notion that frozen meat is inferior to fresh. They may not be familiar with the process that is involved in flash-freezing meat at a processing facility. Carman Ranch conducted some blind taste tests for chefs, comparing fresh beef against frozen. Most chefs could not discern the difference. When Greg Higgins of the famous Higgins Restaurant in Portland stepped up to the plate to buy frozen Carman Ranch grassfed beef, many other chefs followed suit. It takes education and early adopters to create demand for frozen beef.

When meat is cut and packaged (vacuum-sealed is preferable) and then put into a freezer at −10 to −20 degrees F (−23 to −29 degrees C), this meat is frozen quickly, without any freezer burn. Chefs may be used to restaurant freezers that are not as cold and packaging that is not airtight. They often believe that frozen meat becomes degraded, when in reality, once thawed, it is just as fresh as when it was cut. Beef, especially, can be held for a year if handled properly, and still be very high quality. The other aspect of frozen meat is that it must be thawed out properly to maintain its high quality. This takes planning and forethought to remember to pull various cuts of meat from the freezer and place them in the refrigerator to thaw out over 2 to 3 days, something that a hectic restaurant may not be able to do well.

Getting Things Right

Now that you've found your processor, how do you solidify that relationship and keep things working smoothly? Kathleen Harris, founder of the Northeast Livestock Processing Service Co. and a meat-processing "guru," shared her Ten Commandments of working with your meat processor in a recent NMPAN webinar on effective producer–processor relationships. These words of advice include (with our add-on comments in parentheses):

1. Visit the plant. Do they have enough staff? Who takes the cutting instructions? When is the best time to call? (*Do they prefer fax, email, or phone call for cutting instructions?*)
2. Use the plant's own cut sheets. Try to keep it simple and avoid a lot of changes throughout the year or between animals. (*Don't overwhelm your processor, especially at the beginning, with complicated cut sheets. Once you start to regularly bring them more volume, you can get more precise on your cutting instructions. Also, find out if the plant keeps these sheets for a record. If not, ask for a copy to keep on file for next time.*)
3. At delivery of the animal, leave instructions for organ meats, hanger steak, oxtail, and aging time (and any other odds and ends you want to keep like hides, feet, etc.). Make sure your animals are identified. If you want oxtail, your animal's tail must be clean when you deliver it. (*If you want the wool back from your sheep, shear it before you take it in!*)
4. Be punctual about delivery and pickup times. (*If they want your meat out of their freezer in 2 weeks, make sure you pick it up in 2 weeks, and make sure you have the money to pay them at that time.*)
5. Bring a cut sheet when picking up your finished meat to serve as an inventory checklist. Organ meats and other products are often stored in other areas and are more apt to be misplaced. Make sure you pick them up when you are there. (*It is your responsibility to check your meat inventory before you drive off, as well as checking if the meat is properly packaged and frozen.*)
6. Never be demanding.
7. Be sensitive to their time and avoid taking the butcher "off the block."

8. Compliment them on good work—this includes line staff too.
9. Be grateful and appreciative when they get your work done on time or to your specifications.
10. Respect their work, understand their constraints. (*Including the slim margins under which they operate, too.*)

Where Did All My Meat Go?

As you know, there is a difference between the live weight of the animal, the hanging weight after it has been slaughtered, and the final cut-and-wrapped weight of the meat. There is loss and shrink along the way. There are also factors of genetics of the animal, how full its gut was when you brought it in, the weight of its hide or fleece, and other reasons why weights will vary even between animals of the same herd. Very rarely, if ever, is the meat processor "stealing your meat." If anything, they may have inadvertently left a box of something sitting in one of their coolers. If this happens, don't turn into a raving lunatic and accuse them of theft. Kindly ask if you can go through your invoice and do product reconciliation together. Ideally, you should have done this before you left the plant with your meat; that way you don't drive all the way home without a box or two.

To be clear about expected cutting weights, do you know the typical yields of your animals? Do you know the live weights when you bring the animal in? Some slaughterhouses have certified scales that they will be willing to weigh your animals with. There may be a small $5 to $10 fee associated with that procedure, but it's useful information to have. Say you think you are bringing in a 125-pound lamb and you get back fabricated meats that weigh only 40 pounds. Your initial reaction might be to think your abattoir stole your meat! However, what if that live lamb only weighed 100 pounds? Hanging weights on lamb are around 50 percent of live weight, so 50 pounds is what you might expect on the carcass, breaking down to around 40 or so pounds of finished meats. Your live animal estimates were off by 25 pounds and you took out your ignorance on the processor. That is a bad idea—know the live weights instead! Lambs are especially deceiving with their wool. They may look big, but they are hiding a small frame under their fleece. Also, mud and feces caked onto a hide will make that animal weigh a considerable amount more than a clean animal.

A fatter animal, unless you leave all of that fat on the meat, will have lower cutability. This is where yield grades factor in. A yield grade of 1 has the highest cutability, meaning you will get back the most boneless lean meat from that animal. However, a grade 1 carcass could be too lean for your customer base. On the opposite extreme, a yield grade of 5 (likely impossible for grassfed producers) would have a lot of marbling and fat—to an extreme—so much of it would be trimmed off. Unless you have a market for that fat, it would be like throwing away money. Farmers should try to obtain their yield grade for quality control and to make future feeding and breeding decisions. If you want to obtain the highest yields of boneless lean meat from your beeves while maintaining some fat cover for eating quality and aging, you may strive for a yield grade of around 2 or 3.

There are a number of places where meat loss can occur during processing. You first need to understand the factors that affect dressing percentage and cutting yields. Adapted from Danforth (2014a), these meat-loss factors include:

For Dressing Percentage (Carcass Yield):

- **gut fill:** the more full the gut, the lower the dressing percentage. Ruminants have larger gut fill than non-ruminants.
- **muscling:** the more muscle, the higher the dressing percentage
- **fatness:** the more fat, the higher the dressing percentage
- **mud on animal:** the more mud, the lower the dressing percentage because your processor has to clean that off, making the animal weigh less (that's a good thing!)
- **wool:** the more wool, the lower the dressing percentage—more stuff to take off
- **old injuries, abscesses, and other abnormalities:** these defects can severely influence your yield. If there is a part of the animal that is unfit for consumption, it could cost you a whole leg, section of ribs, or shoulder. Ask your processor, up front, if they can photograph the bad meat before disposing of it so you can understand what may have happened and how to prevent this in future stock.

For Final Cutting Yield:

- **fatness:** the leaner the animal, the higher the edible meat yield. In addition, the more fat you leave on the meat, the higher the yield. But, again, make sure you're appealing to your customer base with things like fat caps—some folks like fat, others don't.
- **muscling:** heavy muscling equals more meat
- **bone-in vs. boneless:** keeping the bones in the cuts will dramatically increase the cutting yield and the final weight of the products. You have to determine if taking the bones out will allow you to sell the cut for a higher price to make up for the loss of the bone (weight = money).
- **trimming:** the closer you trim, the less yield you will have; leaving more of the fat layer will increase your cutting yield
- **ground leanness:** the leaner the ground products, the lower the cutting yield (because your processor isn't using the animal's fat in the grind)

Now that you understand that there is loss when your live animals get slaughtered and butchered and nobody is "stealing your meat," refer to the following table, which is an example of general industry standards for dressing percentages and cutting yields. Keep in mind that the factors discussed above, as well as how closely you work with your processor, will influence these numbers. If you develop markets for things like heads, blood, bones, organ meats, and even hides or skins, you very well could improve your profitability per animal unit by increasing your per animal yields. Keep in mind that every ounce of that animal was paid for by your labor and farming inputs. Each ounce you can sell is vital to your business's success. You grew it; you should try very hard to sell it. Even if it means hiring someone on a commission basis to sell more obscure cuts, perhaps that's better than not selling it at all.

Using a goat as an example, say you begin with a 100-pound live weight animal when you slaughter it, resulting in a 48-pound hot hanging carcass weight before it goes into the cooler (which is 48 percent of the original live weight). Then, after butchering and taking out most of the bones in the shoulder and leg, there is 33.75 pounds of final meat yield (which is 70 percent of the hanging weight). So your final yield percentage is just

TABLE 9.2. Typical Dressing Percentages and Cutting Yields

Species	Starting live weight (example)	Dressing %/hot hanging weight in lbs	Cutting yield/final meat yield in lbs (boneless, with trimming)	Final yield as a % of live weight
Beef	1,000 lbs	62%/620 lbs	70%/434 lbs	43.4%
Lamb	100 lbs	50%/50 lbs	70%/35 lbs	35%
Goat	100 lbs	48%/48 lbs	70%/33.75 lbs	33.75%
Pig	300 lbs	74%/222 lbs	70%/156 lbs	52%
Chicken	6 lbs	70%/4.2 lbs	58%/2.5 lbs cut up	41.6%
Turkey	20 lbs	77%/15.4 lbs	58%/9 lbs cut up	45%

Adapted from Danforth, 2014b

33.75 percent of the living animal. You can improve upon the final yields by saving and selling the bones, fat, trimmings, offal, and head.

From a pure meat efficiency standpoint, nothing else, it would appear that pigs have the best meat yields, then poultry, then ruminants. Anatomically, this is because ruminants have a large digestive system to process forages that those other species don't have. The rumen makes up a large percentage of the live weight of these species and is inedible. But of course, on the flipside, the bacteria-filled rumen allows ruminants to process forages that monogastrics can't make as good a use of. If your goal is to extract as much usable meat from a small area of land, then poultry and pigs are the best way to go. If your goal is to convert extensive grass and shrublands into meat, then ruminants would be best for you. What is your land base and what is your market?

Other Places for Meat Loss

AT THE PROCESSOR:
- Are any animals condemned at slaughter?
- Dry aging losses
- Fabrication damage
- Cutting wastes or grinding loss
- Not giving instructions for excess fat and bones (thus you don't get them back)

AFTER THE PROCESSOR:
- Packaging damaged in transport, cold storage, or marketing
- Quality incidents, money-back guarantees
- Spoilage and expiration
- Trouble selling certain cuts (This should never happen. Get creative, find ethnic markets, turn into value-added products, provide recipes, sample it, cook it, etc.) **Note:** Most processors will accept meat that is in its original packaging and has been kept frozen to re-form, that is, turn into ground beef, sausage, jerky, BBQ, and so on, if you are having a hard time selling it in its original form.

A Little Meat Science (Goes a Long Way)

You don't have to be a meat scientist to be a successful meat marketer, but having an understanding of some of the basics of meat and butchery will be helpful in your decisions about how to have your meat handled and processed. The better informed you are, the better you can maximize carcass values, meat quality, and customer satisfaction.

What is meat? Meat is mostly the muscle tissue of an animal. Most animal muscle is roughly 75 percent water, 20 percent protein, and 5 percent fat, carbohydrates, and assorted proteins. Muscles are made of bundles of cells called fibers. Muscle fibers are also organized into bundles that are stacked together in one direction and bound by sheaths of connective tissue. The grain of the meat is those bundles oriented in one direction.

What determines meat tenderness? The main factor affecting muscle tenderness is the volume and strength of the cross-links between collagen fibers. Collagen makes up most of the connective tissue material. An animal develops more cross-links as it gets older, or if it gets more exercise; breed and nutritional factors are at play as well. Depending on the cut, meat will have larger or smaller groupings of muscle fibers. The smaller the grouping, the more tender. The less work a muscle gets, the more tender. That is why leg and shoulder cuts are not as tender as cuts along the back. Four-legged animals work their legs and shoulders much more than their backs.

What role does fat play in meat quality? Fat cells store energy in the form of fatty acids but also act as a repository of any organic substance that is fat-soluble. These fat-soluble compounds are derived from the food the animal eats, and the quantities are affected by age and species. An animal with a more diverse diet of forages will have more varied organic compounds and fatty acids, creating stronger and more varied flavors.

This is why grassfed and pasture-raised animals have stronger meat flavors. Some eaters may refer to this as a taste of "gaminess," when really the taste is a mark of more diverse organic compounds and different fatty acid profiles than they are used to eating. Fat contributes more to flavor than to tenderness. A lean animal can still be tender, but the flavor will be different.

How does animal stress affect meat quality? Animals exposed to stress (extreme temperatures, fighting, loud noises, long drives, inhumane handling at slaughterhouse, etc.) become exhausted. An exhausted animal will not produce enough lactic acid in their muscles after they have been slaughtered, and thus the pH of the meat will not drop enough. High pH meat creates dry, tough meat that spoils quickly. On the flipside, a stressed animal can result in adrenaline-filled meat whose pH drops too suddenly. Warm meat with an acidic pH level will be pale, soft, and express too much liquid. In the middle is what you want: a calm animal at slaughter that produces a meat pH that drops to between 6.0 and 5.6 after rigor mortis completes.

Cooling the Carcasses

An animal carcass must be properly chilled in order to protect the meat quality, keep it safe, and enable fabrication. Carcasses can be quite hot when they first go into chilling. The larger the animal, the longer it takes to cool down. The ultimate goal is to get the internal carcass temperature down to around 39 to 41 degrees F (4 to 5 degrees C). In the United States, unlike some other countries, no specific regulatory requirements exist for how rapidly or to what extent this initial chilling is conducted. Many processing plants that have chilling as a critical control point in their Hazard Analysis and Critical Control Points (HACCP) plans often use the critical limit of less than or equal to 39 degrees F (4 degrees C) surface temperature within 24 hours. This measurement is taken on or just beneath the surface of the carcass: Controlling the surface temperature is key to preventing pathogen growth, which is the most frequent biological hazard addressed by a HACCP plan.

In a literature review conducted by Jeffrey Savell, which was funded by the Beef Checkoff, Jeffrey found studies that indicated that rapid chilling caused beef cold shortening and other studies that showed the opposite effect. What is basically understood is that if the process of rigor mortis has not completely set in and the pH of the meat has not dropped below at least 6.2, rapid chilling can be detrimental to meat tenderness (Savell, 2012).

There are a variety of chilling methods used in different plants, but most plants, especially those that process large beef animals, use blast chilling. Cold air of between 28 and 30 degrees F (-2 and -1 degrees C) is blown on the carcasses while they are hanging on rails. Another method that some producers prefer is a two-stage chilling in which the carcass goes into a slightly warmer cooler, maybe 50 to 55 degrees F (10 to 13 degrees C), for 12 to 24 hours and is then moved into a colder cooler at 35 to 40 degrees F (2 to 4 degrees C). This may be more appropriate for smaller animals like sheep or goats that don't retain as much internal heat as a beef carcass. A third method that is sometimes used in combination with blast chilling is electrical stimulation (of primarily beef). This has been shown to help maintain the tenderness of beef through the chilling process (Savell, 2012). However, the equipment is large and expensive; therefore, small plants usually don't do electrical stimulation.

Jamison Farm of Pennsylvania found out that rapid chilling caused their grassfed lamb to be too tough (at least according to chef Julia Child, who told them so). Since they own their slaughterhouse, they were able to add a cooler so that they can maintain a two-step chilling process with their lambs now. The first cooler is at 55 degrees F (13 degrees C), where the carcasses go for one day while their rigor mortis subsides. The second day they move the lambs into a cooler held at 35 degrees F (2 degrees C). Some processors may have this ability, but most will not. Again, their HACCP plans will probably spell out how they are supposed to do their chilling.

Poultry are typically cooled in ice water baths, but can also be air-chilled. Air chilling results in less water uptake in the meat, which means consumers will get more of what they pay for—meat. Some say it is more sanitary to air-chill in that the carcass is not sitting in potentially contaminated water. The final product will be more desirable too, because there won't be a bunch of water pooled up in the bottom of the bag.

Aging

Aging times at the processor are often dictated by the amount of cooler space they have. While they might be fine with longer 21- to 28-day aging, they may simply not have the space available to have that many beef carcasses in the cooler for that long. If this is a "make-or-break" factor for you, then you will have to look elsewhere. Perhaps you can compromise with a 14-day age, which is pretty standard for beef. Another idea might be to just do long aging of the loin primal while the rest of the animal ages for a shorter time, such as 10 days. Adam Danforth in his 2014 book series *Butchering* suggests that older animals benefit from longer aging in order to break down the connective tissues that are more prevalent in older animals and to allow the meat to become more tender. Older, dry-aged animals would probably be a very specialty product that you could sell to discriminating chefs or charcuterie makers—the flavor potential is outstanding.

According to University of Kentucky's Bob Perry (an avid cured-pork maker) ruminants can age for longer due to their fat makeup; they are less susceptible to rancidity. The higher the level of unsaturated fats, such as with pigs finished on nuts, the faster you need to process that animal. Eat it, freeze it, or get it under salt. Otherwise it will go rancid.

Aging halves is preferred to aging quarters or primals because it results in less meat being exposed to the air and thus less trim loss. If you choose not to dry-age because you don't want the shrink loss or the logistics, make sure you let your animals rest for at least 24 hours before cutting and packaging them in order for the rigor mortis to be reduced through enzyme activity. This is true for all animal species.

Value-Added Products

You may want to have some of your meat made into value-added products such as cured whole muscle cuts, sausages, salamis, or smoked meats. Indeed there is great demand for many of these products and "artisan" meats seem to be all the rage. You can also charge a great deal more for these products, hence the term *value-added*. They are much more costly to make and usually incur more trim loss and shrink, so you won't necessarily make more profit on them. These products should be viewed as a way to use parts of the animal that you are not able to sell as easily when fresh: things like hams, shoulder cuts, and trim. By making various processed products, you will naturally have a wider variety that may appeal to a broader range of customers. Many of these products don't require cooking and are thus easier to sell to the convenience-oriented customer, tourist without a kitchen, or simply busy parents who want good meats for their own (and kids') lunches.

Because most inspected facilities that do fee-for-service work for farmers don't have the equipment or approved HACCP plans to make many of these products, we won't focus too heavily on these foods in this chapter. If you find a processor that does have this capacity, work closely with them to develop a recipe that you like. There are literally thousands of options when you consider all of the sausage recipes, varieties of bacons, and other cured and smoked meats.

The following discussion on value-added products is adapted from the "Resource Guide to Direct Marketing Livestock and Poultry" written by Goodsell and Stanton and published in 2010. This is an excellent resource that can be found online and that should be read cover to cover in addition to this book.

Fresh sausages: A fresh sausage is ground meat (often made from trimmings) combined with fat or lard or other binding agents and seasonings such as herbs and spices. Fresh sausage can be packed in bulk, formed into patties, or put into casings to be sold as links. Fresh sausage must be kept under refrigeration and be cooked before being eaten.

Cooked sausages: These include products such as hotdogs and bologna. Although they have been partially cooked, these products still must be kept under refrigeration and sometimes heated up before being eaten. The label will say "Ready to Eat" if heating is not required.

Fermented sausages (also called salami or salumi): A class of chopped or ground meat products that, as a result of microbial fermentation of a sugar, have reached a pH of 5.3 and have undergone a drying and aging process to remove up to 25 percent of the moisture. These products are typically cured but are not necessarily cooked or smoked. The USDA regulates the moisture-to-protein ratio but does not formally define semi-dry or dry sausages. Semi-dry sausages such as summer sausage and Landjaegar have a higher moisture content and should be refrigerated. They are generally cooked or smoked prior to sale or consumption. Dry sausages such as pepperoni or salami are generally shelf-stable and may be consumed without additional heating. Again, the label will tell the consumer if the product must be refrigerated or cooked before consuming.

Cured whole muscle cuts: These include hams, loins, bellies, jowls, breasts (lamb breasts, duck breasts, goose breasts), bresaola, and many others. They are typically packed in salt and spices to begin with, then rinsed off and hung in a cool, semi-humid environment for a while. Sometimes they are smoked to finish them. Many small meat processors do not provide "dry curing," as it's generally called. The USDA has made it very difficult to produce these products using the old-style recipes, many of which originated in Europe centuries ago. A more commercial-style curing is done now in most inspected plants. Instead of using a long natural curing process, meats are cured using accelerated methods in an oven (many times processors use their smokehouse if the temperature and humidity can be controlled digitally).

Smoked meats: Smoking adds desirable color, flavor, and aroma to fresh or fermented meats. Smoking may also be used as a method of preserving meat, but it should not be the only method employed, as any disruption to the smoked surface will destroy the preservation. Approved woods for smoking include hickory, oak, apple, cherry (and other fruitwoods), mesquite, redwood, and even corncobs. Liquid smoke may be substituted for the actual smoking process. Products can be hot-smoked or cold-smoked. Meats are partially cooked during the early stages of a hot-smoke process. Cold smoking at temperatures below 41 degrees F (5 degrees C) is generally reserved for hams. Smoked meats should be kept refrigerated and thoroughly cooked before being consumed.

Jerkies: Jerky is often prepared using marinades and spice rubs. In general, jerky is prepared using lean muscle meats, cut with the grain of the muscle fiber. Some recipes will use a ground and re-formed mix to make the jerky. All visible fat should be removed to prevent rancidity. Jerky is a great product to make from lean beef cuts that don't sell well, such as eye of round. Jerky has a very low moisture level and may be cured, non-cured, smoked, non-smoked, rubbed, or marinated. It is a great product to add to your mix because it is ready-to-eat and many times it is shelf-stable (no refrigeration needed).

CHAPTER 10
On-Farm and Mobile Processing

According to Arion Thiboumery of Lorentz Meats and the Niche Meat Processor Assistance Network (NMPAN), only when your animal production volumes exceed four hundred head of beef, twelve hundred hogs, or two thousand lambs or goats per year should you start considering building your own meat-processing plant (NMPAN, 2014). This will undoubtedly vary according to your own production and processing costs. Even if you are at the volumes Thiboumery mentions, you obviously will need to have a strong market and should conduct a good feasibility study just like you would for any other expensive business venture you are considering.

Building Your Own Facility

Despite the challenges, many farmers are tempted to bring the animal processing a little closer to home, and presumably under their control, by building an on-farm slaughter or processing plant (or both). However, the regulations for on-farm abattoirs are the same as for off-farm facilities. If you want to sell meat, the animal must be processed in either a state- or a federally-inspected plant. If you want to sell live animals to people and custom-process them for the customer, you may be able to get your abattoir approved by your county or state department of health as a custom-exempt processing facility. As a custom-exempt facility, you are not allowed to sell meat unless you buy in USDA- or state-stamped carcasses or fabricated meats. Just because you are putting something on your farm does not change the rules about it; it does not exempt you from any of the laws that other processors face. There is no "right to farm" law that exempts you from federal or state laws regarding meat safety.

Additionally, you must find out what your local land zoning allows you to do. In many areas you will not be able to build a processing plant because county or state zoning laws don't consider that an appropriate use for your type of land. This is especially true for folks raising animals in urban or semi-urban areas on land that is not deemed "agricultural." But even on agricultural land, the laws may restrict slaughterhouses. It's sadly ironic: Your county may allow you to build a 20,000-square-foot fruit-processing building, 2 acres of grain storage silos, or a 10-acre hog manure "lagoon," but not an animal slaughterhouse. Apparently a slaughterhouse is too close to "food manufacturing" or something like that. An example of this is Gunthorp Farms of Indiana, profiled in the pig chapter, was only allowed to build out their on-farm abattoir to a maximum of 10,000 square feet according to county rules, yet neighbors can build much larger hog barns if they want. Find out your county's rules before you even embark on a feasibility study or business plan for an on-farm abattoir. They may stop you in your tracks or require you to apply for a zoning variance, which may be a lengthy and expensive process in your county.

As for the logistics of doing on-farm slaughter or butchering (or both), there are several farming operations around the country that have built on-farm slaughter or butchering facilities to process their own meat and often the meat of neighboring farmers with some success. Others have purchased nearby existing abattoirs to take control of the processing aspect of their business without having to start from scratch. Still others have teamed up with neighboring farmers to cooperatively own the slaughterhouse or the cut-and-wrap operation. Ownership of the meat-processing infrastructure may be a viable option for some farmers if planned thoroughly and executed correctly (just like everything else in business).

Running a meat-processing business involves a completely different set of skills and risks than farming does. Indeed, many dedicated meat processors struggle with issues like insurance, finding skilled labor, high overhead costs, and keeping their plants busy year-round. As a farmer and meat processor, you will have to take on an ever-increasing list of responsibilities and develop even more skill sets than you already possess.

Several crucial factors must be met for the slaughtering and meat-fabrication side of your business to succeed. This research is based on the excellent online resources and processor expertise of the NMPAN as well as our speaking directly with a few farmer-processors. This is good advice when considering a meat business even if you don't plan to ever take on the processing yourself. Factors to research and consider before you open and operate an abattoir include:

1. **Established markets.** It helps to have established markets before you take on the meat-processing end of things. You can build your markets using other processors, for better or for worse, and create a name for yourself (for more on building a brand see chapter 12). Once you have that market established, then maybe you can embark into processing. But to make the most efficient, profitable use of your facility, you will have to increase volume, probably by a large degree. Will you be able to find markets for that new projected volume too?
2. **Ownership vs. partnership.** Can you invest in an existing meat-processing business, or become an owner or a partner such that you can have more control over the processing without taking on the full responsibility (and liability) of running a facility? Have you ever broached that subject with your favorite processor? Would you be willing to finance their expansion or purchase of a new piece of equipment in exchange for guaranteed slaughter dates, better quality control of your meat, or maybe a seat on their board?
3. **Existing vs. new.** If you decide you must have your own abattoir, can you buy an existing custom, state-inspected, or USDA-inspected

building nearby? An already built facility is usually cheaper to buy than a from-scratch facility, with considerably less permitting needed too. You also won't have to deal with all the NIMBY (not-in-my-backyard) attitudes that new plants run up against (although existing plants may have hostile neighbors too; see the discussion of Black Earth Meats in chapter 8). Or, could you go in on the purchase of an abattoir with several other producers, to reduce your cash outlay and risk, and ensure better throughput? On the other hand, an existing building may not have a good layout or it may require considerable investment to get up to USDA standards. Buying a bad floorplan may not be worth it, regardless of how cheap the asking price might be.

4. **Good planning.** Just as in farming, develop a good business plan for your abattoir, not an "if we build it, they will come" mentality, something that plant owner Mike Lorentz found out when he first opened his USDA-inspected abattoir Lorentz Meats in Minnesota without the sufficient volume to cash-flow it properly. He admitted that he hemorrhaged money the first 3 years and barely made it to the profitable stage he now enjoys (NMPAN, 2013). Even a good business plan might overestimate volume: Expect a few lean years to build up the animal volume and client base that you need for the size of your facility. Assume that everything will take twice as long, cost twice as much, and that you will have half the volume you need to get things started. Next, make a plan for survival in those first couple years when you are undercapacity.

5. **Volume and throughput.** Just like any other abattoir, sufficient volume of animals is necessary to keep the facility running and keep employees on payroll. If you have no employees and are completely family-run, you may have more flexibility to slow down or close for a few months out of the year. However, keep in mind the overhead costs that must be paid even when your plant is seasonally closed. How are you going to ensure the volume you need to run through the plant? Can you produce animals year-round yourself or are there enough other farmers in the area that you can work with to bring you animals (or buy them in)? How deep is your relationship with those other producers? (Hopefully you haven't ostracized any of them in the past!) Will increasing your animal volume to keep your abattoir running put strain on your land, people, and other critical resources? If you will rely on other local producers for your throughput, have they already begun to increase production or will you have to wait one or two years before they have sufficient numbers of animals? If you can, start talking to them well ahead of time before you open your doors. Will Harris of White Oak Pastures is a great example of someone who is working with several nearby producers to have the throughput he needs for his plant. Because they follow his production protocols, he buys their cattle and sheep and markets them under his brand too.

6. **Anchor customers.** You normally can't support a facility and all the skilled labor you need to operate it effectively with only the volume that small-scale direct marketers or backyard enthusiasts might bring you. You

need a couple of "anchor" customers that will be there every week with real volume (another Mike Lorentz pearl of wisdom). Can you be your own anchor customer? Also, offering fee-for-service processing in addition to processing your own animals can smooth out throughput issues. You can later drop this service as you scale up your own animal production, if you so choose.

7. **Butcher first, slaughter later.** Consider adding a butchering facility before you add a slaughter component. Build up your market first before taking on slaughtering, which requires considerably more infrastructure, permitting, inspections, and so on. Also, good, quality meat fabrication is normally the bottleneck in different regions anyway, so it's the most logical place to focus. Likewise, unless you plan on selling across state lines, the butchering does not have to be USDA-inspected; your facility can instead be inspected by your state or your county departments of health (or food safety)—though you will still have to develop HACCP plans and run a clean facility. And if you think you may want to add a slaughter component in the future, check into your local zoning to see if that will even be possible. You don't want to plan for something that will be legally impossible to do in the future.

8. **Flexibility.** You must build in flexibility and adaptability to market changes and the ever-changing tastes of customers. For example, can you add value-added products, ready-to-eat meats, a retail shop, or a deli? Keep in mind, though, that just because a meat product is considered "value-added" does not mean you will make more money on it. They require more processing, labor, ingredients, and specialized equipment. Also, you might be able to do both custom-exempt work and inspected work in the same building, which would broaden your set of potential processing customers.

9. **Cash flow.** How will you keep your plant busy year-round and pay the bills? How can you find enough customers who will pay on time? Wholesale accounts like grocers and restaurants often take 30 to 60 days to pay: How will you provide cash flow in the intervening time? Striking a balance between cash buyers and wholesale buyers will probably be crucial. Likewise, keep in mind that under the Packers and Stockyards Act, if you buy animals from other producers, you have to pay them within 48 hours of receiving the animal (actually by the close of the next business day). Yet the buyers of that meat (restaurants, institutions, grocers) may take 60 days to pay you for that meat. What will you do during those cash-flow gaps?

10. **Retail component.** Many experts suggest adding a retail component to your abattoir so you can test out your market, build your consumer base, get regular feedback from customers, and sell all the parts of the carcass. Can you add a small butcher counter or retail freezer case to the front of the building and allow customers to come in and buy direct? Will this require more labor or can customers be served by existing butchering staff or via a self-serve system? Having a retail case might also smooth out labor hours for your employees that you want to keep on year-round by having something to keep them occupied with.

11. **Scale.** You have to have the right number of animals going through your plant and the

right number of customers buying that meat. Finding that "sweet spot" is challenging and has led to many plant closures. Consider this question posed by a failed North Dakota cooperative processing plant: Are you too big for local markets but too small for national markets? Can you identify a regional customer or set of larger producers that will keep you running? Additionally, you will not have the economies of scale of larger processing facilities. You cannot spread your capital costs over a high volume of animals. How then will you cover your operational, overhead, and capital costs? Can you add value to each carcass to earn more money per carcass? How will you make your "packer margins" (just like the big guys) for the hides, offal, feet, and blood?

12. **Cost.** Building an inspected facility is expensive. Add at least one or two more zeroes to your estimate. Where will you come up with that money? How will you stay afloat in the first couple of years before you break even? White Oak Pastures said they spent $2.2 million to build their red meat plant and $1.5 million to build their poultry plant (both USDA-inspected) in Georgia. Little Farms LLC spent $110,000 to build their custom-exempt poultry plant in Washington. Acre Station Meat Farm of North Carolina said if they had to build today what their parents built in the 1970s, it would easily cost around $2 million. Do you have access to that kind of money? Can you afford the high debt load that will come along with those figures? If you have to bring in a private investor(s), are you willing to give up equity and probably some control?

13. **Labor.** Will you be able to find the skilled labor to operate the facility (including yourself)? Just as other abattoirs struggle to find good, skilled labor, your operation may run into the same shortage of butchers. Many meat processors say they have a hard time finding trained butchers who will stick around, so they end up just training local people who will be with them for the long term. Likewise, do you want to manage a bunch of employees (yet another skill set)?

14. **Cold-chain logistics.** How will you manage the cold storage and transportation logistics of the meat? Can you afford to take that on as well, or will you need to contract that out? Is there anybody in your area that already does this, or will you have to incubate a new business to do your cold storage and transportation?

15. **Value-added products.** Making value-added meat products can be very expensive. Not only do they usually require specialized equipment, individual HACCP plans for each process, and significantly more labor, but the price people expect to pay for those products might not cover those added costs. For example, consumers are used to paying very little for hot dogs, a product that comes from scrap in the big meatpacking facilities. In a small plant, a buffalo-chopper may be expensive relative to the volume of meat run through it. Hot dogs are labor-intensive too; thus many small plants choose not to do them. This is true for a smokehouse as well: According to small-plant owner Greg Gunthorp, it takes one or two people running it all the time. You don't turn on a smokehouse for a couple of hours to run a batch of 20 pounds of bacon. You need to fill that smokehouse to its capacity and run it all day, with more like 400 pounds of bacon. That may take 20

pigs' worth of bellies to fill the smokehouse properly and run it to capacity. Do you have that many pigs?

❦ ❦ ❦

Examples of several unique on-farm abattoirs that appear to be on the track to financial success include:

White Oak Pastures (Bluffton, Georgia)

White Oak Pastures includes two relatively new plants, one for red meat and the other for poultry. The red meat plant processes five species (cattle, pigs, goats, sheep, and rabbits) while the poultry plant also does five species (chickens, turkeys, geese, ducks, and guinea fowl) with around 110 employees and growing. White Oak Pastures (WOP) meat plants are located on a fifth-generation farm owned by the Harris family. The Harrises were conventional cattle people for nearly 150 years. Then fifth-generation son Will Harris decided to switch to a grassfed beef operation and began direct-marketing some of his meat. He started out using outside processors at first, but there was always a glitch—quality issues, scheduling, following cutting instructions, or something else. Will even offered to finance his local processor to expand in order to be able to process more of Will's animals, but the processor was not interested. Consequently, after a lot of deliberation and planning, Will Harris opened a full-service, USDA-inspected red meat slaughterhouse on his farm. He added a poultry plant a couple of years ago to diversify even further. It should be noted that the poultry plant is a fully inspected USDA facility, well beyond the twenty-thousand-bird federal exemption level, therefore the equipment and infrastructure was more costly. When interviewed in 2014 WOP was processing five thousand birds a week of various species in the poultry plant. They may increase that level in the future.

In addition to slaughtering and processing ten different animal species, WOP has a retail butcher shop, a restaurant, a vegetable CSA, and is starting an agritourism initiative. WOP raises the majority of their own animals, but they also are helping over a dozen local farmers market their meats under the WOP label (following strict protocols). All of WOP meat is Animal Welfare Approved, with the beef and lamb certified as 100 percent grassfed too. The land itself is certified organic, along with all the vegetables they also produce for their CSA. Will participates in three other third-party certifications as well.

Will says the only reason he was able to pull off the construction of these two abattoirs was their family's strong financial position after 150 years of farming. He was able to access credit through a local bank, obtain a Whole Foods loan, and draw on personal savings accumulated from good business practices. It was not easy at first, as he states: "We hemorrhaged money the first one to two years while getting to our breakeven point of about 100 head of beef a week." He now sells 90 percent of his meat through wholesale channels such as Whole Foods stores and food service companies and the rest direct via his retail store, chef-direct, and Internet sales. Will's two daughters have joined him as the sixth generation to be involved in the farm, now handling the sales and building the agritourism component to the farm.

For cold-chain logistics and transportation, WOP owns refrigerated trucks for regional delivery. WOP also excels in their commitment to sustainability. They compost all solid wastes on site and use an anaerobic digester for liquid wastes and blood from the processing plants. This is one of the advantages of on-farm abattoirs—they can close the waste gap by getting the nutrient-rich slaughter waste back onto the fields that grow the forage that feeds the animals.

Gunthorp Farms (La Grange, Indiana)

The Gunthorp family operates a poultry and swine plant, one of the only facilities in the country that does both. They kill birds in the morning and pigs in the afternoon. This requires a considerable cleanup midday during the one-hour lunch break to switch from poultry to pigs, especially because their pathogens are similar. They have HACCP plans for slaughter, cut and wrap, ground meats, partially cooked meat, rendered fats, ready-to-eat meats, cured meats, and smoked meats. They use real hickory wood for all their smoking.

The Gunthorps started with a custom plant for 3 years and then went on to get USDA inspection to broaden their marketplace. For over 10 years they have been under USDA inspection with no problems thus far—it helps that Greg has a photographic memory of every USDA FSIS regulatory code! The Gunthorps process mainly their own animals but will also do some processing for other farmers. Especially popular is their real bacon curing and smoking process; they are one of the few USDA processors in their region that will do it. Gunthorp Farms largely processes to order fresh for restaurants, but does sell some frozen meat in the winter. Main sales channels are restaurants, food co-ops, home delivery services, and they have a mobile BBQ that they take to a few special events and food fairs. Their main core value is raising animals on pasture.

Jamison Farm (Latrobe, Pennsylvania)

The Jamisons started their farm in the 1970s and bought the slaughter plant located 5 miles from their farm in 1994. Technically it is not an on-farm plant, but it is a farmer-owned one. The plant had been a downer cow slaughterhouse that the owner did not want to run anymore. So the owner financed John and Sukey Jamison to take over the plant until they were at the point where they could qualify for a bank loan. To build a plant of that size today the Jamisons estimate would cost around $1 million. The capacity of the slaughter plant is around sixty to seventy lambs a day or fifteen steers. Buying the plant and doing their own meat fabrication allowed the Jamisons to work better with chefs by cutting to their specifications, offering fresh meat, and controlling for quality. It also allows the Jamisons to make better use of the whole animal. What the chefs don't typically want (trimmings, grind, and bones) unless they are buying whole animals, Jamison Farm makes into sausages, stews, lamb pies, and stock. John oversees the animal slaughter and Sukey oversees the butchering side of the business. In fact, she does a lot of the value-added products herself. Having the commercial kitchen at the plant allows them to do these value-added things that otherwise they could not do if they had to use an outside USDA-inspected processor.

Humane animal handling is important to the Jamisons. They bring five lambs at a time into the kill chute so the animals aren't stressed. They try to keep the whole process as calm as possible. They also custom-kill cattle for other farmers in their region in order to even out the volume and throughput in their plant. Jamison Farm slaughters between three thousand and six thousand lambs a year, either ones that they produced or

Northstar Bison

Northstar Bison of Rice Lake, Wisconsin, got its start 20 years ago when Lee and Mary Graese got their first two bison. Coming from a background running a nutritional products company, the Graeses were keenly interested in the health benefits of grassfed meats, especially the low-fat nature of bison. Having recently purchased a farm, they had to find a farming pursuit that would actually pay the mortgage. Bison (also known as buffalo) meat was beginning to be recognized in many places as a sweet, lower-fat alternative to beef. The Graeses quickly built their herd to meet that increasing demand and generate income for their family.

Like many animal producers, they worked with an outside meat processor for many years while they built their business and established a brand. The wilder nature of bison proved to be difficult to round up and transport to the slaughterhouse; the stress they exhibited during the slaughter process was not good for meat quality and tenderness. The state of Wisconsin actually allows the Graeses to field-kill their bison in the presence of an inspector, but they have to get the carcass on the rail at the processing plant within an hour. This short timeline was stressful and unattainable most times when they were transporting the meat to a processor. Due to a number of factors, including their desire to scale up and their processor not being willing to fit in more of their animals, as well as their desire to make better use of the whole bison carcass, the Graeses went on the search to find a meat plant to purchase themselves. Via word of mouth they found a plant in Conrath, Wisconsin, that was shortly going to close its doors if a buyer did not come along. Even though it is an hour away from their home farm, they snatched it up for a fair price. The control of the processing has transformed their business in many positive ways.

Since Lee had always accompanied his animals to the processor and even helped with the butchery, he learned the butchery trade without having to go to school for it or apprentice for years. This made the transition into operating a meat-processing plant that much easier for them. They also kept on the most skilled and committed staff that were already working at the plant, so it was relatively straightforward to get things back up and running. The Graeses have made a few changes to the plant, particularly after they received their Animal Welfare Approved audit, which suggested a couple of minor changes in the knock box and holding pens for other animals that they process such as pigs and lambs. But in general the plant was in good working order. The Animal Welfare Approved staff person who conducted the audit found the field-kill method of harvest for the bison to be superior to other methods of slaughter, stating, "This was my first experience with field harvest of bison and it totally changed my thinking about all other bison harvest experiences. There is no doubt that the meat produced by Northstar Bison will be higher-quality and have less stress associated with the entire process."

Because bison are considered a non-amenable species, they are not required to undergo USDA inspection—it is voluntary. This is why Northstar is able to do field-kill with their bison and also with their elk. All the amenable species go to the plant for slaughter and undergo standard ante- and post-mortem inspection just like all meats. Northstar Bison uses state inspection by the Wisconsin Department of Agriculture Trade and Consumer Protection, Division of Food Safety for the field-kill process and their processing plant. The Conrath plant has land in pastures next to the buildings, so they transport the bison and elk there in advance of the slaughter and let them acclimatize for several weeks. Then they field-kill them and deliver them straight inside for evisceration.

Owning the plant allows the Graese family to process animals weekly and sell fresh and frozen meats to a wide variety of customers. They have grown to become the largest distributors of grassfed bison in the country and are now supporting several other grassfed bison producers in their region. They can also take advantage of the many "extra" parts of the bison that would otherwise go to the renderer or pet food trade. They make edible tallow and soap out of the extra fat, send the bison hides off to be tanned and turned into beautiful leather products, and even use the bison fur, or "down" as they call it, to spin into yarn and turn into gorgeous scarves and other knitwear. This has added an average of $150 in profit to each head of bison, a significant amount for a tight-margin business.

---

ones they bought from neighboring farmers. After years of perfecting rotational grazing on their land, they have come to realize that they can't raise enough sheep on their land alone without degrading the pasture. To keep the stocking density down, the Jamisons also buy in weanlings and finish them for a couple months on their grass instead of having to care for as many ewes over the winter.

What About Mobile Processing Units?

There is a lot of interest around the country in mobile slaughtering and mobile processing units (called MSUs or MPUs) for both red meat species and poultry. An MSU is for red meat species and can only do slaughter, scalding and skinning, and evisceration to a hanging carcass, but then the carcass has to quickly be transported to a fabrication facility nearby. An MPU is for poultry species (and sometimes rabbit too) and can slaughter and do some meat fabrication as well because the size of the animals is much smaller and they don't require walk-in coolers or rails, which take up a lot of space on a red-meat MSU. MPUs for poultry have been much easier to get off the ground: They are significantly cheaper and smaller than MSUs, and usually don't require a docking station (that is, power, water, septic, and concrete pad) at the farms they travel to.

The reasons some people advocate for mobile slaughter is that some producers want to slaughter animals on their own farm to reduce animal stress, ease transportation logistics, and exhibit more "control" over the animal harvest in some way without opening their own facility. Others feel that they are too far away from the nearest inspected slaughterhouse or that the ones that are close to them don't provide the quality of service the farmer is looking for. This is all relative, of course. One farmer may feel that a 2-hour drive is too much, while others make an 8-hour round trip without too much grumbling if the price is right and the service good. Undoubtedly, there are places in this country that are deficient in slaughterhouses and very remote, such as island communities, which could benefit from mobile processing. One of the most successful MSUs in the country resides in the San Juan Islands of Washington (Island Grown Farmers Cooperative). It works because farmers own it and are committed to it, and the only other option is to ferry live animals to the mainland for slaughter—a very expensive and stressful option for most folks and their animals. Similarly, an MPU for poultry is off to a good start on Martha's Vineyard, an upscale island community off the coast of Massachusetts with no existing slaughter facilities. MSUs may also work in regions where high land prices and wealthy residents take a NIMBY attitude toward building a slaughterhouse. Places like the Hudson Valley of New York or around the Puget Sound and Seattle area of Washington are an example of where an MSU might work more easily than a brick-and-mortar slaughterhouse.

An MSU or MPU can also be used to test out a market before building a stationary facility. Farmers often outgrow them after a few years as they scale up their operations to run more animals. The cost per animal can also be quite high when compared to a fixed building that has higher throughput. The higher prices may work for somebody doing a few animals a year, but once you make your operation a full-fledged business where cost control is essential, high slaughter costs usually don't work. If a cheaper facility is a couple of hours away and you can take four times the number of animals to them, that might be a better option.

For animals that are highly stressed during handling and transport, mobile processing probably makes the most sense. Range-reared cattle that spend little time around humans, primitive sheep breeds, bison, elk, deer, and wild boar may all be hard to round up, send through chutes, and load onto a trailer. Of course this depends a lot on how often they are handled and whether the breeding program is selecting for calmer animals (or not). While loaded onto a livestock trailer, wilder animals may fight or even cause damage to other animals with their horns, antlers, or tusks. This damage will translate to damaged meat too. If you go the custom-exempt sales route, you can field-harvest many of these species using a regular custom mobile slaughter truck or, if available, you can use a USDA-inspected MSU. The state of Texas actually allows field harvesting of wild game by rifle with an inspector in the truck (Broken Arrow Ranch pioneered this). After the kill, they haul the carcass to a mobile slaughter truck where they eviscerate and cut the animal in half. Once it's cleaned and hanging on a rail, they take the hanging carcasses to a USDA-inspected plant for further processing. If you raise wild game or wild-type breeds, mobile processing may be to your advantage.

States may require extensive infrastructure just to have the MSU located on your farm on slaughter day. They may require a concrete pad for parking the MSU, a dedicated septic system, potable (tested) cold and hot water, 220-volt power, and a flush-toilet bathroom for the inspector. They may also require you to pay a rendering company to haul off all of the viscera and blood.

All of these costs add up. Poultry MPUs seem to have an advantage in that they are much less expensive to build and often don't require any special infrastructure on a farm. However, if your state or county does not allow on-farm composting of offal, blood, and feathers, you may have to pay a rendering service to come and haul those materials away. That, of course, is a shame, since those materials are valuable fertility sources. From a biosecurity standpoint, however, they can transmit pathogens such as salmonella or campylobacter to your poultry pastures unless they are properly composted (Trimble et al., 2013). It is important for anyone who is on-farm poultry processing to consider how you will handle the waste and compost it properly.

CHAPTER 11
Packaging, Labeling, and Cold Storage

Now that you have thought about the logistics of meat processing, you have to consider the logistics of packaging, cold storage, and whether or not you will deal in fresh or frozen meat or both.

Packaging and Labeling

What may seem like a minor detail—how your meat gets packaged and labeled—is actually one of the most important aspects for maintaining meat quality, keeping in line with the law, and increasing your sales. Well-packaged meat will last for a long time, especially if deeply frozen. We have seen all kinds of timetables published about how long properly packaged frozen meat will last. Some say 6 months, others say 12. Yet we've had pork roasts that were frozen for 2 years and suffered no reduction in quality. Ideally you will be selling through your meat inventory much faster than that. But if you have a seasonal production model in which you slaughter animals for 6 months out of the year and then sell frozen meat for the other 6 months, you will want your product to last at least 6 months in cold storage. If the consumer is going to take that meat home and store it for another few months, your meat packaging will need to hold up for even longer.

Well-packaged meat is more attractive for customers and helps them to better trust the product. It maintains a higher level of food safety too, which is absolutely paramount to being in the meat business. Properly sealed packaged meat is easier to thaw out correctly because it won't leak, and the meat will taste better as a result. Good labeling is not only required to sell meat, but will help differentiate your product and encourage sales. If you don't package your meat for retail sales and instead sell boxed subprimals to restaurants or custom halves and wholes, the labeling is less important. However, to be in line with the regulations you still have some basic labeling requirements.

Packaging

The packaging of your meat needs to meet several standards. These include:

1. **Airtightness.** To prevent outside air from entering and to hold in the meat moisture in order to prevent freezer burn and rancidity of meat.
2. **Oxygen-free environment.** Free of as much air as possible within the package itself, which will prevent oxidation and ice crystals from forming.
3. **Durable.** So as to keep from puncturing or tearing to the extent possible, withstand some handling, and prevent transmission of juices or flavors from other frozen foods stored in the same space.
4. **Food-grade material.** Must be able to touch the meat without imparting any odors or flavors to it, and be able to withstand freezing and warming up when thawing out without changing chemically.
5. **Affordable.** Packaging can't be so expensive that it reduces your profitability or costs too much for the consumer. Likewise it shouldn't be too labor-intensive for the butchers, such as having to insert a bone guard on every bone-in cut. Your processor will likely charge more for that extra work.
6. **Labeled properly.** Your label has to be able to stick to the package even while it's sitting in boxes with other meat or in coolers, being handled to some extent. If you are packaging your own meat, such as with exempt poultry, make sure your labels will stick to the slippery plastic package and the ink will not smear if the labels get a little wet. All packaged meat for sale has to be labeled in some way. Custom-processed meat in which the buyer of the animal pays for the cut and wrap can get away with a basic stamp or handwritten label of what the meat is, along with a "not for sale" phrase on it. More on this later.

Types of Meat Packaging

The main kinds of meat packaging offered up in US cut-and-wrap facilities are vacuum sealing, shrink wrapping, and wax-lined butcher paper or freezer paper. There are advantages and disadvantages to each method. You may need to experiment on a small scale to find out which material works best for upholding your standards and meeting consumer expectations. Also, ask your customers what kind of packaging they prefer.

Vacuum Sealing

This is the best option, as it pulls out the air, creates a tight seal, and allows the meat to be seen by the potential buyer because it is transparent. Bones may puncture and ruin the seal, therefore you may have to use bone guards. Frequent rough handling, like customers rifling through meat coolers at the farmers market, may also break or puncture the seal. The meat will rapidly develop freezer burn after that and become unsellable. We ate a lot of ruined vacuum-sealed meat, but on the other hand, it gave us an excuse to eat and get to know our meat really well. We also offered up this meat for a discount to customers who were going to use it right away, or we used it to trade for produce, cheese, and other things we didn't grow ourselves.

Shrink Wrapping

Shrink wrapping is economical and works well for whole carcasses like poultry, but can puncture more easily unless you use two layers and does not always present favorably, especially on portioned cuts. Shrink wrapping is achieved by wrapping the meat in one layer, then dipping in hot water, then wrapping with another layer and dipping in hot water again. That outer layer has a tiny puncture hole in it to suck the air out. Usually the label is placed over that hole. If customers are going to thaw out the meat in water, the hole will let in water between the two layers. Shrink wrapping is also a more labor-intensive process because of the two layers and two hot water dips, so it may cost more at the processor. Poultry shrink wrapping uses a single Cryovac bag that is then dipped briefly in hot water. It pulls most of the air out and presents really well. Use a thick enough Cryovac bag to reduce the possibility of puncturing.

Paper Wrapping

Paper wrapping uses wax-lined butcher paper, also called freezer paper, along with special tape that can withstand freezing temperatures while still maintaining its seal. Usually there are two layers of paper, the first one pulled really tightly up against the meat. Sometimes the inner layer is a plastic cling-wrap film and the outer layer is paper. An all-paper wrap can present a challenge to creating an airtight package. It is more difficult for inexperienced processing employees to do it well, especially when they are pressed for time. Also, because the paper is opaque, it is better for non-retail portioned cuts as when custom-processing halves and wholes. Thawing out can be more challenging with paper wrap because fluids can leak and the package can't be placed in water to thaw out (as some customers like to do). But it is cheaper, and many farmers have no problem selling USDA-inspected meat in paper-wrapped packaging.

All packaging materials must come with a letter of guarantee that they are food-grade. The processor has to keep these letters on file in case an inspector wants to see them. If you have any doubts about a packaging material, you can ask to see those letters yourself. Although it is the processor's responsibility to verify that the material is food-grade, you may just want to double-check yourself and get a copy for your records. You never know when a customer is going to ask about a material—people especially have concerns about plasticizers. Sometimes you might have to customize your packaging, such as using a thicker-grade plastic for vacuum sealing because you know the wear and tear your meat packages go through or when a retail store makes a certain requirement. Customized packaging materials will likely cost more, both in unit terms and potentially in labor costs at the processor.

All labels should be slapped on the meat packages before they are frozen or go into boxes. Make sure the labels are made from a durable material that can withstand freezing, thawing, and condensation, such as a polyester- or vinyl-based material. If you have deep pockets or are at a large enough scale, you could consider investing in preprinted packages, such as you often see with ground beef. This is a very presentable packaging and the labels won't come off—a good idea for those focusing on retail sales.

Boxing

Placing your meat into boxes will also be an additional charge. A 40- to 50-pound capacity box, which is what fabricated meat is typically put in, might cost an additional $2 to $3 per box (this includes some labor time too). You are not allowed to reuse boxes according to USDA FSIS regulations, so you have to pay for brand new boxes each time. However, some farmers drop off their own plastic tote containers or freezer baskets for their processor to place the packaged meat into. This may work for your processor and save you money as they can be reused. Freezer baskets also have the advantage of improving cold air circulation and, thus, producing more uniform freezing results.

Make sure all meat is thoroughly frozen before going into boxes. Your processor will probably put cuts onto sheet trays or into freezer baskets for a day or two first before placing into boxes. You may want to double-check what their standard freezing process is. According to Heritage Meats plant owner Tracy Smarciarz, neither you nor your processor should stack your meat more than three boxes high to ensure good cold airflow. Keep this in mind for your post-processing cold storage too. One beef producer we know lost a bunch of ground beef when his processor boxed and palletized it while it was unfrozen and it never fully froze, even in cold storage. The beef in the center of the boxes where the air was less cold ended up rotting, and the farmer lost thousands of dollars.

Label Claims and Approval Process

Meat labels are a special breed and are much more regulated than other food labels. You can keep it simple and get fairly quick generic approval from the USDA, or you can make all sorts of claims on your label that will have to be individually verified: That process will take much longer. Unless your meat will be sitting packaged on a retail shelf, it is probably best to keep your labels simple. If you direct-market meat where you can talk to your customers, save the label claims for your conversation or a brochure instead of placing them on your meat label. Simple labels will be approved much quicker, and they will be cheaper and often printable at the plant itself. If you sell custom meat or meat by the carcass, primal, or subprimal, you don't need to worry about labels at all. The slaughterhouse will stamp the carcass or label the boxes of primals for you. Custom exempt meat packages will just be stamped "Not for Sale."

If you process in a state-inspected facility and not a USDA-inspected one, you will need to submit your label for approval to your state meat and poultry inspection program.

If you are having any processed meat products made for you, all ingredients must be listed on the label. This is especially important if any potential food allergens are in the list of ingredients. Every last ingredient, from the salt to the sodium nitrate, must be listed on the label. If it's just fresh meat, then no ingredients list is required.

As of January 6, 2014, a long list of special statements or claims on your label will now require approval by the USDA FSIS. You must back them up with documentation, usually a narrative description of how you define and meet the claim, along with a signed affidavit. These label claims include:

- those relating to nutrient content, health, or prevention of disease conditions
- those saying "organic" or "containing organic ingredients"
- those regarding the raising of animals (e.g., vegetarian-fed, acorn-fed, antibiotic-free, grassfed, humanely raised, etc.)
- instructions concerning pathogen elimination, such as "cook to 165 degrees"
- statements identifying products as "natural"
- any third-party claims
- breed claims
- "certified-[something]" claims
- gluten-free
- negative or "free-from" claims (hormone-free, antibiotic-free, cage-free, etc.)

- undefined nutrition claims (e.g., "heart-healthy")
- geographic styles (e.g., "Creole-style," "French-style," etc.)
- in the case of poultry, the claim of "free range" or "free roaming" access to the outside must be demonstrated

"Natural" label claims must include a statement about what this means: either no artificial ingredients, no artificial colors, or minimally processed. So if you include this claim on your label, it should say somewhere else on the package what "natural" means.

To obtain meat label approval, go to the USDA FSIS online label submission program, called LSAS (Label Submission and Approval System). You can still submit label applications by paper, but the online program is supposedly faster.

The steps for submission are:

1. First talk to your processor. They may or may not submit your label request themselves. They are ultimately responsible for the label since it has their stamp on it. You cannot submit a label request without having a plant that goes along with it.
2. Set up an LSAS account (or if your plant has their own, they may do all the rest of this for you, usually for a fee).
3. Fill out the application online, supplying all supplementary documentation required.
4. Make a PDF copy of the entire application and supplementary documents for your records.
5. You should receive a confirmation message within 7 days. Make sure you receive it and respond with a confirmation yourself.
6. Check in on the online dashboard to see where your application is in the process.
7. Provide any additional materials that they request in a timely manner. There will almost always be a little back and forth, especially if you have any claims on the label. You may simply be required to provide more documentation to back up your claim.

Getting federal USDA FSIS approval for your label can take from 8 up to 16 weeks, so plan well ahead of time. This used to be really hard if you changed processors, but now (as of January 2014) you can just change the plant stamp and not the label without getting FSIS approval, but your processor must do this for you. If your initial label approval process is taking forever or causing you too much grief, you can actually hire a label expeditor at a cost of $100 to $300 to potentially reduce the wait time. If you plan to make any label claims, such as "all natural" or "grassfed" or "antibiotic-free" your label will take even longer to approve. If you try to make any nutritional claims, those actually have to be supported by lab data. This is very expensive, but may be worth doing once, to figure out if your grassfed beef or pastured pork has higher levels of omega 3s or conjugated linoleic acids (CLAs), for example.

If you have third-party certifications like organic or Animal Welfare Approved, those certifying bodies will also normally have to approve your label if you are placing their logo or name on your label.

Additionally, if you plan to sell branded, packaged meats in retail outlets, you will need to purchase a retail-ready UPC code. These can cost anywhere from $1 to $8 each. Each product will need a different UPC; even different-sized packages need different UPCs (such as a four-hot-dog package versus an eight-hot-dog package). However, if a retailer is going to display your meat fresh in the meat counter or cut and package it themselves, you won't need your own UPC codes.

Most label-printing companies will require an initial art setup fee and a minimum order size of five hundred to one thousand labels at a time. This can be a large investment and it also takes time. Don't start working on your labels just 4 weeks before you plan to take your meat to market.

Lastly, here is the most unfortunate aspect of the whole label-approval and label-making process: Your label goes with the plant. If you decide to change plants, for whatever reason, those labels you paid for don't come with you. The old plant will keep them. You have to get new ones made with the new plant's stamp. That could be leaving a lot of investment on the table. Try to settle in with a processor before you get custom labels made.

Safe Handling Instructions

All retail cuts of red meat and poultry, even exempt poultry, must bear a label with safe-handling instructions on it. This is the *exact text* that is required by USDA FSIS (as of March, 2015):

- Keep refrigerated or frozen. Thaw in refrigerator or microwave.
- Keep raw meat and poultry separate from other foods. Wash working surfaces (including cutting boards), utensils, and hands after touching raw meat or poultry.
- Cook thoroughly.
- Keep hot foods hot. Refrigerate leftovers immediately or discard.

Nutrition Labeling

As of 2011, nutrition labels are now required for all major cuts of single-ingredient raw meat or poultry. The only exemptions to this rule are for poultry processed under the P.L. 90-492 exemption (under twenty thousand birds) and for ground and chopped meats processed in small plants. You can obtain generic nutrition information off of the USDA National Nutrient Database for Standard Reference website to insert into your printed labels or you can get nutritional testing done on your own meat if you think it has a better nutritional profile than the average meat. Many meat processors will have the standard reference nutrient profiles already built into their label-making services. If you want to use the generic information that the USDA has already gathered via testing, ask your meat processor if they can assist you with that. Like all labeling, there is often a one-time setup fee by the meat processor, but once you have done this, you are good for awhile, unless you change processors or decide to get your own nutritional testing done.

Sarver Heritage Farm of West Virginia, a third-generation family farm, decided to go ahead and pay a nutrition lab to analyze a cut of their grassfed beef. They had a hunch their beef was going to have a different nutrient profile from the generic one, and it turned out the differences were stark. For around $800, they got a rib roast analyzed and the paperwork submitted to USDA FSIS for label approval with the East Coast lab Eurofins. On the next page is a summary table of some of the differences they found between their grassfed beef and the standard reference beef.

When we asked the owners of Sarver Heritage Farm, the Doering family, about the decision to

TABLE 11.1. Nutritional Comparison of Sarver Heritage Farm Grassfed Beef vs. Standard Grainfed Beef

Nutritional parameter on a 4-ounce serving	Sarver rib roast	Commodity rib roast (according to USDA testing)
Calories	190	360
Calories from fat	90	280
Saturated fat (grams)	5	13
Unsaturated fats (grams)—both mono and poly	3.5	Not tested for
Vitamin A (of daily value)	2%	0%
Omega 3: omega 6 fatty acid ratio	1: 1.57	1: 10 (not tested for, this is a normal ratio for corn-fed beef)

do their own nutritional testing, they had the following insights to share: "Having nutrition labels for our grassfed beef does provide a direct comparison between what appear to be two completely different products [as evidenced by the table above]. We chose to have the testing done, mainly, because we were not comfortable with using the FSIS-provided label, as it did not represent our product. We were leery, prior to getting the testing done, to market the 'health' benefits of grassfed beef. We let others make those claims or referred folks to [the] Eat Wild [website]. Having the testing done allows us to speak confidently about our product and present its nutritional value honestly. Customers seem surprised at the differences."

It would seem from the Doerings' experience that paying to have your own meat nutrition testing done is a worthy expense that will help you to differentiate your product. Keep in mind that you cannot put this label on every muscle cut unless you get each cut analyzed. Yet the Doerings can put the label on their rib eyes and rib roasts since that is where the testing was done, and they also display a copy of this on their website. Apart from the meat label itself, you could create point-of-purchase marketing materials with this nutrition analysis to display near your meat, such as a brochure, a printout, or a laminated poster for all to see.

For ground and chopped meats processed in small plants, you only have to state the lean-to-fat ratio of the meat but not the full nutrition content on the label. The lean/fat ratio must be phrased like this: "80% lean/20% fat" or "80% lean to 20% fat." You cannot just write "80/20" on the label because consumers may not know what you are talking about.

Fresh vs. Frozen Meat

While particular markets are looking for fresh meat, many direct markets are just fine with frozen meat. Some meat processors will freeze all packaged cuts immediately, while others are willing to store meat fresh-chilled if the producer picks it up right away. Farmers that sell to restaurants, food service companies, and meat counters will usually get their meat cut into primals or subprimals, bagged, and boxed for fresh delivery that week, unless some of it is being dry-aged. You will have to decide for yourself what makes the most sense for your markets and your access to meat processors.

In our particular case three reasons dictated that we worked with only frozen meat:

1. our customers preferred frozen meat because they could safely get it home without it getting warm;
2. frozen meat worked better for farmers market sales because it would stay properly chilled over the course of the 4- to 5-hour market; and
3. our slaughterhouse was located too far away to consider fresh meat sales.

We weren't going to take three animals a week on an 8-hour round trip (actually it would be double the drive when factoring in the return trip to pick up the packaged meat). The transportation and labor costs per head would have been too high to justify fresh meat sales. Obviously, if you scaled up such that you had orders for a dozen or more animals each week, then a weekly trip to the processor might make more sense. What are some of the issues and costs you need to factor in when thinking about fresh meat versus frozen?

Inventory Management

Adapted from the University of Maryland's "Meat Marketing Planner" guide (an excellent resource by the way): Develop a "first in, first out" (FIFO) system for your meats. Not even frozen meat has an indefinite shelf life. Have guidelines on how long you will retain the product before getting rid of it through sales or donations; make the guidelines clear to your employees too. Having some organization to your inventory will also assist in packaging and order fulfillment. Knowing which cuts sell the quickest will help you develop your cut sheet orders to match consumer demand more accurately.

If you don't have the processing date written on your meat packages, how will you know which are the oldest so you can practice FIFO? You could group them into freezer crates or boxes with the processing date written on the side. Or use a dry erase marker and write it directly onto the side of the freezer. If you do print the processing date on your label, keep in mind that this may affect purchases. Even though meat that is 6 to 12 months old is perfectly fine if frozen well, consumers may deem it too old and not want to buy it. We used to print the processing date on our meat labels until too many people got turned off by meat that was as little as 3 months old. After that we took the processing date off the labels and just stuck to a good FIFO inventory system with dates written on the side of the boxes, not the packages themselves. Anything older than 12 months we ate ourselves, even though the quality was perfectly fine. You never know how long a customer is going to store the meat in their freezers after they buy it from you.

Cold Storage Considerations

Frozen meat is easier to deal with in many ways, but it requires good freezer storage. You might have a few options in that department depending on your scale:

1. buy a bunch of large, energy-efficient chest freezers;
2. bring a box freezer or container freezer onto your property or build your own walk-in freezer on the farm; or
3. rent freezer space in town.

Some states require that on-farm freezer storage be inspected annually, often with a small permitting or licensing fee. Probably the best place to ask is your state Department of Agriculture meat or poultry division. No need to panic about this: As long as your freezers are in working order, kept clean, and inside some sort of building, you should be fine. We would wipe down the outside of our freezers once a month and empty, defrost, and clean the insides once a year. Probably a bigger concern is walk-in-type freezers that could harbor pests such as rodents.

If you are dealing in small quantities of meat, you can probably get away with a couple of chest freezers. We purchased three 6-foot-long Energy Star–rated chest freezers for around $800 each that held up for 5 years, at which time we sold them for $500 used. They sipped electricity—our bill went up only $10 a month—so it was affordable.

Once you scale up beyond the chest freezers you have a couple of options: rent cold storage space nearby (for many farmers "nearby" may be 1 or 2 hours away) or build your own. Many farmers purchase used freezer container boxes, haul them to their farms, wire them up, and use them. Costs run between $15,000 and $18,000 for a 40-foot-long freezer container delivered and wired up or $5,000 to $8,000 for a 10 foot by 10 foot walk-in. Twenty-foot-long containers, which cost between $8,000 and $10,000, are more ideal but are really hard to come by. All freezer containers typically require three-phase power, which you may have to pay to have installed on your property if it's not already there. Some folks have installed electronic inverters that convert the three-phase power from the unit to standard one-phase at the electric junction box: That will run you another $1,000 or so to purchase. Freezer units can be huge energy hogs too, and the compressors are notorious for failure. The repair costs can be frightening. You could also build a super-insulated box yourself and then install a freezer compressor on it. Several farmers have done this and have been very pleased with the results and with the lower costs—provided you are a good carpenter and you use some scrap materials. If you had to pay somebody to build it from brand-new materials, it might not be any cheaper than bringing in a container unit. If that kind of investment is too much for you right now, in many parts of the country you can lease a reefer (refrigerator) container instead.

When storing your freezer container outside, keep in mind that humidity may become an issue. You might have to install a dehumidifier or air exchange unit on it so condensation does not build up inside. For all freezers, consider owning a backup generator in case of power failures. You don't want to lose thousands of dollars' worth of meat during a 5-day power outage.

Having an on-farm walk-in freezer gives you much more flexibility as a meat producer. You can sort inside the cooler, build mixed boxes, pack special orders, make CSA boxes, and check inventory. When renting cold-storage space elsewhere, often you won't have the same options for sorting through your inventory. Luckily the cold-storage facility we rented space in would allow us to pick off of our pallets when we went in once a week. We were required to call ahead of time, and they charged us a small $3 "picking fee," which covered their cost of an employee opening up the door and monitoring that we picked from our pallet. When we needed to do more extensive sorting and boxing, they would let us pull out a bunch of boxes onto a rolling cart and bring them over to some stainless steel tables they weren't using. We could take our time sorting and making mixed boxes for a buying club we had partnered with. When we were done, we would load the remaining meat boxes back onto the cart and the employee would once again open up the rolling freezer doors for us to reload the pallet.

Not all cold-storage places are as affordable or amenable to the needs of direct-market meat producers as the one we rented in California. In many parts of the country cold-storage rental space just does not exist. You may have to dig a little to find it: Try calling fish processors, fruit processors, meat processors, or even food banks in your region to see if they have underutilized freezer space. Maybe they have never thought of renting out space and would be open to making a little money from some redundant space they have. Just be clear about your needs, how often you will be entering and exiting the freezers, how much space you need, what temperature you need it kept at, food safety considerations, etc. Your rented cold-storage space may also need to be inspected by your county or state health department in order for you to obtain a license to store meat. Ask if that would be a problem for the folks you are talking to. Make sure you add your meat-storage facility onto your general liability insurance policy too.

You should consider the power usage and energy efficiency of your freezer choices. We once met a farmer who was spending over $1,000 a month on power for two large walk-in freezer units that were underutilized and had big air gaps around the doors. That is probably on the high side: Others have said that they pay around $200 a month in additional energy costs for their freezer units. In comparison, we spent just $30 a pallet each month for walk-in freezer space at a nearby fish-processing facility that was just 10 minutes down the road. Since it was a rental space, we had no capital or maintenance costs for it. Spending $30 a month versus $1,000 a month is huge (that doesn't even factor in depreciation and repair costs for the $1,000 a month freezer units). You can see how these kinds of costs add up quickly. If building your own, add more insulation than you originally planned and extra sealing around the doors.

In doing research for this book we found that costs for freezer space rental varied between $15 and $40 a month for a pallet. If it's convenient, renting is probably much cheaper than purchasing or building your own walk-in freezer, factoring in things like repairs, depreciation, and electricity. There is a new nonprofit model in New York called the Meat Locker (described in the accompanying profile) that is creating a novel system for cold storage. While it is designed to provide freezer storage for customers who buy meat in bulk direct from farmers, local farmers can also rent space if it is available.

Frozen meat sold at farmers markets, roadside stands, and butcher shops needs to be kept at 20 degrees F (–7 degrees C) or less. Fresh chilled meat must be kept between 26 and 41 degrees F (between –3 and 5 degrees C). If it drops below 26 degrees F, it is no longer considered fresh but rather frozen. Although these are the typical temperature ranges, check with your state or farmers market management for any specific temperature rules they may have. Keep a working thermometer in at least one of your coolers so you can periodically check the temps throughout the day at market. Although when we sold at farmers markets we would often forget to bring the thermometer, the one and only time a health inspector spot-checked our meat coolers we happened to have a thermometer with us that day, thankfully, and we were within the correct temperature range.

What about fresh meat sales? For cold storage, not frozen storage, you may consider installing a Coolbot, which is a low-cost refrigeration unit that can bring a well-insulated space down

The Meat Locker

The Meat Locker and Meat Suite are two intertwined projects developed and run by Cornell University Cooperative Extension in the Finger Lakes region of west-central New York. They were developed to help farmers find more consumers and develop the freezer trade market for custom quarters, halves, and whole animals. Cornell surveyed over two hundred consumers and forty farmers, and they found that freezer meat was the most profitable market channel for farmers and created the most affordable meat for consumers: a classic win–win market. However, most surveyed consumers did not have chest freezers nor a desire to invest in them. For students and apartment dwellers, it simply was not an option.

The first component, Meat Suite, is a simple farmer–consumer linking website where you can find central New York regional farmers producing the kinds of meat you may be looking for. You plug in the distance range you are willing to consider from your zip code and the meat you are looking for. A number of farms will pop up, each one describing the species they raise, the practices they use, their location and contact info, and their pricing by the (hot carcass weight)/pound. The website also explains how buying meat in bulk works, along with definitions for terms such as hot carcass weight, final weight, dry aging, and so on. Once you have identified a farmer you want to buy from and have made the arrangements for slaughter and butchering, the next component is cold storage.

The Meat Locker is the cold-storage component of this freezer trade arrangement. Currently, it comprises a 10 foot by 14 foot walk-in freezer housed within a semi-industrial building in Ithaca. The freezer is open 2 days a week for consumers to obtain their meat out of freezer bins. There are two sizes of plastic bins they can rent by the month: a 25-gallon bin for $5 or an 18-gallon bin for $3. When you show up, a Cornell employee fetches your bin and brings it out. You can fish out whatever meat you want to take home for the week or month—you can come to the Meat Locker as often as you choose. Any kind of meat can go into the bins, including personally hunted game, custom-slaughtered animals, or USDA-inspected meat. If space allows, farmers will also be able to store their meat in the freezer as well. Up to seventy spaces are available.

To begin this project, the prices were kept at affordable levels because Cornell wanted to prove the concept before considering raising prices or opening other spaces. They also received a USDA grant for the initial costs, around $20,000, which included a fire-suppression system and a new electrical panel in addition to the freezer itself. It is debatable if this could be done as a for-profit model in other places, but it is certainly something that nonprofits could do with minimal grant funding. A group of farmers could also cooperate to build a similar facility or a butcher shop could do the same.

to 35 degrees F (2 degrees C). For infrequent refrigerated storage, such as your weekly poultry slaughter that you take fresh to market, this might be a good option. For diversified farms, Coolbots also work well for produce cooling.

Running fresh meat logistics and dealing with all those buyer orders can become another part-time job. At least with whole carcass restaurant or butcher shop sales you don't have to get the meat cut up, so maybe you can eliminate one trip to the processor if your meat does not have to hang for long at the slaughterhouse or if the restaurant can do aging itself. The farmers that have been most successful with restaurant sales either have on-farm abattoirs (for example, Gunthorp Farm, White Oak Pastures, and Jamison Farm) or they are able to move large numbers of animals each week making the transportation costs lower (like Carman Ranch does). They all have full-time sales staffs as well. These are certainly factors to consider for fresh meat markets.

Ice

In addition to cold storage, ice for keeping meat cold during processing and marketing can also become a significant expense. If you are processing poultry on your farm, you will need ice to cool down the carcasses. If you are transporting fresh or frozen meat without a refrigeration unit, you will need some sort of ice. If you are displaying your meat at a farmers market, fair, or other event, you will also need a lot of ice. You have a few options that vary considerably in price. You can buy an ice machine: These can often be purchased used from restaurant supply warehouses. A reliable one could cost from $1,500 to $3,500. Then you have to power it and supply it with water, both of which cost money. Ice must come from potable water, so you may have to install a water filter to get it to a potable level. Your state or county Department of Public Health may even require that you get your ice water tested. You also may have the option of buying ice from town, either in small bags from a grocery store or in bulk from ice companies. This of course depends on where you live and how far you are willing to drive.

After buying ice in small bags from our nearest mini-mart and paying retail prices for our weekly poultry processing and market ice, we decided to save some money by trying the bulk ice. Paying retail for bagged ice is a great way to enrich the grocery store or mini-mart and deplete your pockets. We were spending around $320 a month on bagged ice. So we cut down on the number of bags by buying potable water ice just for the chill tanks and began buying bulk ice for the coolers from an ice company that supplied the vast Salinas Valley produce industry. We would position two large plastic totes in the back of our pickup truck and pull under a chute that dumped ice into the totes. We then brought the bulk ice home and used it on poultry processing days to fill our coolers that we packed fresh whole birds into for market days. Although it didn't cost very much, it did require an extra 1-hour trip into town and dealing with large totes of ice was logistically challenging without the right equipment. We know of another farming family that spends $600 a month for ice to fill their coolers for farmers market sales. We are sure they could save money if they found a different solution.

You also need ice to make a cold-water bath for small animal chilling like poultry or rabbits: approximately 1 pound of ice per 2 pounds of meat. If you are processing your own poultry under exemption, you need to factor that ice cost into your equation—it can add up. Air-chilling poultry may be a great way to save money on ice.

We discovered that you could keep meat cold by freezing plastic water bottles and placing the packaged frozen meat on top of the frozen bottles. Once a year we bought water bottles and threw them in our chest freezer. On market days we would pull out a bunch of them to line the bottom of the coolers. When we returned from market we plopped them back into the freezers. Every once in a while we gave out a frozen bottle to a customer who was going to have the meat sitting in the car for a period of time. We probably had to buy two cases of bottles a year. This new system cost us around $50 a year, as compared to the roughly $2,000 a year we were previously spending on ice just for markets. That was a really good cost savings for us. We have seen other farmers use a similar practice with refreezable gel packs in lieu of ice. The only difference is those gel packs are expensive and you can't drink them on a hot day, nor will you want to freely give them out to your customers (as you could with water bottles) due to their cost.

CHAPTER 12
Principled Marketing

Marketing is truly the heart and soul of the new meat supply chain. You may, like many people, have an aversion to the word "marketing," so try to think of it more as building relationships of mutual trust and cooperation. To advance from being a good livestock or poultry farmer to running an actual meat business, you have to become an expert marketer as well. This road can sometimes be rocky, but we're sure you are used to that in the world of farming.

Basic principles of marketing apply to every business; a meat business is no different. We will draw upon those principles in the context of a farmer-driven, ethical meat business. But some marketing tenets are unique to selling meat, a perishable food product that requires a high degree of trust on the part of the buyer. Trust is a nuanced emotion that marketing may or may not draw out. The number one reason people will do business with you is trust. You can build a trusting relationship by sticking to the following principles:

Commitment. Keep your promises. Offer guarantees to demonstrate that commitment to your customers. This will also reduce their hesitation or perceived risk in making that first purchase. If you are in the good meat business and sell only quality product, you will almost never need to provide money back.

Professionalism. Be professional and always honest. Dress professionally too. That doesn't mean you can't wear the cute wranglers and cowboy hat (people often *love* that look) but take off the manure-caked muck boots and your stained, ripped jeans. Also, clean your truck, van, or trailer when you are delivering meat; that's part of your professional look too, and helps demonstrate your commitment to food safety. Join and announce your memberships in professional organizations, such as the American Grassfed Association, American Pastured Poultry Producers Association, your local Chamber of Commerce, and the Better Business Bureau. This shows that you are striving for continuous learning and that you run an honest business. It also shows that you know what you are talking about and have a certain level of expertise.

Know your markets. Research your customer demographics, your sales channels, your competition, and trends in the marketplace. Ideally, don't start with any animals until you have your market in mind and understand the needs and wants of those buyers. For example, don't raise lard-type pigs if your market prefers lean, or raise lean Texas Longhorn cattle if your market likes good marbling. Also, market research is not foolproof. Stated consumer demand and real demand are different. Lower your volume projections and your revenue expectations to begin with until you see how real demand plays out.

Customers have the final say. You may love your product, but ultimately your customers will have the final say. If they don't care for your products and you aren't willing to change, you will not be successful. For example, if you insist on finishing your cattle on grain because you think it tastes better, but the majority of your customers keep asking for a grassfed product, they will eventually move on and seek out another brand that is in line with their values. If you are not customer-centric, then direct marketing meat may not be the business for you. That said, don't chase your tail around customizing products for a small share of customers. If your most loyal customers aren't asking for it, put it on the back burner or don't consider it at all.

Commit to being first in the markets you serve. Try to get there first, build the largest market share first, and have the products that people are seeking first. If you can't do this, look for other places where you might be able to be first in that market or for that particular segment. For example, maybe you can't be "the first" to bring pastured chicken to your marketplace, but maybe you can be first with pastured duck and geese.

Deliver total quality. Offer only quality products. If you ever doubt the quality of a product, keep it for yourself or feed it to your dogs.

An approach to figure out if you are providing "total quality" is to determine where your business is positioned on the three axes of Strategies for Leadership in the diagram below. Will you be a leader in operational excellence, customer intimacy, and performance superiority? Can you be the best in at least one axis and pretty good at the other two? *Operational excellence* is producing the product efficiently, affordably, and—we would add—in a socially and environmentally responsible way that is also convenient enough for the customer. *Customer intimacy* is segmenting and targeting markets precisely and then tailoring offerings to match exactly the demands of those niches. It involves understanding your customers and solving for their needs in a friendly, proactive way. *Performance superiority* is producing a differentiated product that is high-quality, unique, and, in the case of meat, tastes great, making the

Strategies for Leadership

Operational Excellence

Operational Competence

Product Differentiation

Customer Responsiveness

Performance Superiority

Customer Intimacy

FIGURE 12.1. Strategies for Leadership. *Courtesy of the Wharton School*

competition obsolete. Plot where your company is on each axis, then figure out where the competition is on the axes too. Which axis are you the best on? Do you at least meet fair value on the other axes you aren't the best at?

Which of these two examples below do you think will do better in the three strategic leadership categories mentioned above?

> **Meat Business A.** They produce a very tasty, high-Choice grade grassfed beef that is only available in June of every year. Customers must prepay for that meat 1 year in advance and pick up their meat bundles from the farm the fourth weekend in June. Only twenty animals are available each year and the price is four times that of average grocery store meat.
>
> **Meat Business B.** They produce a high-Select/low-Choice grade grassfed beef 7 months out of the year fresh and 5 months out of the year as frozen. Customers can buy in bulk or can purchase by the retail cut. Prices are generally 1.5 times that of grocery store meats. Customers can pick up from the farmstand, which is open 5 days a week year-round, or it can be delivered to their door via a monthly buying club for a small fee.

Business A may score high on performance superiority because they can produce a high-grading grassfed beef. They may be meeting the needs of a select group of their customers, but are probably leaving out a whole lot of other people. They are not scoring high on operational excellence because they can't produce very much for more than 1 month out of the year. This is an extremely niche product, probably too much so. Business B scores higher on customer intimacy because they are addressing a wider range of customers' needs; they are doing well on operational excellence because they can produce more efficiently, more year-round, and are making the product available more conveniently to their customers. While they may not produce "the best" grassfed beef, grading high Select or low Choice is pretty good for grassfed beef. They are differentiating themselves from grainfed beef and from the lower grading grassfed beef that is prevalent.

This is not to say that Business A will not survive; however, they are unlikely to be able to scale up that very niche model and would not compete well if more players moved into their marketplace. If Business A had to compete against Business B, Business B is likely to capture more market share, have a more scalable model, and probably ultimately be more profitable.

Customer-Centricity

What is your philosophy on customers? Are they empty vessels that you fill up with meat in exchange for their money, or are they partners in your business, possibly even friends? There are two dominant views on selling: One is just helping a customer part with their money. This is the normal way of doing business, and this is the "product-centric" approach. The other way is helping to meet a customer's need; this is the "customer-centric" approach (adapted from Moyer, 2003). Put yourself into the customer's shoes: What do they need? Even if you think you understand what their needs are, it's always a good idea to periodically survey some existing and potential customers about what their meat consumption practices are.

To aid you in thinking about this approach of considering your customers' needs, we will discuss the concept of customer-centricity. Modern customers now want and indeed expect full-scale solution providers for everything they buy. Instead of focusing solely on the product and expecting customers to flock to you, a customer-centric business focuses on *fulfilling the needs of their core customers.* Not all customers are created equal, contrary to the whole "the customer is always right" school of thought. Some are more valuable than others. A customer-centric business focuses on those customers with the highest "lifetime value." In the meat business, and indeed many businesses, *80 percent of your business will come from 20 percent of your customers.* Therefore, you can spin your wheels trying to get the other 80 percent of your modest customers to become more loyal to you and buy more meat from you, or you can make the loyal 20 percent even happier. They have the highest "lifetime value," so focus on them. Ways to do this include:

- loyalty programs (rewarding customers for sticking with you)
- referral programs (rewarding customers for telling their friends about you)
- cross-selling (getting the same customers to buy more from you)

Marketing will be a constant, ongoing activity for any direct-market meat producer. A mistake that many young businesses make is thinking that they can step back from marketing after they get their initial customer base established. You need to understand that no matter how wonderful your products are, you will have attrition. You may lose between 20 percent and 50 percent of your customers each year, regardless of the great loyalty programs you have established. People move out of town, their eating habits change, they get busy with new jobs, go back to school, and they lose their jobs too. Therefore, you will need to pay constant attention to marketing and continually adapt it to the changing demographics of your targeted segments.

Finding New Customers

When you are just starting out or you have developed a new enterprise, or if your current sales are stagnant, it may be time to pound the pavement a little and find new customers. There are many tactics that other successful farmers have drawn upon.

Sampling: Sample your products and give away freebies on occasion to enthusiastic customers; participate in or conduct meat tastings with potential or existing buyers (especially important with chefs). As mentioned before, Carman Ranch of Oregon conducted some blind taste tests among chefs to find out if there was a difference between their frozen grassfed beef or fresh. Most chefs found no difference, and this motivated them to keep Carman Ranch on their menus even in the wintertime when they could only get frozen meat. Try it, then buy it (or sample it, then sell it).

Education: Provide education about what you do and give presentations to as many community groups as you can find. Also, generate newspaper stories. Give the local media something interesting to write and talk about. To build awareness, offer your customers what is called "conversion literature," which will open their eyes to the problems of CAFO meats and keep them coming back to you (Schafer, 2005): books like *The Omnivore's Dilemma* by Michael Pollan, *Pasture Perfect* by Jo Robinson, *Holy Cows and Hog Heaven* by Joel Salatin, and *The Grassfed Gourmet* by Shannon Hayes. Our best piece of conversion literature by far was *The Omnivore's Dilemma*: Thank you Michael Pollan for driving so much business to pastured livestock producers! When customers read these books, it will also reinforce their emotional commitment to you and turn them into a "convert" who will also help with word-of-mouth marketing.

WOMM: Encourage WOMM (word-of-mouth marketing) from patrons (via a rewards system). Turn your loyal customers into cheerleaders and evangelists on your behalf. One of the best ways to do this is through occasional farm tours. People who have a fantastic, visceral experience on your farm will tell everybody they know about it.

Network: Half of all consumer first purchases of a product or service are influenced by social capital (JFK School of Government, 2006), meaning the more interaction you have with your customers and the more trust you build, the more likely your customer will tell other people, who will then be more likely to try your meats. Don't be a hermit farmer. Join groups and be active in your community as much as possible to foster those social networks.

Build a Good Team

Now that you have a good handle on marketing principles, your next task is to build a good team to do marketing right. If you don't enjoy marketing or feel you don't possess the proper skills, find somebody who does and add them to your team.

Your good meat alone and your stellar personal reputation will not sell the meat for you; you have to employ the right people and create a well-crafted brand as well. On the other hand, having a product you can be proud of *will* help sell those products for you. That enthusiasm will transfer over to your future customers. Are you proud of what you produce?

You also need a marketing plan, and you should actually implement it. A bunch of papers in a binder does not market your meat for you; steps need to be taken to put it into action. You can hire contractors to help you with specific aspects of marketing, such as developing your logo and graphics or setting up a spiffy website. We worked with a graphic design firm to develop all of our egg-branding materials: logo, egg carton design, retail signage, and graphics for our website. They did it all in exchange for a little money and meat. We're pretty sure bacon was involved. The quality of their work far exceeded what we might have been able to create on our own or could have coaxed friends to do for free. Sometimes (most times) you get what you pay for when working with qualified experts. Look at other local food brands and marketing materials that you like and find out who designed them. That's a good place to start when looking for a designer you'd like to work with.

Also develop a budget for what you are willing to spend on your whole marketing campaign and don't penny-pinch on this. Remember, good branding materials such as a logo, taglines, color scheme, and website should serve your business for many years, so they are worth a decent investment of capital. For example, if you think you will update your marketing materials every 5 years, the value of that work essentially has a 5-year depreciation schedule (although you don't account for it like that on your books since it is a "professional service"). That makes the one-time $4,000 to $8,000 investment not seem so bad—that's only $800 to $1,600 a year if the materials will last for 5 years without any major updates. Some bigger companies spend much more than that, but it's all relative to what your business is earning. You could go much longer with those graphics if you don't think they need refreshing. Nevertheless, almost every company updates their branding materials periodically to better reflect the times and to avoid becoming dull or stale. At the very least keep your website up to date.

As part of your marketing plan, you need to spend some time researching meat market trends and consumer demographics of your target markets.

Meat Market Trends

So what exactly are buyers (consumers, chefs, retailers, etc.) looking for these days, in no particular order?

- Consistent quality (for example: all grade Choice beef)
- Values (for example: predator-friendly lamb)
- Cuts and styles available (for example: while many beef producers grind a large portion of their beef carcasses thinking everyone just wants ground beef, perhaps you can be the rancher that has all sorts of interesting steaks, roasts, sausages, salami, oxtail, etc.)
- Specialty breeds (for example: chefs seem loony for Berkshire pork right now)
- Feeding program (for example: soy-free chicken)
- Lean meat or well-marbled meat (there are buyers for both ends of this spectrum)
- Location, region, state (for example: Piedmont North Carolina pork, Driftless Region Wisconsin grassfed beef)
- Is it a family farm? Has it been worked for multiple generations?
- Humane attributes (for example: no tail-docking sheep, no sow gestation crates)
- Clean and safe meat (for example: no fillers, tenderizer injections, colorants, preservatives, MSG, etc.)
- Healthy (for example: lower in saturated fat, lower in cholesterol, higher in omega 3 fatty acids)
- Paleo, primal, meat-centric eating (low carbs, high protein)

According to the National Restaurant Association (the other NRA!), **locally sourced meat** was the number 1 food trend for restaurants in 2014. Number 3 on the list is **environmental sustainability**—can you as an ethical meat producer play up those values to your customers in a better way? Number 6 is **hyper-local sources**, such as a restaurant garden. Could you partner with restaurants to grow animals specifically for them, such as The Maple Grille model in Michigan? Perhaps they could even pay you ahead of time or contract with you to grow animals for them. We have seen farmers do this with fruits and vegetables, so meat seems to be a logical extension of that practice. Finally, number 10 on the top 10 list is **farm-branded items on the menu**. If you do sell meat to restaurants (or any farm products, for that matter), can you ask them to put your name on the menu? Such as "Rancho Real grassfed beef tenderloin with balsamic reduction" or "Porcine Paradise milk-fed pork chop over polenta and creamed rainbow chard." It is smart to include not only the name of your brand on the menu but also your main production practice, such as grassfed or pasture-raised. That will give just a little more information to the consumer and will encourage them to seek you out for other meat purchases. Farm-branded ingredients not only boost the credibility of a restaurant that says they do "farm to table" but it also acts as advertising for your farm.

Other consumer trends for buying "niche meats" from our research include: personal issues come first like **taste and healthfulness,** then **supporting local, the environment,** and **humane animal care,** in that order. Based on this information, your marketing should focus first and foremost on how your meat tastes, the health attributes of your meat, and how it will complement or make an awesome meal for your customer. Then you want to focus on your proximity to the customer: Are you hyper-local, within the same town? Same region, or same state? Since the word local is being used on everything from Idaho-grown potato chips to pork coming from Canada, we ask that you use honesty when using this word and don't jump on the "local-washing" bandwagon. Let's keep some integrity in the word *local*. If your pig was born and raised up to slaughter weight in Canada and then trucked down to Oregon for slaughter, does that make it an Oregon product? Not in our opinion. Would you call that northwest-grown? How about beef that you finish that was born and raised in Saskatchewan until they were yearlings, then transported to Montana for grass finishing? We suppose if you actually raise the animal for a good portion of its life, say the last 8 months of a steer's life, then you can call it Montana-grown. There is, of course, no clear definition of *local*. Just don't lie. Don't call it your pork when you just bought the finished hog at auction. Don't market it as your beef when your neighbor actually raised it. If you are a group of farmers marketing together or a branded meat company that buys from multiple producers that uphold your quality standards, why not let your customers know that you are a "family of farmers" instead of pretending it all comes from one farm? Customers want to hear the truth and they will usually support the idea of several producers working together.

The "Red Meat Market" sidebar describes a virtual "meat market" that takes its understanding of customer demographics combined with a thorough understanding of emerging meat trends to create a new "community" around regional meats.

Meat Buyer Demographics

Did you know that nearly all hog and cattle producers send money to mandatory programs called the Pork and Beef Checkoff? Despite what you might think of these programs and their relative usefulness for marketing your niche meats, they do actually crank out a great deal of useful information, including market trend analyses. We only recently discovered these sites while doing research for this book—we wish we had known about them earlier! For example, in a recent Pork Board survey of five hundred randomly selected people, consumers were asked about their niche pork purchases (niche being defined as organic, pasture-raised, or "natural"). The top three reasons given for buying niche pork were freshness of the meat (66 percent, even if that meat was in fact frozen at time of purchase), healthfulness (62 percent), and flavor (56 percent). As a producer of these meats, you may be tempted to highlight the environmental benefits of how you raise your pigs or the humane treatment of the animals. But if you only have a small window of time to catch people's

Red Meat Market

Red Meat Market, a social media meat community and brand operating in Chicago, Illinois, is approaching meat marketing a little differently. Even though their experiential model is completely unique, there is a lot to learn from the way they approach building community around a brand. Mark Wilhelms, the founder, had an extensive background developing social media campaigns for large clients and wanted to use his expertise to drive positive change in the meat system.

In our interview Mark described how you couldn't launch a brand in this day and age without both demonstrating authority in the marketplace and building customer affinity. This allows your customers to influence others. So Red Meat Market first focused on building an online community around local meat, appealing to the "meat curious" and the more typical male meat-lover types. They proceeded to add an experiential component, furthering people's education and attachment to the good meat movement, through things like on-farm dinners, cattle round-ups, and butchering classes. The events are especially "dude-friendly," appealing to the twenty- to forty-something male meat-lovers out there (who also happen to be the largest consumers of red meat) by providing beer, food, and a casual, laid-back atmosphere. Red Meat Market sells branded merchandise like hats, T-shirts, and aprons that also solidify brand awareness and support people feeling that they are part of a culture. Once Red Meat Market did all this, they worked with some co-packers to develop and roll out some locally raised, antibiotic- and hormone-free meat products, starting with products that appealed to the same target segment. Products such as easy-to-cook, tasty items like sausage and jerkys. All of the products have irreverent names, like a bratwurst called Herman the German, named after Mark's great-grandfather, and Killbazza, a fiery hot kielbasa sausage.

Mark stresses the importance of how successful brands of today have to be "felt in the gut." You have to create a brand that is not only real and believable, but also fun and engaging. Because Red Meat Market has figured out their target segment so well, they can cater best to that segment and create that emotional connection. They are not trying to be everything for everybody. Although Red Meat Market does target women as well, since women do most of the shopping, female buyers generally want to make their husbands and family happy with the products they buy. Men are the influencers and decision-makers when it comes to meat, so Red Meat Market offers things like a Father's Day sausage-making class and a Cinco de Mayo bull breakdown class to appeal to them. Every participant goes home with a bunch of meat and an experience they can tell their friends about. There is a high "gloat factor" that men love.

According to Mark and the Red Meat Market model, once you have engaged consumers, the next step is to activate them. Ask them to do something, such as sign up for a newsletter, share information with their networks, buy products, attend events, and be influencers in their communities. Then they feel like they are part of something greater then themselves. This is no longer the classical "product-centric" model of buy, buy, buy. This is building a brand that caters to the needs of people by offering them high-quality products they can trust and engaging them in a larger movement for a more transparent, regional meat supply chain. As an individual farmer with a farm brand or as a group of farmers, you can do the same thing. When we added farm tours and butchering classes to our farm "brand," we too saw an increase in the connection and loyalty our customers felt toward our farm.

attention, either through packaging, signage, or a conversation you have, you may be better off highlighting your pork's quality, flavor, freshness, and health benefits. That is what resonates most with consumers. The same study identified that the top two barriers for consumers to purchasing niche pork were the lack of availability where they shop and the inability to find the product at all. The third most mentioned barrier was price, not as high up on the list of challenges as one might imagine. Keep in mind that consumers often will *say* one thing in a survey, but actually *do* another. So they may say that certain kinds of meat aren't available and that is why they don't buy them. But if you were to go through the hoops to get your ethical meats into the grocery stores where most people shop, would they actually buy them? That would make for an interesting experiment.

Another fascinating study by the Pork Board is called the Pork Intake Development Study, conducted by Texas A&M University, compiling 6 years of data on pork consumption across the country during the last decade. This study analyzed multiple demographic variables, including region, season, age, gender, ethnicity, income, presence of children, and more about how much pork people eat, what cuts, and other questions regarding pork.

For example, the study found that the area that we live in, the Pacific states, has the lowest consumption of fresh pork of any region, at around 79 pounds per person per year. On the other side of the spectrum, the north-central states of Missouri, Iowa, Minnesota, and the Dakotas have the highest pork consumption, with 119 pounds per person. Since these are some big pork-producing states, that sort of makes sense. But say you were trying to market pork in the Pacific Northwest: You would have to work a little harder to get pork to the top of mind of consumers. However, folks in this region eat a little more chicken and seafood than the central states, so you might have better luck selling those meats. The study also pointed out that Asians, African-Americans, and Latinos consume more pork per capita than Caucasians. When we lived in a very ethnically diverse region of California, we could sell any and all parts of the pig. Where we now live in Oregon, it would be more challenging due to the lack of diversity, and it might encourage us to grind more of the "harder-to-sell cuts" for sausages, or consider value-added products like hot dogs.

Keep these types of demographics in mind when trying to understand your particular market. To find out more about the particular demographics of your town or region, try searching a website called PRIZM, developed by Nielsen Ratings Group. By inputting your zip code you can find out the dominant market segments in your area, median income levels, ages, and ethnicities. For example, in our little zip code of 97040, we found one of the dominant groups is called "Shotguns & Pickups," a lower-middle-class group that predominately lives in manufactured housing, drives trucks, likes hunting and riding ATVs, and so on. The neighboring zip code to the west of us is called "Greenbelt Sports," and they are older and upper middle class without kids (most have moved out), and they like outdoor sports and vacationing in the tropics. They shop on eBay as opposed to Walmart. Also, the median income over there is $52,000, whereas in our zip code it is $50,000 (and the "Shotguns & Pickups" households make even less—around $42,000). If we were to try marketing pasture-raised organic pork, which segment should we go after? One demographic might be interested in coming out to our property to shoot wild pigs, while the other probably wants portion-cut meat neatly wrapped up and frozen. Use this kind of information to help you decide where to market and how. Rural demographics might be more interested in sides of meat—they often have chest freezers and some even know how to break an animal down themselves. Urban folks might be looking for smaller quantities of portion-cut meats.

Along these lines, consider how your customer segments may differ in their interest and ability to buy your meat. Some simplifications about rural versus urban or suburban demographics will point out these subtle differences.

RURAL CUSTOMERS MAY INCLUDE:

- people who are used to hunting or buying meat by the side; they often have a chest freezer
- survivalist types, homesteaders, do-it-yourselfers, or barterers
- people who want to save money or trips to a grocery store that might be far away
- second-home owners and retirees
- vacationers
- farmworkers, often Latinos or other ethnic minorities, many of whom know how to slaughter and butcher animals and sometimes prefer live animals

SUBURBAN CUSTOMERS MAY INCLUDE:

- families with children
- "soccer moms" and busy parents who may not make time to cook
- home gardeners
- men who like to BBQ, tailgate, and run the smoker
- baby boomers without children (or their kids have moved out) and retirees

URBAN CUSTOMERS MAY INCLUDE:

- professional couples without children
- foodies and people who eat out a lot
- families with children that choose to live closer to everything
- "weekend warriors"—people who like to BBQ, tailgate, and get out to the country on weekends
- ethnic minorities of all income levels

Additionally, keep in mind that middle-aged women (often moms) are the group that does the most food shopping (Goodman, 2008). They spend the most money on food of any demographic. However, men eat more meat, according to the Pork Checkoff and other studies. So women do the shopping (still) and men do more of the eating (is this fair??). How will you position your marketing to target those shoppers and eaters? Will you give it a masculine or feminine feel or can you find some neutral style that appeals to all genders?

Beef Marketing in Focus

The Beef Checkoff also maintains a variety of websites: from consumer-oriented ones that help people decide which cuts of meat to buy and how to best cook them, to ones for farmers and industry professionals on how to produce, cut, and market beef. On www.beefretail.org you can find updated quarterly lists of the top 10 beef cuts sold for each region of the United States. If we were raising cattle and trying to figure out how to process our animals, we would look at this list for a good idea of what might sell in our region. For example, a boneless tri-tip roast is the number 1 cut of beef sold in California, while in the Northeast region it is the boneless top round first steak. The averaged top 10 muscle cuts for the whole country, according to the last quarter of 2013, were:

1. rib eye steak (boneless)
2. chuck center roast (boneless)
3. strip steak (boneless)
4. subprimals (boneless, not specific)—used mostly in food service, restaurants, butcher counters; not home use
5. stew meat
6. top round first steak (boneless)
7. rib eye steak (bone-in)
8. top sirloin steak (boneless)
9. cubed steak
10. sirloin tip steak (boneless)

Despite the popularity of those beef cuts, ground beef makes up the majority of beef sales in the United States—56 percent of all sales in 2013. It is estimated that by 2020 75 percent of all beef sales will be ground beef: Convenience and versatility in cooking beef is the wave of the future. The United States is actually importing trim from foreign countries just to satisfy our current ground beef demand in this country. Steaks and middle meats used to reign supreme, but high beef prices and a population that lacks cooking skills and eats out a lot has led to the demise of the Sunday pot roast or even the steak dinner.

Interestingly, steak prices overall are coming down while ground beef prices are climbing. The ratio of steak prices to ground beef is now closer to 1.5/1, according to economist Bill Helming of the National Cattlemen's Beef Association. A key takeaway from this data is that there is an overemphasis in the industry on producing high-grading carcasses, while cattle destined for ground beef do not need to grade as high. We are living in a "ground beef world" now. Some suggest ranchers select out their high-grading animals early and manage them separately for middle meat quality while managing the lower gaining and grading animals on the less choice forages and pastures for sales primarily as ground beef. This also points out the opportunity to perhaps buy in cull cows for your ground beef trade. We're not suggesting you go get a bunch of unhealthy downer cows, but perhaps forge an alliance with a nearby organic dairy to buy their cull cows for your ground beef sales. Use animals for what they are best at—don't try to force a lean animal into a steak world. Likewise, don't use a nicely marbled animal for a ground beef market—that's bad math. If you better match your animal to what market it's destined for, you will be more profitable and you will make your customers happier.

Another trend is the uptick in grassfed beef consumption. In 2010, grassfed beef was 3 percent of the beef market and it has risen to as high as 6 percent in major metropolitan areas (Mintel Group Ltd., 2012), even as overall beef consumption has continued its decades-long downward trajectory.

Willingness to Pay

Research on consumers' "willingness-to-pay" looks at the average price premium that consumers will pay for certain food attributes. This research is helpful when designing your marketing strategy and messaging. Umberger et al. (2009) demonstrated that consumers would pay price premiums if they were given information about production practices and nutritional benefits of grassfed and organic meats, as well as information about the treatment of the animals and environmental impacts associated with their production. Johnson et al. (2012) showed that mainstream food shoppers (those that mostly buy from regular grocery stores or big box stores) were willing to pay an additional $0.94 per pound for grassfed beef. This is probably not enough of a premium to market grassfed beef through this channel. However, if a farmer can get grassfed beef directly into the freezers of those shoppers without having to pay the distribution costs, then it could make sense. One way to improve farmer profitability and reduce consumer prices is to sell meat in bulk, such as by the quarter, half, or whole animal. In the same study, 24 percent of the consumers surveyed had purchased beef in bulk before. Of the other 76 percent that had not, 69 percent of them said they would consider purchasing bulk beef if they either knew the producer or if a friend recommended a particular farmer (more on how to increase word-of-mouth marketing later in this chapter). In the study 51 percent of consumers had switched to "natural" or organic beef due to food safety concerns. This study was conducted in Portland, Oregon, a well-educated, food-oriented town. The location may skew the results, but in general, more and more consumers are hearing about the benefits of pasture-raised and grassfed meats. It's much easier to enter this market today than it was even 10 years ago due to more pervasive consumer awareness. Another hopeful bit of information gleaned from the same study was that in blind taste tests of grassfed versus grainfed burgers cooked exactly the same way with only a little salt and pepper, an equal amount of consumers liked the taste of each type. That also might indicate that the tide is turning with regard to the taste preference of American consumers.

It is interesting that every willingness-to-pay study finds that consumers will pay different premiums based on diverse product attributes. There is no magic formula, such as farming in one way will increase your price premium by X dollars. But these studies do point out the essential importance of telling your story as a farmer and being transparent about what you do. Too many websites that we researched for this book did the exact opposite. Many are unclear on what the operation is, such as if it's a family farm or an investor-owned corporation; if it's organic or not; if it's 100 percent grassfed or not; if the poultry range outside each day or not; if they use conventional feeds or not. The point is: Be clear and your customers should reward you with an appropriate premium that covers your costs and provides a profit too.

Strategic Messaging and Branding

Before you develop your brand, you need to start with a statement of who you are and the key message you will be making. Start by developing a positioning statement as described below. This is like stating where you are on the three axes of Strategies of Leadership mentioned earlier in this chapter. A positioning statement in the world of marketing looks like this:

1. **Target Segment(s).** For whom am I creating these meat products?
2. **Point of Difference.** What will be their reason to buy my meats?
3. **Frame of Reference.** To whom will I be compared? Who is my competition?

This position should both (a) be defensible and (b) require you to make choices. *You can't be everything to everybody.* Here's an example in the context of an integrated farm and meat business:

> *Southern Breezes Ranch produces the highest-quality grassfed meats in the state of South Carolina. Our meats are not only flavorful and tender, but they are also naturally low in fat and high in vitamins. We transform our grassfed meats into oven-ready marinated roasts, sauced ribs, and seasoned burgers for the busy family and active adults alike.*

In this statement above, the target segment is busy families and active adults in South Carolina. The point of difference is the flavor, tenderness, low-fat, high-vitamin grassfed meat and also the more ready-to-eat preparations of the meat. The frame of reference is other grassfed meats that are available in South Carolina (which may not really be available). You can see how this statement is defensible (as long as the meat is indeed flavorful, tender, low-fat, high-vitamin, and grassfed) and that the farmer had to make a choice about how to word it, particularly with regard to the target segment. Young college students may not identify themselves in that statement, but they purchase and prepare less meat than older demographics, so it might be okay to make that type of distinction in your target audience. Your goal is to make your key segments feel welcome and a part of your brand.

Once you have a positioning statement, you need to develop a brand. A brand has many elements that need to work together. It is not only the outward appearance of your business, it is also how you *do* business and how you *treat others.* The design aspects of a brand include:

Name. This needs to be memorable, recognizable, meaningful, persuasive, appealing, fun, aesthetic, protectable (legally, hard to copy), adaptable, and transferable to several products and services. For example, some friends of ours started a business called Surfside Chickens. They proceeded to add pigs and vegetables to their operation, which don't really fall under the "chickens" category. So they decided to change their name to Fiesta Farm, which is broad enough they could grow just about anything. It also is fun because fiesta means party. We started with TLC Ranch, which stood for "Tastes Like Chicken" because we raised meat birds and layers. Luckily, TLC can mean a lot of things so it just became "Tender Loving Care" Ranch for all the animals we raised. Make your name broad enough to incorporate the different enterprises that you may add over time. Not doing so can result in expensive changes to marketing materials and packaging.

Logo. Colors affect purchases. Red stimulates appetite (good for food products) while blue curbs it. Green conveys grass, outdoors, and environmental responsibility. Black and white logos reproduce well, but they don't often look good next to meat; they are a bit too washed out. However, if you don't want to spend a fortune on meat labels, you may want to stick to black-and-white printing unless your product will sit on a store shelf and needs color to pop out.

Symbols. Use animals, hills, trees, sunrises, kids frolicking—you get the picture. However, consumers have started suing companies portraying certain symbols on their packaging if they confuse the customer or portray an image that is not true. Petaluma Poultry of California was recently sued for portraying children outside feeding laying hens on grass. Because the picture portrays outside birds and their hens don't actually have outdoor access, some consumers felt like this wasn't "truth in advertising," setting the basis for a lawsuit. Save yourself a few bucks on potential lawsuits and instead use truthful symbols, pictures, words, and phrases on your packaging that reflect the reality of your farming system.

Packaging. Does your packaging keep the meat fresh and safe, does it show the meat off (is it transparent?), and is it consistent with the rest of your brand? Depending on where you market, the outward appearance of the packaging and labeling may be less important. Some farmers do just fine bringing paper-wrapped meat to the farmers market even though customers can't see what the meat looks like inside. We always vote for the plastic vacuum seal so that the meat is visible and customers have a better sense of what they are getting: One chop or two? Six hotdogs or eight? A well-marbled cut of beef or a lean one? Likewise, your packaging, as much as feasible, should be consistent with your brand values. If you value keeping trash out of the landfill but your packaging has three different layers of plastic to it, that is inconsistent. If you pack your meat into little Styrofoam boxes or hand a plastic bag to every customer that comes by, that is incongruous. An example of getting it right is a lamb producer near where we live that gives out little insulated, reusable bags to customers who buy their meat. Also, their logo is on the side. They encourage you to bring it back and give you a small discount for doing so. This encourages not only reuse but also customer loyalty. That's precisely the kind of creative marketing we are talking about.

Slogan or Jingle. Be memorable and concise. "Nice to meat you"; "Bacon: the gateway drug"; "Good for your tummy and good for the land"; "Putting the good back in meat." Steal any of these (as long as they are not already trademark-protected). You're welcome.

Experiential. Strong brands of today are experiential, meaning they are multi-sensory, emotional, intellectual or cognitive, inspire behavior or action, and are relational or cultural. How do all of these things apply to a farm and meat business? You can create a one-dimensional brand or a three-dimensional brand, meaning there will be multiple senses working when people learn about and interact with your brand.

For example, are you the meat seller at the farmers market who always has a little Coleman stove going with meat samples sizzling away, wafting a glorious odor of meaty goodness around the market? Or are you the guy with frozen meat hidden in coolers with a salesperson hiding behind a hat, sunglasses, and a smirk on their face? Does your farm open itself up now and again for on-farm tours so people can see, smell, and hear firsthand how you raise your animals? Or do you have a strict "biosecurity" protocol where nobody can come to your farm and you have "No Trespassing" signs posted on every corner? How will that affect how potential customers perceive your brand? Biosecurity is important, but it doesn't mean that you can't let customers ever visit your farm. Do you put pictures of baby lambs on your website, or just have a couple photos of meat in packages? All of these things will affect the perception of your brand. Likewise, you, the farmer, are your brand. How do you carry yourself? If you come across as distrustful of visitors, the government, and other groups of people, your customers may have a hard time trusting you. Trust is a two-way street.

Market Differentiation

You are different from the publically traded, multinational, vertically integrated meatpackers. Whether you agree with that model or not, you are different. How do you highlight that difference? How do you tell that story? *You are selling a story, not a product.* Emphasize the benefits of your products over its features—what it *does* for people, the planet, rather than what it *is* (Savory et al., 1998). Are you conveying that story throughout your branding materials, your customer service, and your commitment to quality? Do customers know the advantages of your products? Do you state them clearly? Here is an example of how *not* to do it and how you might want to rephrase things:

Abundance Acres doesn't use hormones, antibiotics, GMOs, chemical wormers, or body modifications on our animals. They spend all day outside.

How is this statement conveying the advantages of this farm? How does this statement convey how it helps the consumer meet their needs? How about this instead?

Abundance Acres supports health and community by producing the finest-tasting healthy meats that will bring an element of fine dining into your home kitchen. Our animals are raised in a way that respects our local environment, enhances biodiversity, employs the highest animal welfare standards, and creates living-wage jobs for our fantastic employees. When you buy from Abundance Acres, you are supporting your own health and the health of our community.

This statement conveys the advantages clearly: health, community, fine-tasting, improving your home cooking, good for environment and animals, and creating jobs. It conveys meeting the customer's needs of health, having good-tasting ingredients for their home cooking, and supporting a local business that is community-minded. Notice how it is not negative, but focuses on the positive attributes of the farming business.

We do think it is okay to say somewhere else in your marketing materials something about what is not in your meat, such as synthetic hormones, antibiotics, conventionally grown feeds, and so on. But your consumer-facing materials—such as the label and the front page of your website—should be focused on the positive attributes and what sets you apart.

Boasting Nutrition

There is copious research demonstrating the various nutritional qualities of grassfed and pasture-raised meats (Kim et al., 2009; Pugliese et al., 2008; Raes et al., 2004). Are you utilizing this information throughout your marketing efforts to boast about the nutritional values of your meats? According to the research, animals that are either fully grassfed or raised with plentiful access to forage exhibit the following tendencies in their meat:

- lower levels of saturated fats
- higher levels of monounsaturated fats (MUFA) or polyunsaturated (PUFA) fats
- higher ratio of omega 3 fatty acids to omega 6 fatty acids
- higher levels of conjugated linoleic acids (CLA), a form of PUFA
- higher levels of dososahexaenoic acids (DHA), a form of PUFA
- higher levels of tocopherols (vitamin E)
- higher levels of beta-carotene, the precursor to vitamin A
- often higher levels of protein
- overall, higher nutrient density

People are increasingly concerned about their health, especially given the rise in obesity, heart disease, stroke, diabetes, and cancers. Although there will probably never be a scientific consensus on exactly how much and what kind of fats, vitamins, and minerals we all need to be in peak health, the nutritional characteristics above are fairly consistent in the research. If your customers care about what kind of fats they are consuming, their vitamin intake, and protein intake, you should inform them of the research. If you don't want to get into the specific fatty acid and vitamin details, emphasize the overall nutrient density of your meats. *More and more people are eating for heightened nutrient density.*

You won't be able to make most of these nutrient claims on your label unless you commission some sort of scientific study, but you can conduct your own nutritional testing and submit your nutrient label for approval to USDA FSIS. As described in chapter 11, you are required to include a nutrition label on your meats, and USDA FSIS provides some generic nutrient labels for different kinds of meat. However, as explained above, grassfed meats do vary in nutrition and you may want to demonstrate your meat's higher values. For $800 to $1,500 you can pay a lab to test your meat. If you don't want to bother with that, you can still use your other marketing formats such as your website, Facebook page, word of mouth, and other places to get the word out about the nutritional benefits of your grass-based meats.

Reduce Friction (Make It Easy to Buy)

The ultimate goal of a sales-oriented business is to sell things—either products or services. First you need to figure out what is keeping people from buying from you in the first place, or keeping them from coming back regularly. Typically this is what business professors call "friction." *Friction is anything that slows the customer down from making a purchase or completing any kind of transaction.* Customers will always follow the path of least resistance. If you want to generate more sales, you have to identify the friction points in your selling system and eliminate them. In terms of a direct-market meat business, friction often includes issues of accessibility: Can a customer get hold of you? Where can your products be found, and are they always stocked? Does a customer have to travel far and wide to get to your meat? If a customer goes out of their way to a place where they expect to find your meat, let's say a Saturday farmers market or their local food co-op, and your product is not there or the selection is low, they may not come back. They don't know if they can trust the accessibility of your meat or if they can depend on you. Think of how frustrating it is when you go to find something at your local store and it is out of stock. No wonder so many Americans love their big box stores, where they can dependably find so many of the things they are looking for.

Another form of friction is price. If they don't think they can get your meat at a price they can afford, they will often look elsewhere. For example, there is one small grocer in my town that regularly stocks a decent selection of "localish, naturalish" meats. But because it is a small store that moves limited volume, we are hesitant to go there because we know that the prices are around 15 to 20 percent higher than we expect for those brands. We could make a once-a-month trip into Portland where we would have fifteen to twenty grocers, food co-ops, and farmers markets to choose from that would have decent meat at prices we could more easily digest. The perceived price barrier is a source of friction for customers.

Other friction points for customers, especially with regard to buying local meat, include:

- **Payment method.** Get yourself a Square or Paypal account now—they are so easy to use and allow you to take credit cards anywhere.
- **Cash flow.** If you require all the money up front for a meat CSA or a side of beef, that may squash many sales because people often don't have that kind of cash all at once. Can you provide more flexibility in a payment plan, such as 50 percent up front and 50 percent upon delivery, or a monthly fee instead of one lump sum?
- **Cold storage.** If you require the customer to own a large chest freezer, that may also kill the sale. Think of other ways that you could store the meat for them or help them divide the meat up among more people so that they only have to buy an amount that fits in their regular freezer. Also, if your customers are hesitant to buy something perishable like meat, you can provide ice or insulated bags (with your logo on them!) for them to safely take your meat home.
- **Seasonality.** You can't change the fact that the weather and grass growth doesn't allow you to finish animals at certain times of the year. We have described earlier in this book a myriad of ways to extend the harvest season for your animals. First work on that aspect. Second, many of the ways to market meat are also seasonal—things like farmers markets and farmstands that are usually just open during the summer and maybe early fall. How can people still access your meat at other times of the year? If you just ignore them for 6 months, will they come back to you when you are selling again?

They have to find somewhere to buy meat during those intervening months, and they may come to decide that they would rather choose the dependable, year-round source over you. It's hard to lure back customers year after year.

- **Customer service and ease of checkout.** This is especially important when you are selling meat directly to customers. How long do they have to wait in line? Is your staff friendly and knowledgeable? What if you added a second person at your farmers market booth who walked around with a credit card reader and could help the other folks waiting in line? Would the additional sales surpass the added labor costs? Or could you create more of a self-serve atmosphere that lets customers choose and pay for their products right away? Do you offer any sort of follow-up support or service, such as recipes, cooking instructions, or money-back guarantees if they don't like the meat? This is an important aspect to consider. A grassfed beef producer that we bought meat from would guarantee his product. In all our years of selling his beef, we only had one customer bring a package back. It was a package of filets that had unfortunately been cut up into medallions—they were looking for a whole filet. We exchanged the product for free and made our customer happy in the process.

Networking and Cooperation

Finally, and hopefully this should be obvious, you should build cooperation with other meat producers in your region to strengthen your market. You may think of them as competition, but having strong social networks with your competitors can help "lift all boats." When you run out of a product, you can refer customers to them, helping meet your customer's need. When that other farmer runs out of product or doesn't have enough animals, they can refer their customers to you. Again, this is the customer-centric approach. Your customers will greatly appreciate this and you will build trust among your "competitors." Next thing you know, you all could be going in on bulk feed together, doing group marketing, building a cooperatively run slaughterhouse—the sky's the limit! Also, if people were to come out to your farm to pick up something, wouldn't it be nice if there were several farms they could visit on the same trip? Build that farmer network—your "competitors" may come in handy for advice, bartering, borrowing equipment, lending a hand, creating an agritourism network, and making your customers pleased.

CHAPTER 13
Financial Management, Pricing, and Other Business Essentials

"Farmers feel slighted, processors overworked, and the consumer feels ripped off by $10 per pound meat," said meat consultant Chris Fuller in our interview with him. Why doesn't anybody feel like they are making money or getting a fair deal on meat, especially when prices keep escalating? Where is all the money going? You need to understand the costs of production, transportation, processing, cold storage, marketing, shrink, and other aspects of the business to recognize where the money is going and what (if anything) you can do about it. At the very least, know your side of the equation.

Financial Management and Cost Control

Starting at the beginning, do you know how much it costs your business to raise up an animal to slaughter weight? This is called "cost of goods sold." Factored into that, do you know how your animals are performing, their feed conversion ratios, and live animal loss rates? These are performance indicators that will make a difference on how expensive each animal is to raise. How much edible meat is each carcass yielding ("cut outs") and what is your typical shrink loss during processing? If you don't have a clear handle on these things, you need to spend more time doing recordkeeping in order to identify that lost revenue. You don't have to keep track of every animal, but at least take data on each lot or group of animals that you sell and market to get herd performance information. In the first half of the book we discussed some of the ideal herd and flock performance indicators that you can compare your animals to. This can point out issues such as poor genetics (poor performance of the breeding stock), a certain feeding regime did or did not work as you expected, raising animals to a certain size was a good or bad idea, raising animals during colder or hotter parts of the year was foolish, and so on.

Likewise, do you know which animals to cull and which breeding animals are performing to your expectations? Do you have an identification system for each breeding animal and a spreadsheet of performance data for each breeding individual in the herd? Do you have benchmarks for your breeding stock, such as number of piglets weaned or calf weaning weights, that help you to determine which to cull?

Cost control is key to building a successful business. Can you list your top five costs of production and roughly what percentage of your overall expenses they are? If you can't list these off the top of your head, you need to spend some more time with your numbers. Of your greatest costs, why do you spend what you do on them, and are there any alternatives? For example, if you spend a lot of money on hay each fall to get your animals through winter, why are you doing that? Could you make your own hay or contract it out? Could you buy hay at a different time of the year when there is less demand and it's less expensive? Would it make sense to invest in hay-storage infrastructure to be able to stockpile hay when prices are low and protect your investment? Don't forget that "bought in" hay robbed someone else's soil of nutrients, not yours, and that by using it you are increasing the biomass of your soil in addition to feeding your herd. Do you have too many animals to get through winter? Could you better stockpile forage in your pastures so the animals could self-harvest some of it over the winter? As Jim Gerrish emphasizes in his ranching books, what are you doing to "kick the hay habit" out of your system so you can become more profitable? Hay is just one example of many where producers are learning to cut costs without sacrificing quality or animal welfare.

In Rebecca's last book, *Farms with a Future*, we dove into all the financial reports you should be keeping—things like profit and loss statements, balance sheets (statements of net worth), cash flow projections, and budgets. We won't go into them again here. Suffice it to say, you need to spend some time with your bookkeeper or accountant to set up these forms and develop a schedule to regularly update them over time. Just because you have money in your wallet or the bank does not mean you are making a profit. Farming is one of the most difficult yet rewarding endeavors you can undertake. Don't do it for free! If you don't know your numbers, you may very well be giving away your labor for free; we see this all the time. These financial reports will show you where you are making money and where you are losing money, so you can take corrective action before things get out of hand. Don't throw good money at bad enterprises and don't be that person who waits until tax time to figure out how your business is doing.

We waited until the second year in business to start paying better attention to the bookkeeping. That is when we discovered that the profitability of broiler chickens was not where we wanted it. Not only were the profit margins slim compared to other animal enterprises (less than 10 percent for broilers) but also the profits per hour of labor invested were quite low since pastured broiler production is so labor-intensive. Armed with that information, it was easy for us to eliminate the broiler enterprise to focus on more profitable ones going into our third year. Empower yourself to make better decisions by understanding the costs of production, labor hours, and profit margins of each of your enterprises. You will be glad you did.

Seeking Profitability

There are four ways to make more money in any business—farms are no different.

1. **Decrease cost of goods sold (COGS).** These are costs specific to the production of the animal, including purchased stock, feed, minerals, veterinary care, labor, a portion of overhead (equipment, infrastructure, fencing), death loss, and others. Start out with your biggest costs of production to see if there are any alternatives. Pick one or two significant ones each year to research and test out substitutes. For example, feed is usually a livestock producer's biggest expense. You could look for other feed suppliers, try to buy in bulk, purchase whole grains and mill them yourself, or cut the feed with cheaper hay or mill-run grains, among other options.
2. **Decrease overhead.** These are costs that you must pay to have your farming business, including insurance, accounting fees, licenses and permits, mortgages, interest expenses, property taxes, some depreciation, marketing, website, guard or herding animal care, and others. Again, pick a couple of items each year (it's too overwhelming to do them all at once) and do some research. Call an insurance broker, for example, who will compare different policies for you. Find an accountant who will work for a meat share instead of cash. Refinance your property at a lower interest rate. Keep going.
3. **Increase weights and volumes produced.** Do this through heavier animals, higher-yielding animals, more animals, reducing death loss, utilizing more of the animal, and more. Say you finish one hundred lambs a year for market. Could you put an extra 20 pounds on each of them before you take them to slaughter? Could you find a rabbit breed that has a better meat-to-bone ratio? If predators are taking around 9 percent of your broiler chickens each season, could an investment in another livestock guard dog get that down to 3 percent instead? These changes all result in more meat to sell.
4. **Increase prices.** Set a higher cost per pound or unit, or decrease portion size while keeping the same price as before. Maybe you could bump up your prices 5 percent across the board without anybody noticing, or just increase the price of your high-end cuts but keep your modest-priced cuts the same. You could also decrease your package size by a small amount but keep your price the same, meaning you earn more per pound. If any of these seem like devious marketing tactics to you, then don't do them.

We have observed that many small-scale producers focus mainly on the fourth tactic to increase profitability by continually raising prices. While raising prices is the most effective way to make more money overall, it is not the most "fair" to your customers, nor will it work over the long term if it is your only tactic. Eventually your customers will feel they are no longer getting good value and will look elsewhere. For example, we knew an unprofitable egg producer who was charging $8 for a dozen eggs from conventionally fed outdoor hens. We were charging the same price for organically certified eggs from truly pastured hens that got moved to fresh pasture every other day. The other egg farmer's chickens were not well cared for, his pastures were heavily degraded, the egg quality was poor, and yet he was earning twice the profit margin as us due to the lower priced conventional feed costs. Even with this profit margin, he still managed to go out of business. We feel that all businesses should focus on internal cost management and efficiencies instead of continually passing on their lack of business skills in the form of higher prices onto their unsuspecting customers.

That said, if you are always sold out (overall, or of a particular cut or animal), your prices are probably too low. A way to solve this is to charge more when demand is higher and less when demand is lower.

Instead of always falling back on price increases to make your business more profitable, you need to tackle the trickier aspects of your business, such as analyzing your costs of production, looking for efficiencies, testing out alternatives, culling poor-performing animals, and other tactics we mentioned above. Overhead can be reduced by spreading it out over more volume or over multiple enterprises. You can work with your processor to get back more of your animals and find decent markets for the whole animal, increasing your profits per head. There are a myriad of strategies short of raising your prices all the time.

Find your bottlenecks that limit throughput and then seek to address them. Some of these examples could be your limiting factors to profitability:

- **Land base or feed quality:** Low soil fertility, poor pasture regrowth, wrong type of browse, lack of shade, using poor-quality hay, or finishing animals on brown grass, for example.
- **Genetics or number of animals to weaning:** Wrong animal for the climate or forage base, trying to grass-finish high maintenance animals, poor mothers, not culling enough, breeding too young, wrong body condition for breeding, birthing at the wrong time of year, not providing shelter during partition, slow-to-gain animals, or poor feed-conversion genetics, for example.
- **Logistics of feed, water handling, and fencing:** Buying feed in small, more expensive quantities due to lack of storage or cash flow; hand feeding and hand water hauling; can't get water to all pastures; water is too hot or too foul; fencing is too expensive to set up properly; non-portable fencing; animals keep escaping; or high predation losses, for example.
- **Animal transport:** Lack of facilities to sort and load animals well, lack of the right truck or trailer for hauling, lack of crates for poultry, or the slaughterhouse is too far away.
- **Processor's capacity:** Can't get the slaughter dates you need, can't bring in larger lots of animals, don't have the space for dry aging, don't do cut and wrap, and so on.
- **Sales volume:** Takes a long time to move the volume of meat you have or the markets you are in are too slow. We should have given this issue more thought with our own business. We now wonder: If it took 12 months to sell three hundred pigs' worth of meat and we had to pay all of the cold storage and marketing costs for that time, what would have happened to our profitability if it took only 6 months to move the equivalent amount of meat?

Another way to increase prices is to turn a product into a service, such as a meat-of-the-month club, monthly mixed-meat box, weekly meat dish (ready-to-eat), year-round classes, and other events or services. When products are bundled like this or emulate a service, you can increase the prices without it being obvious.

Seeking Financing

In *Farms with a Future* there is also an entire chapter devoted to the topic of finding money to start or scale up your farming business, thus we won't repeat that information here. However, it is important to identify some of the ways that farmers have obtained the financing to build out their direct-marketing programs, develop their own processing facilities, enhance their brand awareness, and other strategic business moves. These include:

1. Self-funding either through savings, off-farm income, or retained business earnings
2. Traditional debt financing: credit cards, term loans, mortgages, lines of credit, equity loans, USDA farm loans
3. Non-traditional financing: friends and family loans, consumer loans and prepayments, buyer loans and prepayments, microlending (e.g., Kiva Zip), crowd-funding (e.g., Kickstarter or Indiegogo), county or municipality loans, bond funds
4. Equity financing: Slow Money, angel investors, institutional investors, direct public offerings
5. Grants: private foundation grants, state and federal grants (e.g., a Value-Added Producer Grant from the USDA could fund a feasibility study for a new slaughterhouse or farmers cooperative)
6. Tax deferments, tax credits, and tax exemptions

Cash Flow

Cash flow is a struggle for most businesses; livestock farms struggle with it even more due to the grow-out period of animals. You pay for your operating expenses now to care for the animal and must wait potentially up to a year or two to get paid for the sale of the animal. Obviously, the smaller the animal, the quicker the payment cycle. Indeed this is why many beginning livestock producers start with rabbits, chickens, or pigs: They are easier on the cash flow.

Money comes into your business when your customers pay you for the product they receive. If they don't pay you on the spot (COD, or cash on delivery) for their meat, then you have to invoice them and they pay later—just like you pay the bills you have stacked up on your desk (hopefully) once a month. How many outstanding payments you have at any given time is called "accounts receivable" in accounting lingo. Joel Salatin, in his book *Salad Bar Beef*, stresses the importance of keeping accounts receivable low: Try not to be waiting on a bunch of income that people owe you.

There is a temptation for small businesses or those just starting out to extend credit because they're dying for a sale, but this is an unhealthy business practice. We learned this the hard way ourselves when we sold eggs for a short time to restaurants. Restaurants are notorious for being slow payers. Not only do they take 45 to 60 days to pay an invoice, but you often have to hound them a couple of times for payment. They could even go out of business before you collect payment, since restaurant failure hovers around 25 percent in the first year. And, how much labor will you have to spend to hound your customers to pay you? As a small or even mid-scale farmer, you don't have the cash flow yourself to be the bank for slow-to-pay buyers. Be aware that most restaurants or retailers will not do a COD payment, but you should still be firm with them about what your payment terms are. Perhaps begin with a 14-day payment policy. When we sold eggs to Whole Foods Markets, they would direct-deposit the payment within 7 days of delivery. That was a reasonable amount of time to wait and didn't require hounding them for a payment (or trying to find that check in your pile of mail). See what you can work out with your buyers.

Online payment systems are making transactions a lot quicker and more efficient, and they can reduce your administrative costs. For example, you could accept deposits and final payments for freezer meat through PayPal or some other online system. Just make sure you create a system for recording those payments and for moving that money into your business bank account on a regular basis.

Another way to address cash flow is to develop some sort of prepayment system. If you collect a portion of the price early on, this could help even out your cash flow and provide you the funds for livestock production costs such as feed. This works well for CSA-type arrangements or selling meat by the side. When we sold halves and wholes, we usually collected half of the fee several months in advance. This served as a deposit for people to reserve their animals and also provided the income to buy the feed to finish our pigs.

A way to address cash flow over the long term is to diversify your business. If you raise one livestock species, perhaps you can spread the sales out over the year with fresh and frozen meat sales. Or have multiple animal species that finish at different times of the year. Raise some fast-growing species like rabbits or poultry while you are raising up slow-growing species like cattle where you won't get paid until the following year. Or develop a complementary enterprise to run in the "off-season" to bring cash in at that time of the year, such as butchering classes or value-added products.

Pricing Strategies

Now that you hopefully know how to figure out the costs of production for the entire enterprise and the per head price, let's figure out how you might go about pricing your meat cuts. Take your pricing seriously—getting your prices wrong can destroy your profitability and create a weak foundation for future growth. There are several theories on how to price products. These include:

Cost-plus accounting. You take the cost of production per head plus an arbitrary profit percentage you want to make on top of that. So, for example, if an animal costs you $3.50 a pound to produce and you wanted to make a 25 percent profit margin, your per pound price should average out to $4.38 per pound.

Demand-side pricing. If you follow the law of supply and demand, you price your products to the point that you move the needed volume of meat in the time period you have to move it. This price can change seasonally based on supply. So if all your high-end cuts are always sold out at your market booth, you raise the price of those cuts to help meter them out a bit more slowly over your market season. If you have 10,000 pounds of beef to sell in a 5-month farmers market season and you want to have a decent selection for all 5 months, you price your cuts up to a point that they don't sell out immediately and stick around a bit longer. Say of that 10,000 pounds of beef you

Financial Management, Pricing, and Other Business Essentials

have only 200 pounds of filet mignon. If you priced your filet at $10 per pound, it would probably be sold out in the first month. The remaining 4 months at market would get pretty tedious having to tell every customer that inquires about filet that you are sold out until next year. However, if you priced the filet at $24 per pound, it could last 3 or 4 months of the market season instead. So you will have made more money and kept a better selection at market for a little longer.

We priced our eggs this way. If we had a frenzied chaos of pushing and shoving to get our eggs at the farmers market and sold out the first hour of market, our prices were too low. If we raised our prices up to the point where we sold out of eggs in the last hour of the 4-hour market day, that was better. We made better income for our time standing at market, we controlled some of the chaos, and we did not have to tell people for 3 of the 4 hours that we were sold out. We preferred that a few potential customers walked away due to our high prices than that many more walked away disappointed that we were sold out of eggs. Demand-side pricing might seem ruthless to you, but it is how most of American businesses price their products or services.

To effectively do demand-side pricing, it helps to keep track of numbers of customers, volumes sold, and cuts sold, so you can see how changing your pricing affects those variables. If you sell at farmers markets, for example, create a market load list that shows how much of each cut you brought to market, the retail price, and how much you sold by the end of the day. You can also write a little check mark for every customer that purchases from you. We like to look at this data later to compare different markets and the total number of customers versus income earned at each market. You may have a high-volume market that only brings in a moderate income for your time, while another market may have fewer people but each person buys a larger share of your products. That kind of data is invaluable for determining whether a market is worth your time or whether you need to focus on either (1) attracting more customers or (2) getting each customer to buy a bigger share from you. From a marketing perspective, it is cheaper to get regular customers to buy more things from you than it is to attract brand-new customers.

Competition pricing. You base your pricing on what your competitors or local stores are charging for each cut of meat. You could also index your price to the commodity prices provided weekly by the USDA Agricultural Marketing Service. This is often recommended in farming books, but we don't encourage it. Yes, it is a good idea, on a regular basis, to find out what others are charging and keep that in mind. But you should not be competing solely based on price: You are doing something different using an animal production system based on ecological and animal welfare principles and producing a superior, nutrient-dense food. When you make your brand or image about price, you are downplaying those values. You will never be able to compete with the big guys on price, so don't even try it. That said, if you are selling freezer meat, you can and should advertise that buying meat in bulk is cheaper than buying individual cuts at a retail store. For example if you added up all the cuts offered on a side of pork from a retail store, it might come to $600, but buying it on a hanging weight price of $4 per pound plus processing may only cost $500. Let your customers know they will be saving $100 over store prices, getting infinitely better product, and supporting a local farmer in the process.

In our case when we were selling TLC Ranch pork, we never even looked at the retail price of pork in nearby stores. Why? The pork sold in stores was not even close to being a comparable product. There was no local, pasture-raised, organic heritage breed pork in the stores; they only had CAFO conventional pork raised in the Midwest. Comparing prices would have served no purpose for us. Instead, we would first figure out our costs of production for each animal. Then we would add a percent profit margin on top of that (usually 30 percent, which also helped cover loss and input cost variability) and then test out those prices in the market and do a little demand-side price adjustments. We think our process worked well.

Target return-on-investment (ROI) pricing. This means that your profit margin for a product will be determined by how quickly you want to be paid back for your investment. For example, if you invested $100,000 of your own money into a meat business through a combination of savings and taking money out of your retirement, perhaps you want to get that money back within 3 years (so you can replenish those accounts). Say the loss in interest you would have accrued had you kept that money where it was amounts to another 5 percent annually. So you want to earn $115,000 in profit within those 3 years, plus cover your costs of production and personal labor. Simplified, that would mean you need to make $38,333 annually in net profit after taxes, or about $48,000 in net profit before taxes (assuming a 25 percent overall tax rate). If you sold two hundred beeves a year, you would then need to clear $240 in profit on each animal to meet your target ROI. You then determine your retail prices based on that profit margin. Phew! That's a complicated process, but it is one that profit-motivated businesses often use. Rather than draw down your savings and retirement investments forever, as many farmers do, how about building it back up and then some over time? That is the key to long-term financial sustainability.

❋ ❋ ❋

Other pricing strategies (sorry, but a lot of marketing psychology will be used here):

Don't price all your meat cuts the same. This actually makes your customers *less* inclined to buy your meat. They want to know there are differences, that some pieces are better than others. We once met a lamb producer selling frozen meat cuts at a small farmers market who priced all of their lamb cuts at $11 per pound. We asked why all their meat was the same price. They said it just "made things easier for us and the customer." We then asked if that meant their high-end cuts would sell out quickly and their low-end cuts would not sell as easily and they said "yep." So we proceeded to buy up a bunch of center-cut lamb chops because at $11 per pound that is a great deal. Yet we weren't about to buy ground lamb or lamb shanks for that price—it was just too high for those economy cuts. Two weeks later we went back asking for lamb chops and, lo and behold, they were already sold out of them for the rest of the year. So yes, it's a bad idea to price everything the same. These farmers probably lost out on a hundred or so dollars per carcass, plus they lost the confidence of customers when they were sold out of all the tender cuts on the animal and had only overpriced economy cuts to sell over the course of the market season.

Price increases should be no more than 10 percent at a time. That seems to be the magic amount in which customers don't notice the increase or don't complain, as shown in countless marketing studies. If you get your pricing right from the beginning and update it regularly (at least once a year), then you hopefully shouldn't have to make any price jumps higher than 10 percent at a time. Just as we think you should generate a profit-and-loss report monthly so you can stay on top of the financial health of your business, you should also update costs on production spreadsheets so you know quickly if you need to raise your prices. We would suggest trying to adjust your prices at least once a year and probably quarterly. Don't feel bad about this: Big businesses change the price of their widgets every single day. Ever notice how gas prices change by the pennies each day? Gas suppliers monitor their supply and demand, as well as the price of a barrel of oil. If your feed prices, for example, go up by $0.02 per pound in a month and you are buying 10 tons a month, that amounts to an increased cost to you of $400 a month. If you don't change your price for the rest of the year, that little $0.02 increase will amount to a loss of $4,400 over the next 11 months.

End your prices with 9s. Research shows that this actually works to make people think an item is more affordable! So instead of $5.10 per pound ground beef, go for $4.99 per pound. Or instead of $20 per pound for lamb chops, try $19 per pound. People think they are getting a bargain even if it amounts to just cents. They focus on the number before the decimal point.

Frame purchases as a gain, not a cost. Your meat is improving your customer's health, giving them pleasure, making cooking easier for them, supporting a stronger local economy, keeping land in agriculture, protecting grasslands, supporting biodiversity, replacing one CAFO animal, and numerous other benefits. Frame their purchase not as a net loss of money in their pocket but as a net gain for their personal well-being and society's benefit.

Price has to be consistent with positioning. Are you the Dollar Store of meats or are you the Bloomingdale's? If your products are cheap, people will expect lower quality and lower values, and will be less loyal to you over time. Make the transaction about more than getting a bargain. You set the perceived customer value of what you produce by your initial pricing. Everyone will be happy if, as you scale up later on, you are able to lower your prices, but they definitely won't like it if you are always raising them. Start at the right price.

High-end cuts should be considerably more expensive because they are considered "luxury, special-occasion" cuts. There is so little of the loin on the animal, therefore by nature they are scarcer. Not only should your prices be much higher on those cuts, but you can also limit how many pounds or packages each customer can buy at one time. This creates even more of an allure of scarcity, allowing you to elevate the price even more. We're talking about $24 per pound pork tenderloin or $32 per pound filet mignon. Even bacon is scarce (sadly they have not engineered a pig to be all belly yet!), so you could probably charge between $12 and $20 per pound for it. We charged $16 per pound (in 2010) and sold them in 10-ounce packages; that way they cost just $10 each. People have less qualms about plunking down a single $10 bill than pulling out $16. This said, don't gouge people, and if the middle meats are not moving fast enough to suit your liking, lower the price. Starting high and lowering prices makes consumers much happier then starting too low and raising your price.

Offering "sales" makes people think they are getting a good deal. Just saying "sale" will make people think they are getting a bargain, even if the price differential is not that significant. Going from $16 per pound rib steaks down to $14.99 per pound is essentially only a $1 difference. But with the words "sale" next to it, it's a huge psychological boost to the buyer. They are getting a deal and they will feel all the better for it. Think about the last time you got a deal on something—didn't it make you feel like a winner?

Setting Your Prices

Armed with the intricate pricing information above, let's go through an exercise to help you determine the price to charge per pound of meat. Using a beef animal as an example:

Costs of production to bring animal to slaughter weight of 1,000 pounds live weight: $1,250
Kill fee ($100) and butchering costs ($450): $550 (this may be on the high end depending on where you live)
Transport costs: $200
Cold storage and marketing costs: $175
Overhead and equipment or building depreciation: $150
Target return on investment: $234
Total cost per animal: $2,559

Now you know how much total income you need to make from that animal. Using a figure of 440 pounds of salable meat from that animal, you need to average at least $5.48 per pound across all your cuts. Looking at a price average based on 2014 grassfed beef prices from farms in Oregon, California, and West Virginia, here is what you could earn on a beef animal (prices will vary by location—these are only examples). Table 13.1 is based on cutting up a beef animal in a very simple, traditional way. This is probably not the best way to cut up a beef; see appendix C for a more profitable breakdown. This is only for sample purposes.

According to the suggested prices, if you sold all of the 440 pounds, you would net $891 dollars per animal ($3,450 minus $2,559). Factor in a reasonable amount of loss in processing and damaged packages—average 10 percent of your original total earnings ($345)—and you would earn more like $546 per head in net profits. You can play with these numbers, adjusting some prices up and other prices down as you test them out in the marketplace. Just make sure they add up to *at least* your target per pound price (in this case, $5.48 per pound) or your overall total that you are aiming for.

Now say you decide to wholesale your meat instead of direct marketing it. Here is what that could look like:

Total Cost of Animal: $2,559
Minus cold storage and marketing costs of $175 that you don't have to pay
Minus cost of butchering of $450 that you don't have to pay since you are selling whole carcass
New cost of hot carcass: $1,934

If the hanging weight was 62 percent of 1,000 pounds, then for a 620 pound hot carcass you would need to charge $3.12 per pound to cover your costs and make your target ROI. As of mid

Financial Management, Pricing, and Other Business Essentials

TABLE 13.1. Beef Price Sheet

Meat cut	Percent of carcass/ lbs of meat	Suggested price	Total for cut
Brisket	2.55%/12.75 lbs	$8.00	$102
Shoulder clod (grind)	5.39%/26.95 lbs	$7.00	$188.65
Chuck (mock) tender	.83%/3.65 lbs	$16.00	$58.40
Pectoral muscle	.43%/1.89 lbs	$7.00	$13.23
Short ribs	1.03%/4.53 lbs	$8.00	$36.24
Chuck roll	3.81%/16.76 lbs	$8.00	$134.08
Skirt steak	.66%/2.9 lbs	$12.00	$34.80
Hanger steak	.29%/1.28 lbs	$14.00	$17.92
Rib cap	.72%/3.17 lbs	$22.00	$69.74
Back ribs	.77%/3.39 lbs	$7.00	$23.73
Rib eye roll	3.87%/17.03 lbs	$20.00	$340.60
Tenderloin/filet	1.72%/7.7 lbs	$26.00	$200.20
Strip loin	3.15%/13.86 lbs	$18.00	$249.48
Top sirloin	3.67%/16.15 lbs	$14.00	$226.10
Bottom sirloin flap	.98%/4.31 lbs	$8.00	$34.48
Tri-tip	.69%/3.04 lbs	$12.00	$36.48
Knuckle	3.24%/14.26 lbs	$5.00	$71.30
Top round	5.44%/23.94 lbs	$10.00	$239.40
Eye of round	1.49%/6.55 lbs	$9.00	$58.95
Bottom round	3.75%/16.5 lbs	$8.50	$140.25
Flank steak	.54%/2.38 lbs	$12.00	$28.56
Trimmings (grind)	20%/88 lbs	$7.00	$616
Bones	16%/70 lbs	$3.00	$210
Fat (½ made into grind)	18.39%/80.9 lbs	$3.50	$283.15
Organ meats	1.5%/6.6 lbs	$5.50	$36.30
		Total	$3,450

2014, according to the USDA Agricultural Marketing Services grassfed beef price report, Select-grade grassfed beef carcasses were going for around $2.95 per pound, a bit under the target ROI for this example by just $0.17 per pound. Some of the ways you could reduce your costs to get closer to the country average (if that is all your buyer is willing to pay) include finding a more affordable kill fee, reducing your transport and overhead costs per head, or adjusting your target ROI a bit.

It may appear that direct marketing is more profitable. Indeed it can be, assuming you get the prices you set out to obtain and don't lose product for any reason. But the risks can be higher in direct marketing than in wholesaling. If you can negotiate a reasonable carcass weight price and avoid a lot of the costs and labor that goes into marketing, wholesaling might be a better option for you. Or you can do a little of both—it is up to you.

Insurance

Raising animals and selling meat come with some heightened risks. One way to manage those risks and protect yourself from legal action is to have the right insurances. Start off with a good business liability policy. It should be comprehensive enough to cover the physical assets of the farm as well as the products you are selling. Make sure it includes full "product liability," covering issues such as pathogens, foreign objects, and any other problems that could happen with the meat. Include coverage for potential meat loss, such as if your power goes out for too many days, thawing out the meat in your freezers, or losses at the retail end if they don't use your product in time and don't want to pay you.

Liability insurance (also known as commercial general business liability) protects a company's assets and pays for obligations—medical costs, for example—incurred if someone gets hurt on your property or when there are property damages or injuries caused by you or your employees. Liability insurance also covers the costs of your legal defense and any settlements or awards should you be successfully sued. Typically these include compensatory damages, nonmonetary losses suffered by the injured party, and punitive damages. General liability insurance can also protect you against any liability as a tenant farmer if you cause damage to a property that you rent, such as accidental fire, breaking the well pump, or other covered accidents. However, it is not the same as property insurance. If a wildfire, flood, tornado, or robbery cause damage to your property, your general liability policy will typically not cover that: Make sure you ask your agent about every last detail. Obtain a separate property insurance policy for those kinds of natural disasters. Lastly, liability insurance can also cover claims of false or misleading advertising, including libel, slander, and copyright infringement.

Many retailers and even farmers markets will require a certain level of insurance. When they say you need a minimum of, say, $500,000 in liability coverage, that specifies the maximum amount the insurance company will pay against a liability claim. So, if your small business gets sued for $350,000 for medical costs associated with an injury caused by an animal getting loose on the road, plus an additional $100,000 in legal fees, then you are not responsible for paying anything out of pocket because you are under your $500,000 limit. There may be a small deductible that is your responsibility, but that's it.

Incorporation

To reduce the risk of losing your farm due to a lawsuit, you may want to consider separating the farm assets from the business assets. You could do that by forming a separate corporation, such as a limited liability corporation (LLC), that operates the business. Your farm business, as a sole proprietorship, partnership, or corporation, then leases the land and farm equipment (and perhaps even the breeding stock) to the separate farm business. You then insure them as two separate entities. Although there is no such thing as an impenetrable wall in the world of law, taking these actions could reduce the possibility that you would ever have to sell your farmland as a consequence of a legal action against your business. Of course, consult with an agricultural attorney if you have questions about this. Incorporation also has certain tax advantages and disadvantages that would be beneficial to discuss with a tax accountant as well.

Labor Management

Being efficient with your labor as the owner of the farm business along with any labor you hire will help you increase your profitability and probably your sanity. Are you spending more than a few hours a day on daily chores? If so, you are probably spending too much time "in the business" and not enough time "on the business." Cattle rancher Cody Holmes penned a book titled *Ranching Full-Time on 3 Hours a Day*, suggesting that you can run a full-fledged ranching business with less than 3 hours of daily chores. As a business owner, there are many other affairs you need to focus on to manage and build your business: tasks such as labor management, equipment repairs and upgrades, conducting on-farm research, managing meat-supply logistics, recordkeeping, bookkeeping, marketing, and sales. If you are always knee-deep in the daily chores such as hauling feed, setting up fences, filling up waterers, assisting animals with birth, tagging, and vaccinating, how will you fit in time to properly manage your business? The business management tasks then get put off for later and just don't get done or are completed hastily. Try to keep track of your own time and look for ways to be more efficient with it. Spend more time on the activities that are worth more money and consider hiring people to do the lower-cost tasks or the ones you are less skilled at (equipment repair comes to mind).

Likewise, you need to manage your hired labor efficiently. How can you increase your hired labor efficiencies without coming across as a "evil overlord" of a boss? Perhaps you can provide production incentives to increase your workers throughput but still provide them a minimum hourly guarantee so they are not discouraged to work more efficiently. For example, provide daily production goals that they have to complete in a timely manner, but also say that you will guarantee at least a 35-hour workweek. We think paying employees with a salary as opposed to an hourly wage also improves efficiency. Why pay workers to be slow or inefficient in their activities, which is precisely what can happen if you pay by the hour? Your hourly employee slowly meandering over to the well to turn on the pump for irrigation, driving back to the shop to get that tool they forgot, or slowly gathering the eggs is not helping your business. Instead, you could offer a base salary and bonuses or commissions based on overall production volumes, quality standards, or sales targets. More piglets make it to weaning age? Provide a small bonus. Beef steers gain weight faster due to good pasture rotation? Give a bonus. You are able to clear out your freezer meat inventory quicker? Write a bigger check. Sales staff are likely to be motivated by sales commissions. You can't dock wages if production goals are not achieved, but you might choose not to pay a bonus or commission if those targets are missed. Along those lines, you could choose to eventually let go of an employee if they consistently miss targets or perform shoddy work. Those are hard decisions to make, but ones you cannot put off if you want your business to be successful. We know good employees are hard to find, but that does not mean you should keep employing mediocre ones.

Set goals for all of your employees, yourself included. Meet weekly to discuss those goals and then follow up with how they are being attained. If targets are not met, discuss why and come up with a plan to overcome the issue(s).

Scaling Up?

A pearl of wisdom from famous grass farmer Joel Salatin in his book *Salad Bar Beef* is that "a well-managed small business can always grow into a profitable large business, but a minimally profitable small business seldom grows into a successful big business." His point is that you must manage for profits from the beginning and also get your pricing and margins correct as you start out. If the farm does not make any profits at small volumes, it's probably not going to get better when it is producing higher volumes. If you price your products to include a comfortable margin from the beginning, this will provide the funds needed to expand in the future. We have seen too many farmers around the country underpricing their products while they are small, part-time, or "hobby"-type operations. If you lose $100 a pig when you are just selling eight pigs a year, you will never have the profits to scale up and your losses will be even greater when you do. Losing $800 a year for eight pigs is not nearly as bad as losing $8,000 a year if you scale up to one hundred pigs a year. "Don't reproduce error in bulk"—one of our favorite quotes from interviewing young livestock producer and entrepreneur Jerica Cadman of Shady Grove Ranch in 2011—sums up perfectly the issue of scaling up a bad business.

Lauren Gwin, co-director of NMPAN, admitted that there are some very practical reasons why the meat industry looks the way it does today. They solved issues of seasonality by moving toward indoor production and feedlot systems. They built large meatpacking plants to reduce the processing costs per animal, something that small plants are disadvantaged by due to their lower volumes. The key is to use elements of the industrial meat model to your advantage while not compromising your values, the animal welfare, or the healthfulness or quality of the resulting meat. If we just took away the antibiotic crutch that the entire model is dependent upon, said Lauren, the playing field might be leveled to some degree.

How can you scale up using some of the "big meat" techniques while keeping your high ethical standards? Moving toward longer seasons, taking on some seasonal indoor production such as hoophouses in the winter, taking in larger lots of animals at a time to the processor, buying feed in bulk, automating the watering and feeding, and finding a few higher-volume purchasers that reduce your marketing expenses are some strategies you could employ.

From Lauren: "A lot of people try to build direct-marketing businesses without enough throughput to support the costs of production and overhead of that business." Scaling up will help you lower those overhead costs per animal, not only allowing you to price your meat more competitively, but allowing you to make more of a profit, and ultimately make a living from your farming. That is our hope. Now, happy farming!

APPENDIX A
Simple Farm Biosecurity Tips

Some simple biosecurity steps you may want to consider include:

1. All visitors should leave their animals at home, both pets and livestock. If you are custom-processing livestock or poultry for people, make sure their animals arrive in a healthy state and that the truck or trailer parks well away from any of your animals.
2. All visitors should understand the rules on your farm. Either verbally express them to all visitors, have the rules written on a legible sign, or provide a handout to everybody.
3. Kids should always be under supervision of a guardian. Make sure their guardians understand that responsibility.
4. All visitors with visible signs of sickness should be asked to leave the farm and come back when they are well. Don't make people feel permanently ostracized—tell them they are welcome to return when they are healthy.
5. Provide hand sanitizer and boot washes with disinfectant for everybody. In lieu of boot washes you can provide disposable booties for everybody to put over their shoes. Crud coming in on shoes is one of the biggest risk factors.
6. Don't let people touch the animals unless it is in some sort of segregated petting-zoo-type area where you encourage it. Or if you are offering a hands-on experience such as chicken or pig slaughter or cow milking, ask that folks either wash their hands beforehand or wear sterile gloves.
7. You can keep people to the outside of all fence lines or ask them to just stick to walking on the farm roads. If you are doing a pasture walk, stick to fields or paddocks where the animals aren't currently grazing and won't be for a couple months from that time.
8. If a visitor is also an animal farmer or hobbyist animal producer themselves, take extra precautions because they could be bringing in pathogens from their own farms. You can ask them to bring a clean pair of shoes to wear while they are on your farm (or at least ones that they don't wear on their farm). Maybe ask them to park their farm vehicle

on your gravel parking area or just beyond your farm gate and walk in. You may think you are coming off as rude, but the tires on their farm rig may have animal manure on them. The vast majority of pathogen transmission events that we hear of on farms come from other people who raise livestock or poultry themselves or who work in the industry, such as at hatcheries or slaughter plants.

9. If your animals are not currently healthy, don't invite folks out until your animals recover. Their immunities will already be compromised and don't need any extra burden. Plus, why would you want to show off sick animals? If you have a few sick animals, isolate them in a recovery ward area that is warm, dry, and where they can recover (hopefully) in relative peace. Keep visitors away from the sick ward.

10. You may want to prohibit your employees from raising poultry or swine at home; otherwise, they could be bringing diseases from home to work with them. If you don't want to be that draconian, another less secure option is to have workboots and coveralls that they change into when they get to your farm every day. Also make them wash their hands thoroughly when they get to your farm each day.

APPENDIX B
Meat and Poultry Processing Rules

TABLE B.1. State-by-State Red Meat Processing Rules

State	State inspection program?	TA plants or CIS program?
Alabama	red meat and poultry	TA plants allow interstate sales
Alaska	none	no
Arizona	red meat and poultry	no
Arkansas	none	no
California	none	no
Colorado	none	no
Connecticut	none	no
Delaware	red meat and poultry	no
Florida	none	no
Georgia	red meat only	TA plants allow interstate sales
Hawaii	none	no
Idaho	none	no
Illinois	red meat and poultry	TA plants allow interstate sales
Indiana	red meat and poultry	some state plants participate in CIS program, allowing interstate sales
Iowa	red meat and poultry	no
Kansas	red meat and poultry	no
Kentucky	none	no
Louisiana	red meat and poultry	no
Maine	red meat and poultry	no
Maryland	none	no
Massachusetts	none	no

State	State inspection program?	TA plants or CIS program?
Michigan	none	no
Minnesota	red meat and poultry	no
Mississippi	red meat and poultry	TA plants allow interstate sales
Missouri	red meat and poultry	no
Montana	red meat and poultry	no
Nebraska	none	no
Nevada	none	no
New Hampshire	none	no
New Jersey	none	no
New Mexico	none	no
New York	none	no
North Carolina	red meat and poultry	TA plants allow interstate sales
North Dakota	red meat and poultry	some state plants participate in CIS program, allowing interstate sales
Ohio	red meat and poultry	some state plants participate in CIS program, allowing interstate sales
Oklahoma	red meat and poultry	TA plants allow interstate sales
Oregon	none	no
Pennsylvania	none	no
Rhode Island	none	no
South Carolina	red meat and poultry	no
South Dakota	red meat only	no
Tennessee	none	no
Texas	red meat and poultry	TA plants allow interstate sales
Utah	red meat and poultry	TA plants allow interstate sales
Vermont	red meat and poultry	no
Virginia	red meat and poultry	TA plants allow interstate sales
Washington	none	no
West Virginia	red meat and poultry	no
Wisconsin	red meat and poultry	some state plants participate in CIS program, allowing interstate sales
Wyoming	red meat and poultry	no

Meat and Poultry Processing Rules

TABLE B.2. State-by-State Poultry Processing Rules

State	Poultry rule	1,000/20,000 bird rules?	Sales restrictions?	Permit, license, or fee
Alabama—has state PIP	follows USDA 20,000-bird exemption—P.L. 90-492		same as federal	no license but periodic inspections for 1–20,000 bird plants
Alaska—no state PIP	follows USDA 20,000-bird exemption—P.L. 90-492		on-farm sales or direct to permitted food establishments	annual permit with Food Safety and Sanitation Dept.
Arizona—has state PIP	follows USDA 20,000-bird exemption—P.L. 90-492		same as federal	no, but there may be county-level rules and initial inspection by PIP program
Arkansas—no state PIP	follows USDA 20,000-bird exemption—P.L. 90-492		same as federal	manufactured food permit from Health Dept.
California—no state PIP	follow USDA 20,000-bird exemption—P.L. 90-492. Includes rabbits, too, under exemption	if sold off-farm, only family labor allowed for processing	direct to consumers on farm, farmstand, or farmers markets	license required and county health departments may impose stricter rules
Colorado—no state PIP	no exemption—all poultry sold must be processed under USDA inspection		cannot sell exempt poultry	license, sanitation inspections, and USDA inspection
Connecticut—no state PIP	follows USDA 20,000-bird exemption—P.L. 90-492		same as federal	requires facility inspection
Delaware—has state PIP	follows USDA 20,000-bird exemption—P.L. 90-492		same as federal	license and plant inspection by Dept. of Ag
Florida—no state PIP	follows USDA 20,000-bird exemption—P.L. 90-492		same as federal	initial facility inspection and annual food processors permit from Dept. of Ag
Georgia—no state PIP	does not accept federal exemption—however, very small (under 1,000 chickens) farmers can get state exemption	has a special small producer program for under 1,000 birds	on-farm sales only	register with Food Safety Division, periodic inspections
Hawaii—no state PIP	follows USDA 20,000-bird exemption—P.L. 90-492		same as federal	sanitation requirements and periodic inspections

Note: PIP stands for Poultry Inspection Program

State	Poultry rule	1,000/20,000 bird rules?	Sales restrictions?	Permit, license, or fee
Idaho—no state PIP	follows USDA 20,000-bird exemption—P.L. 90-492		same as federal	license and inspection from Public Health Dept.
Illinois—has state PIP	follows USDA 5,000-bird exemption—P.L. 90-492	only up to 5,000 birds under exemption	on-farm sales only	request written exemption from Dept. of Ag
Indiana—has state PIP	follows USDA 20,000-bird exemption—P.L. 90-492 (as of July 1, 2014)		fresh bird sales on-farm only, frozen bird sales at farmers markets or farmstands	Indiana Board of Health has jurisdiction over exempt poultry processing
Iowa—has state PIP	follows USDA 20,000-bird exemption—P.L. 90-492		direct-to-consumer sales only	exempt plant license by Dept. of Ag, inspections required for 1,000–20,000 bird plants
Kansas—has state PIP	follows USDA 20,000-bird exemption—P.L. 90-492. Includes rabbits, too, under exemption		under 1,000 birds, on-farm sales only, 1,000–20,000 same as federal	registration, inspection, and sanitation requirements by Dept. of Ag (not for rabbits, though)
Kentucky—no state PIP	follows USDA 20,000-bird exemption—P.L. 90-492		same as federal	permit by Dept. of Public Health
Louisiana—has state PIP	follows USDA 20,000-bird exemption—P.L. 90-492	if under 1,000 exemption, equivalent to 250 turkeys. If under 20,000 exemption, equivalent to 5,000 turkeys.	same as federal	exemption letter by Dept. of Ag
Maine—has state PIP	follows USDA producer/grower exemptions only—P.L. 90-492	1,000 exemption, whole birds only; 20,000 exemption, whole and cut-up birds allowed	under 1,000 exemption, direct to consumer only; under 20,000 exemption, same as federal	license to process, license to sell from Dept. of Ag
Maryland—no state PIP	follows USDA 20,000-bird exemption—P.L. 90-492. Includes rabbits, too, under exemption		same as federal if you have approved plan and inspection, otherwise on-farm sales only	off-farm sales require training, food safety plan, and inspection along with annual fee
Massachusetts—no state PIP	follows USDA 20,000-bird exemption—P.L. 90-492		same as federal	license by Dept. of Public Health, annual fee; local Board of Health must also approve

Meat and Poultry Processing Rules

State	Poultry rule	1,000/20,000 bird rules?	Sales restrictions?	Permit, license, or fee
Michigan—no state PIP	follows USDA 20,000-bird exemption—P.L. 90-492		same as federal	food establishment license Dept. of Ag
Minnesota—has state PIP	follows USDA 20,000-bird exemption—P.L. 90-492		direct-to-consumer sales only	plant inspection required for 1,000–20,000 birds
Mississippi—has state PIP	follows USDA 20,000-bird exemption—P.L. 90-492	sanitary requirements for 1,000–20,000 plants	on-farm sales only	county health permit required for selling off-farm—usually won't give to uninspected sources
Missouri—has state PIP	follows USDA 20,000-bird exemption—P.L. 90-492		depends on county Health Dept. rules	registration with Dept. of Ag if slaughtering more than 1,000 birds annually
Montana—has state PIP	follows USDA 20,000-bird exemption—P.L. 90-492		same as federal	license by Dept. of Livestock and Dept. of Public Health
Nebraska—no state PIP	follows USDA 20,000-bird exemption—P.L. 90-492		all poultry sold must be processed in a licensed facility	license by Dept. of Ag
Nevada—no state PIP	no exemption—all poultry sold must be processed under USDA inspection			
New Hampshire—no state PIP	follows USDA 1,000-bird exemption—P.L. 90-492	only up to 1,000 birds annually under exemption	direct-to-consumer sales only	license by Public Health Dept.
New Jersey—no state PIP	follows USDA 20,000-bird exemption—P.L. 90-492		same as federal, sold frozen only	local or state Health Dept. inspection
New Mexico—no state PIP	follows USDA 20,000-bird exemption—P.L. 90-492. Includes rabbits, too, under exemption		direct-to-consumer sales only	none
New York—no state PIP	follows USDA 1,000-bird exemption—P.L. 90-492	1,000–20,000 birds require license and inspection	under 1,000 birds direct to consumer only, 1,000–20,000 with A5 license—can include HRI sales	A5 license by Commissioner of Ag if processing over 1,000 birds

State	Poultry rule	1,000/20,000 bird rules?	Sales restrictions?	Permit, license, or fee
North Carolina—has state PIP	follows USDA 20,000-bird exemption—P.L. 90-492		under 1,000 birds, direct-to-consumer sales only, 1,000–20,000 birds same as federal	inspection by Dept. of Ag
North Dakota—has state PIP	follows USDA 20,000-bird exemption—P.L. 90-492		direct-to-consumer sales only, some county Health Depts. may regulate farmers market sales	annual registration with Dept. of Ag—inspection for 1,000–20,000 plants
Ohio—has state PIP	follows USDA 20,000-bird exemption—P.L. 90-492		under 1,000 birds, on-farm sales only	license by Dept. of Ag if processing over 1,000 birds
Oklahoma—has state PIP	does not accept federal exemption—however, very small (under 1,000 chickens) farmers can get state exemption	over 1,000 birds annually requires USDA inspection	on-farm sales only for under 1,000 birds	registration for poultry slaughter with Dept. of Ag
Oregon—no state PIP	follows USDA 1,000-bird exemption—P.L. 90-492. Includes rabbits, too, under exemption	1,000–20,000 birds require Dept. of Ag inspection	under 1,000 birds, on-farm sales only	state licensing by Dept. of Ag if over 1,000 birds
Pennsylvania—no state PIP	follows USDA 20,000-bird exemption—P.L. 90-492	more frequent inspections for 1,000–20,000 bird plants	direct-to-consumer sales only	register as food processor with Dept. of Ag, annual inspections
Rhode Island—no state PIP	follows USDA 20,000-bird exemption—P.L. 90-492		only frozen	license by Dept. of Health
South Carolina—no state PIP	follows USDA 20,000-bird exemption—P.L. 90-492		same as federal	meat and poultry handler registration
South Dakota—no state PIP	follows USDA 20,000-bird exemption—P.L. 90-492		same as federal	
Tennessee—no state PIP	follows USDA 20,000-bird exemption—P.L. 90-492		same as federal, frozen only	license by Dept. of Ag, retail license too
Texas—has state PIP	follows USDA 10,000-bird exemption—P.L. 90-492. Includes rabbits, too, under exemption	10,000 birds only	same as federal	registration with Dept. of Health

Meat and Poultry Processing Rules

State	Poultry rule	1,000/20,000 bird rules?	Sales restrictions?	Permit, license, or fee
Utah—has state PIP	follows USDA 20,000-bird exemption—P.L. 90-492		same as federal	license by Dept. of Ag or Food
Vermont—has state PIP	follows USDA 20,000-bird exemption—P.L. 90-492		direct-to-consumer sales only	license by Agency of Ag
Virginia—has state PIP	follows USDA 20,000-bird exemption—P.L. 90-492		direct-to-consumer sales only	permit from Dept. of Ag for 1,000–20,000 birds, local health depts. might have additional rules
Washington—no state PIP	does not accept federal exemption—however, very small (under 1,000 chickens) farmers can get special permit	Dept. of Ag offers a under 1,000 birds special permit and up to 20,000 birds a food processor license	under 1,000 direct-to-consumer sales only, 1,000–20,000 same as federal	permit or license with Dept. of Ag
West Virginia—has state PIP	follows USDA 1,000-bird exemption—P.L. 90-492	working on final rules for 1,000–20,000 exemption	same as federal	license by Dept. of Ag for over 1,000 birds
Wisconsin—has state PIP	follows USDA 1,000-bird exemption—P.L. 90-492	bird-by-bird inspection for over 1,000 birds	on-farm sales only	
Wyoming—has state PIP	follows USDA 20,000-bird exemption—P.L. 90-492		same as federal	inspections by Dept. of Ag

APPENDIX C
Sample Cutting Instructions

Each animal gives its life to feed us. We should honor that animal by using as much of the carcass as possible, as profitably as possible. To help you make the most money per animal, we have developed two sample cut sheets each for beef, pigs, and lambs or goats designed around the two distinct meat-eating seasons of the summer (BBQ!) and the fall and winter (back-to-school, short days, cold nights, warm roasts). Keep in mind that your clientele may prefer different cuts than this, smaller package sizes, and more processed meats. This is not a one-size-fits-all approach. The key is to work with your processor and your customers to find out the best way to cut up your animals. If you process animals several times a year, it is fairly easy to have some processed in one style and some processed in another. Even each haul to the processor can be split into different cutting groups. Just remember that your processor will likely have minimum volumes to run certain products, like grind or smoked products. Also, the more complicated your instructions are, the more room for error and the more costly it is for the processor. Likewise, just because you have always been cutting up your animals in the same way for years does not mean there might not be more profitable ways to cut up the animal or there aren't ways that will please your customers even more. But the price of getting it wrong can be substantial, with some cuts sitting around for months in your cooler, taking up space, using electricity, and potentially deteriorating in quality. You can opt to have some frozen cuts ground or turned into other products if you are having trouble selling them as is.

In the following sections we offer up the sample cut sheets and notes to take to your butcher or to play with and modify as you wish. Butcher and meat consultant Chris Fuller of www.meatchris.com helped us flesh these out, focusing on the seasonality of cooking, along with some of the hottest cuts and best ways to maximize carcass yields. As a former meat plant manager, he understands how to extract the most value from the animal carcass. How you cut up your animals can have a strong impact on your profits as well as the customers you attract and keep. In addition to how you cut up your animals, you need to think about other important characteristics. These include:

Sample Cutting Instructions

Portioning by weight or by size: Portioning by size is easier and results in less trim waste because muscles are not uniform. Evenness of cooking can be affected by the portioning of the meat. If a steak, for example, is cut thin on one end and thick on the other, it won't cook evenly and may make for an unhappy customer. Also, portioning into pieces that are too small might not appeal to people unless they expect to be buying stew or kabob meat.

Pack size: One per package, two per package, or four per package? Smaller packages work better for single people and the elderly who consume smaller portion sizes. Larger packages will scare off these customers but appeal to larger families. How about a mixture of package sizes to appeal to a diversity of clients?

Weights: Make the 1 pounders, 2 pounders, or 5 pounders? Same issue as package size. Some variety is good, although we stopped doing roasts over 3 pounds, figuring that if somebody had a large family they could just buy two 3-pound roasts. Moving a 5- or 6-pound roast can be tough.

Fat trim: No fat, ¼ inch, ½ inch, 1 inch, 2 inches, and so on. Almost nobody wants zero fat trim, and the same goes for 4 inches of fat, too. Play around between ½ inch to 1½ inches, and collect feedback from your customers.

Boneless or bone-in: Boneless meats are more popular with customers; however, you are not only losing value on the bones, but you are also increasing your processor's labor costs. Will you sell less if you include the bones? Can you charge the same or nearly the same, therefore making more income off the bones? We started leaving the bones in all our chops and many of our roasts—not a single customer ever complained.

If you are new to direct marketing, talk to friends, butchers, chefs, and other "steakholders" to find out what meat fabrication characteristics above will work best given your target segments. Obviously, retailers, restaurants, and distributors may want your meat cut up differently or left in subprimals, so this section will be less relevant for those customers.

BEEF

CHUCK

Cut name	Suggested size in lbs or thickness	Quantity per head
Delmonico Steak	1"–1¼"	4–6
petite chuck tender	whole 0.5 lb	2
flatiron filet	1–1.5 lbs each	4
brisket	3–5 lbs	4–6
Summer cuts—focus on grilling items		
ranch steak	1"	8
chuck eye steak	¾"–1"	18–24
Winter cuts—focus on roasts, holiday items		
cross rib roast	3–4 lbs	4
boneless chuck roast	3–4 lbs	6–8
cross cut shank	2"	8

RIB

Cut name	Suggested size in lbs or thickness	Quantity per head
Summer cuts—focus on grilling items		
rib eye steak	1¼"–1½"	22–28
back ribs	half rack (3–4 rib)	4
Winter cuts—focus on roasts, holiday items		
rib steak	1¼"–1½"	12
rib roast	4–6 lbs	2–3

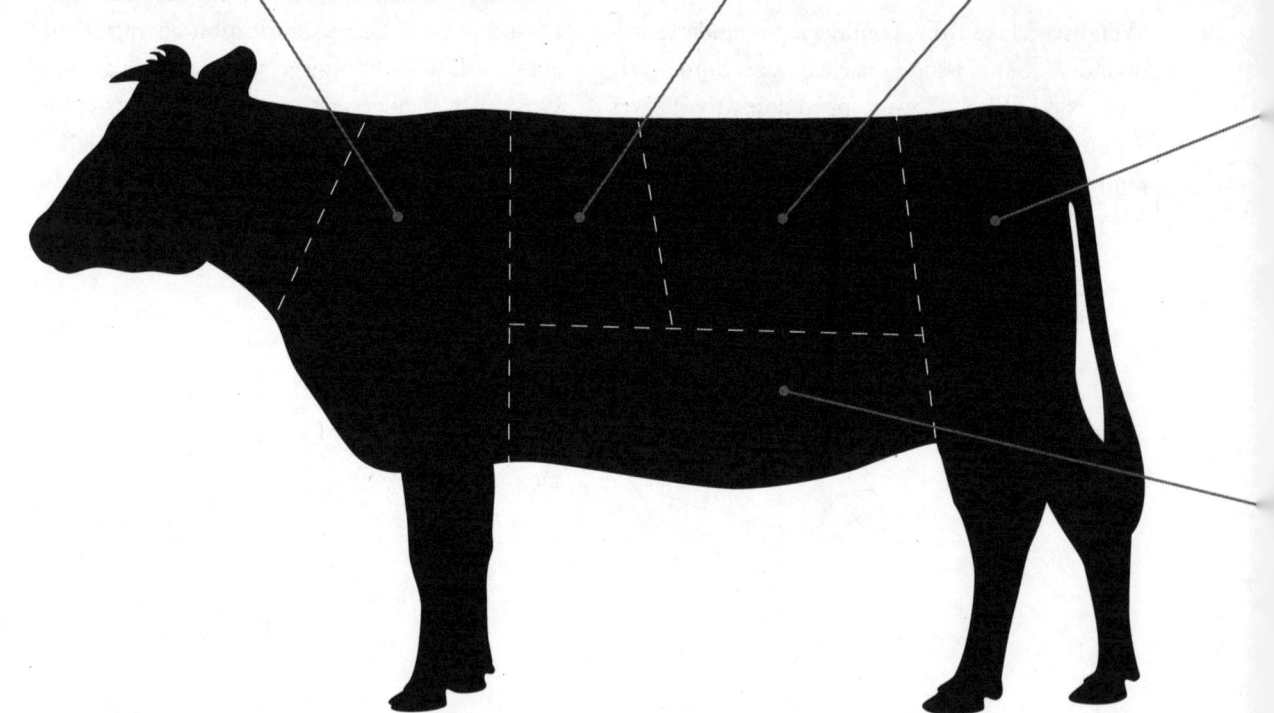

Sample Cutting Instructions

SHORT LOIN AND SIRLOIN

Cut name	Suggested size in lbs or thickness	Quantity per head
T-bone steak	1"–1¼"	12–14
New York strip steak	1"–1¼"	12–14
top sirloin	1"	10–12
baseball sirloin	1¼"–1½"	8
tri tip	whole	2
Summer cuts—focus on grilling items		
filet mignon	1¼"–1½"	6–10
sirloin cap steak	1"–1¼"	10–12
Winter cuts—focus on roasts, holiday items		
tenderloin roast	whole	1 or 2
sirloin cap roast	3 lbs	2

ROUND

Cut name	Suggested size in lbs or thickness	Quantity per head
Summer cuts—focus on grilling items		
sirloin tip kabobs	1"	10 lbs
London broil	1½"	8–10
Western griller steak	¾"–1"	16–20
Winter cuts—focus on roasts, holiday items		
sirloin tip roast	3–4 lbs	4
rump roast	3–4 lbs	8
cubed steak	¾"–1"	16–20
cross cut shank	2"	8

PLATE AND FLANK

Cut name	Suggested size in lbs or thickness	Quantity per head
flank steak	whole	2
skirt steak	1–2 lbs	6
sirloin flap meat	1–2 lbs	6
Summer cuts—focus on grilling items		
flanken cut short ribs—best only	½" thick	20–30
Winter cuts—focus on roasts, holiday items		
English-style short ribs	2–3" thick	10–16

MISCELLANEOUS

Cut name	Suggested size in lbs or thickness	Quantity per head
hanger steak	trimmed, whole	1
cheeks	whole	2

OFFAL

Cut name	Suggested size in lbs or thickness	Quantity per head
heart	trimmed, whole	2 (1 split)
tongue	whole	1
sweetbreads	whole	4–5
liver	1½"	15 +/- lbs
whole head	whole	1
ox tail (segments)	1.25–1.5 lbs per pack	2 packs
marrow bones	6" long, split or whole	8–12

TRIMMINGS

Cut name	Suggested size in lbs or thickness	Quantity per head
sirloin and other steak	steak burger	20–30 lbs
other	regular burger and sausages	180–200 lbs

Summer Beef Cut Sheet Notes:

Chuck: Delmonico steaks are the first few cuts of the chuck eye directly adjacent to where the butcher breaks the rib from the chuck. They are the highest-quality chuck eye steak and can be labeled Delmonico and sold for a good bit more than the chuck eyes.

The way this chuck is broken down maximizes grilling cuts, minimizes roasts (there are none, save for the brisket), and thus maximizes the value of the carcass. Also, less burger typically equals more money.

You must realize these cuts are worth at least double what burger is going for in many markets. If your average grassfed burger price is $7 to $8 a pound, then the flatiron, Delmonico, and petite tender should sell for $15 a pound or more.

Rib: By separating the back ribs from the rib eye steak you're automatically getting more money per pound than if they were sold together as rib steak.

A 16-ounce rib steak ($16 per pound), which includes 14 ounces of meat and 2 ounces of bone, will bring you $16. If you sold this meat as a 14-ounce boneless rib eye steak ($18 per pound) and 2-ounce back ribs ($4 per pound), it would bring you $16.25. While this may not seem like a lot, when calculated over the multiples of pieces and many cuts in the carcass, it adds up. Every penny counts in this slim-margin business of meat sales.

Plate and Flank: Flanken-style short ribs are simply a three- or four-rib plate of short rib run across the band saw at ½ inch thick. They are great marinated and grilled. Also known as "Korean-style" short ribs, they are very popular right now.

Short Loin and Sirloin: Unlike the example with the rib steaks, T-bones usually fetch a good price in comparison to New York and filet, especially considering the bone adds a lot of weight and you can't sell that bone like you can the back ribs. Therefore, if you can sell porterhouse and T-bones, do it. If you can't, then resort to the boneless filet and New York.

The sirloin, broken into its three parts, *must* be sold at a much higher dollar amount per pound in order to make this a sensible option for a producer. However, the steaks are much better. So, if the market allows, Chris Fuller recommends cutting it into these pieces. If not, stick with boneless top sirloin steak—less goes into burger. Also, use any sirloin trim separately to create a "steak burger" or "ground sirloin" blend to get a few more bucks per pound out of that sirloin trim.

Round: Round steaks (London broil, aka top round, and Western grillers, aka bottom round) are all good for marinating and grilling. They are not the most tender cut, but they still can typically be priced higher than ground beef, *except* for the eye of round. This is Chris's least favorite cut, so he recommends having it ground because he doesn't want people eating it and thinking his beef is bad! In our opinion, after having 4 years of experience trying to sell London broil at farmers markets, we would recommend grinding them for sausage instead. We could always sell more beef sausage in the summer than London broil.

Sirloin tip is the best cut to make kabobs because they're consistent muscle, lean, and flavorful.

Miscellaneous and Offal: Hanger steak, cheeks, head, and tongue should all find an easy home in ethnic markets. If you cannot find any ethnicities that will value these cuts, perhaps you can stockpile them and sell them to a butcher shop or restaurant in a more diverse, urban area a few times a year. A good way to lose money is to pay for that stuff and then just dispose of it. We had no problem selling hanger steaks at farmers markets since they are a bit of a rarity and cook similar to a skirt steak. Sometimes they are called "butchers' steaks" because rarely do they make it out of the butcher shop.

For the fabrication techniques for all of these cuts go to www.beefretail.org or the North American Meat Processors (NAMP) guide.

Winter Beef Cut Sheet Notes:

Only the variances from the summer cut sheet will be explained below.

Chuck: This sheet has many more roasts than the summer sheet. Of course, people enjoy roasts and braising much more in cold weather. Pot roast and stew are mostly cold-weather dishes. Therefore, instead of chuck steaks and arm steaks we have chuck roasts and arm roasts. The way this chuck is broken down maximizes the flavor and weight in roasts while still extracting the highest-value steak cuts. And still: Less burger equals more money.

Roasts cannot sell for the same amount as steaks (in general), but there is less labor involved in making them. Potentially there could be cost savings negotiated with the processor here. Maybe the winter beef cut is less expensive than the summer and you can balance out the processing costs over the whole year.

Cross cut shank makes great soup bones, or osso buco for those who like to prepare braised meals in the winter. We found osso buco (both beef and lamb) to be a very popular cut. They come from both the foreshank (chuck) and the hind shank (round). Make sure you get both.

Rib: By *not* separating the back ribs from the rib eye steak you must charge a bit less than boneless steaks, but you can sell weight that otherwise, in winter, would not sell. Or, you can go boneless rib eye (like in the summer sheet) and save the ribs to sell in the summer. Ribs just don't sell that well in winter.

You should have several rib roasts (at least 4 pounds each) to satisfy the holiday orders for "prime rib." Consumers will request this item for Christmas, even if it's not Prime grade beef. Even if your regular customers normally want 2- to 3-pound roasts, for the holiday they

Sample Cutting Instructions

will mostly be serving guests and will want something a little larger. *Tip:* If you can work it out with your processor, consider leaving your rib roasts whole and have the processor cut and re-pack to order once you've gotten holiday requests. A processor can receive frozen meat, in its original package, and re-pack it at a low cost to the producer. You can then market them as "custom order" rib roasts for the holiday.

Plate and Flank: English-style short ribs are simply a three- or four-rib plate of short rib run across the band saw at 2 to 3 inches thick. They are great braised or slow-cooked.

Short Loin and Sirloin: T-bones usually fetch a good price in winter as well as summer. Sometimes people just want to grill. However, New York and filet are a great choice in winter because, again, people want a nice roast for the holidays. Whole tenderloin is one to keep on hand. Sell some T-bones if they're popular, but for winter Chris recommends doing New York strip steaks and saving your whole tenderloins for special order.

The sirloin cap, as opposed to the sirloin cap steak, is a small roast that cooks up nicely—tender and full of flavor. This is a good roast for small families who may want to cook it on the grill, in the smoker, or in the oven for a nice Sunday meal. If your customers aren't going for that, cut it into steaks and sell it as another option for those who really like strip steaks; it's very similar and a good-quality steak at a good value.

Round: These cuts (sirloin tip, London broil—aka top round, and bottom round) are generally all lean, but make great rump roast. Chris prefers top round for rump roast as it slices really nicely with little to no gristle. The bottom round is better suited to cut into small cubed steaks. Many people like to make recipes with cubed steak in the winter—Swiss steak, stroganoff, or similar "stewed" steaks or chicken-fried breakfast steaks can be made with this cut. If you only make cubed steak from the bottom round you'll end up with a fair amount of steaks, but won't be overwhelmed with a freezer full of them.

Miscellaneous and Offal: In order to extract additional value from your beef carcasses, turn offal into dog food; tongue, liver, kidneys, fat, and ground bone can be made into high-end dog food. Hides, tendons, and pizzles of cattle can also be turned into dog chews. Keep in mind that pet food is regulated differently than human food. You may be able to make pet foods in a separate commercial kitchen facility and not have to ask your processor to do them for you. Some processors will make pet food items though.

PIG

SHOULDER

Cut name	Suggested size in lbs or thickness	Quantity per head
bone-in Boston butt roast	3–5 lbs	4–6
front hock	whole	2
Summer cuts—focus on grilling items		
country-style ribs	1¼"–1½"	6
shoulder steaks	¾"–1"	10–12

RIB

Cut name	Suggested size in lbs or thickness	Quantity per head
Summer cuts—focus on grilling items		
boneless rib chop (aka boneless loin chop)	1"–1¼"	20–24
baby back ribs	whole	2
Winter cuts—focus on roasts, holiday items		
crown rib roast	whole	2

Sample Cutting Instructions

LOIN

Cut name	Suggested size in lbs or thickness	Quantity per head
bone-in loin chops	1"–1¼"	20–24
Summer cuts—focus on grilling items		
sirloin kabobs	1 lb per pack	4–6

LEG

Cut name	Suggested size in lbs or thickness	Quantity per head
rear hock	whole	2
Summer cuts—focus on grilling items		
ham steaks	¾"–1"	10–12
sliced deli ham	1 lb per pack	6–8 lbs
Winter cuts—focus on roasts, holiday items		
bone-in ham	whole, cut in half, or quartered	2–8

BELLY

Cut name	Suggested size in lbs or thickness	Quantity per head
spare ribs	whole	2
belly—make bacon	sliced	16–20 lbs

MISCELLANEOUS

Cut name	Suggested size in lbs or thickness	Quantity per head
kidney fat	2–3 lb packs	2–4
back fat	2–3 lb packs	2–4
cheeks	whole	2
jowl	sliced (can be cured and smoked)	2–3 lbs

OFFAL

Cut name	Suggested size in lbs or thickness	Quantity per head
heart	trimmed, whole	1
tongue	whole	1
liver	½"	4–5 lbs
whole head	whole	1

TRIMMINGS

Cut name	Suggested size in lbs or thickness	Quantity per head
Summer cuts—focus on grilling items		
pork trim	sausage	15–20 lbs
Winter cuts—focus on roasts, holiday items		
pork trim	sausage	30–45 lbs

Summer Pork Cut Sheet Notes:

The thing with pork is that everyone loves bacon and sausage. Therefore, if you feel that some of the above products won't sell to your customer base, but sausage will, by all means turn some of the leg, shoulder, or other lower-end cuts into sausage. Chris has had clients who instructed that their pig carcasses be turned into only pork chops and bacon, with the rest going into sausage. This is something you'll need to research in your market.

Chris believes the above cut sheet shows a good amount of cuts for a wide range of customer interests: breakfast meat, BBQ, grilling, and luncheon meat.

Shoulder: Country-style ribs come from the shoulder and are essentially the same cut as the Delmonico (see the beef cut sheet). They are really just a pork chop that is sliced in half to create a "rib" shaped piece of meat. They are great for grilling or pan-frying.

Boston butt is the most popular cut for BBQ pork. It is best slow-cooked, either smoked or braised. The higher fat content makes it ideal for slow cooking as the meat stays moist. Leave the bone in.

Shoulder steaks have good meat, but they are not the tenderest cut. That is why it's recommended to cut them slightly thinner than a pork chop. Also, they are a larger cut than a pork loin chop and therefore portion size becomes a factor in determining the thickness as well. They are great for grilling. If you can't sell these, grind them into sausage.

The front hock is a good piece to sell as an addition to beans, greens, or other dishes that need a little pork flavor. There is not a lot of meat on them, but there is tons of flavor. These can be cured and smoked as well to add a salty and smoky flavor. We used to have our smoked hocks cut into thinner "discs" which our customers loved because they could just throw a small piece into a soup or pot of beans for flavoring instead of a whole hock.

Rib: Just as folks enjoy bacon, there is also a large demand for baby back ribs. Unfortunately, there are only two racks per animal and they are not very big. Just as with the belly, baby back ribs are attached to the rib portion of the pork loin when they are cut from the carcass. And, just as with the belly, you can leave more meat on the ribs, but that means less meat for the boneless rib chops. This may be a trade-off you're willing to live with as you can charge a good bit for baby backs. Chris recommends directly communicating how much meat is to be left on the baby backs with your butcher to make sure you're both clear on the expectations here. After trying to sell baby back ribs ourselves, we recommend cutting them in half since they are fairly large and sometimes intimidating to customers. We also found that we could sell bone-in rib chops for days but baby backs were not requested as often. You will have to test that one out in your market.

Loin: Now that you have your boneless pork chops from the rib section, you can cut bone-in loin chops from the loin section. This is the same cut that is called a T-bone on a beef. It looks just like a T-bone, as it includes a portion of the tenderloin and a portion of the "strip loin." Many people enjoy eating bone-in pork chops on the grill.

The other option with this section is to make more boneless pork chops and save the tenderloin whole. However, commercial pork tenderloin is much larger than your average market hog raised on a small farm. They usually don't quite weigh a pound. Therefore, it makes much more sense to Chris to leave them on the chop, include the bone, and charge a fair amount for a nice, thick, bone-in pork chop.

The sirloin can be boned out and turned into very nice kabob meat. If you have a market for pork kabobs, this is a good option. If you don't have a demand for kabobs, the sirloin can be cut into chops, but a better option in Chris's opinion is sausage. Year-round, you can never have enough sausage.

Leg: Ham steaks are essentially cured and smoked ham, bone-in, sliced into steaks. Many customers enjoy these as a breakfast meat, or baked for dinner. With two legs on a pig, Chris recommends turning one into ham steaks for sale to these customers. With the other leg he recommends boneless ham, sliced thin for luncheon meat. It will be the same flavor as the ham steaks, but for a different use by the customer. In our experience, since none of our processors had a good ham recipe, we preferred to grind our whole legs into both ground pork and sausages.

Keep in mind the leg is a lean cut, so if you're considering this cut for sausage, make sure you ask the processor to add some fat to the mix to keep the proper ratio (70/30, lean/fat). You should have plenty of backfat on the animal for this purpose.

The rear hock can be treated the same as the front hock, left fresh, or cured and smoked—both treatments can be used for a flavoring item in other dishes such as beans or greens. Or you can bone it out and use the meat for sausage.

Belly: The belly is made up of two cuts: belly and spare ribs. They are attached when they are first cut from the carcass, and at this point a decision must be made: By weight, do you want more bacon, or a meatier rib? Most of the spare ribs that are sold these days in the grocery store have belly meat attached to the ribs. That is why much of the commercial bacon is so slender. In Chris's experience people are willing to pay much more for bacon than they are for ribs. Therefore, he suggests leaving as much meat on the belly as possible and making a thinner rib with less meat on it. These spare ribs are very tasty, but do not have as much meat on them as what's generally found at the store. He recommends explaining this to the customer when they're purchasing the ribs so they are not taken aback when they get home with their package. We used to ask our processor to leave a little meat on the spare ribs; we didn't mind losing a little belly to the ribs because we grew out our pigs to such a large size that they had pretty thick bellies.

Of course, you don't have to make bacon from your pork belly. There are many chefs these days who use fresh pork belly in a number of

Sample Cutting Instructions

dishes. However, the overwhelming majority of consumers will purchase bacon before they will purchase fresh belly, and you can get a higher price for bacon. But, depending on your market, you may consider saving a fresh belly or two to have in the event you get a call for it. It also depends on how well your meat processor does bacon. If bacon is not their specialty, you may be better off doing fresh belly or fresh sliced belly. Nothing is worse than selling a customer unsavory bacon—they may never come back.

Miscellaneous and Offal: Kidney fat, also called leaf lard, is the lightest, highest-quality fat on the hog. It is generally kept separate from the other fat and used to make lard for baking. You can ask the processor to coarse-grind it for you if you like to make the larding process easier on the customer. Sometimes it can be challenging for your processor to obtain the leaf lard if the slaughterhouse has removed the kidneys. Most of our pig carcasses had the kidneys still attached so they could be utilized, including their leaf lard.

Backfat, as mentioned above, is not as high quality as kidney fat, although it can be made into lard used for frying or other cooking applications besides baking. It is great to save for blending into your sausages or it can also be sold in bulk bags to home cooks who like to render their own lard. Not all processors have rendering capabilities.

Pork cheeks, although small, make great braised dishes, like tacos. However, if you feel that you will not have enough volume, you can ask that they be included in your sausage trim.

Pork jowl is a fatty cut that comes from between the head and the neck of the pig. It can be treated by the processor the same as pork belly to make a "face bacon." You can sell this item fresh as well. It is a superior cut for charcuterie applications.

Pork heart and tongue are, like cheeks, a very specialty item. They may not yield much, but over a period of time they can be stockpiled and sold to an ethnic customer.

Pork liver is popular for some dishes such as German-style terrines and sausages. It is a very different flavor than beef liver and some customers will only eat pork liver, if they eat liver at all. It is very rich and flavorful.

Pork head is the essential ingredient for head cheese. Not many folks make head cheese these days, but it is coming back into style in restaurants. Consider keeping your pork heads for wholesale customers.

Trimmings: Pork trim is used mainly for making sausage, although you can make fresh ground pork if you have requests. Again, Chris recommends making all the sausage you can because it is a very popular item in most markets. You can make various flavors with either your own proprietary recipe, or by using the processor's pre-blended mixes. We were not happy with our processor's spice mixes due to the addition of some preservatives, so we elected to have ours just made into ground pork. We then told customers exactly how to make their own spice blends to mix into it, making it into whatever flavor of bulk sausage they wanted. Also, you will have to decide if you will do cased sausages or bulk sausage. Sausage links in casings are great for the grill, but many people (ourselves included) are not enthusiastic about the origin of those casings (which in large part are synthetically derived from cow hide collagen). You can get natural lamb or pig casings (from real intestines), but they are more expensive, not organic, and frequently not an option at a cut-and-wrap facility. Because we had so many customers questioning the origin of our casings, we opted to stop making links and do all bulk sausage, which is also less expensive to have made compared to links.

For the fabrication techniques for all of these cuts go to www.porkfoodservice.org or the NAMP guide.

Winter Pork Cut Sheet Notes:
Only the variances from the summer cut sheet will be explained below.

Rib: When it comes to ribs, there are a limited amount compared to how popular they are with customers. Chris recommends stockpiling your ribs from the winter to sell in the summer when everyone wants them for the grill. Ribs don't move much in the winter.

Crown rib roasts are a very popular holiday item. You can leave the roasts whole, or have them cut in half. Or, if you want to commit to a very fancy item, you can have two rib roasts sewn together by the butcher to make a traditional crown roast. Chris likes to market any frenched rib roast as a crown roast (frenched means when the meat is removed to expose the clean rib bone), but a true crown roast is two frenched rib roasts sewn together in the shape of a circle to create a "crown." That is, however, a lot of meat, and many families may not want that much. It is truly a special occasion cut, maybe best left as a "reserved" cut only, where people prepay a deposit to reserve themselves one.

Leg: Ham roasts are very popular for the holidays, even moving into the spring with Easter. Chris recommends making many large and small ham roasts to have a variety for your customers. Cut a few in half, some in quarters, and leave a few whole. Of course, there are only two legs on a hog, so if you're only processing one hog at a time, consider what the most popular sizes will be. Again, if your processor does not make a good ham, then don't have ham made. You could have a small fresh ham roast made, which makes a nice, lean roasting joint, or you could cube some up for stew meat. As in the summer, you could also turn most of it into sausage, which sells well year-round.

LAMB AND GOAT

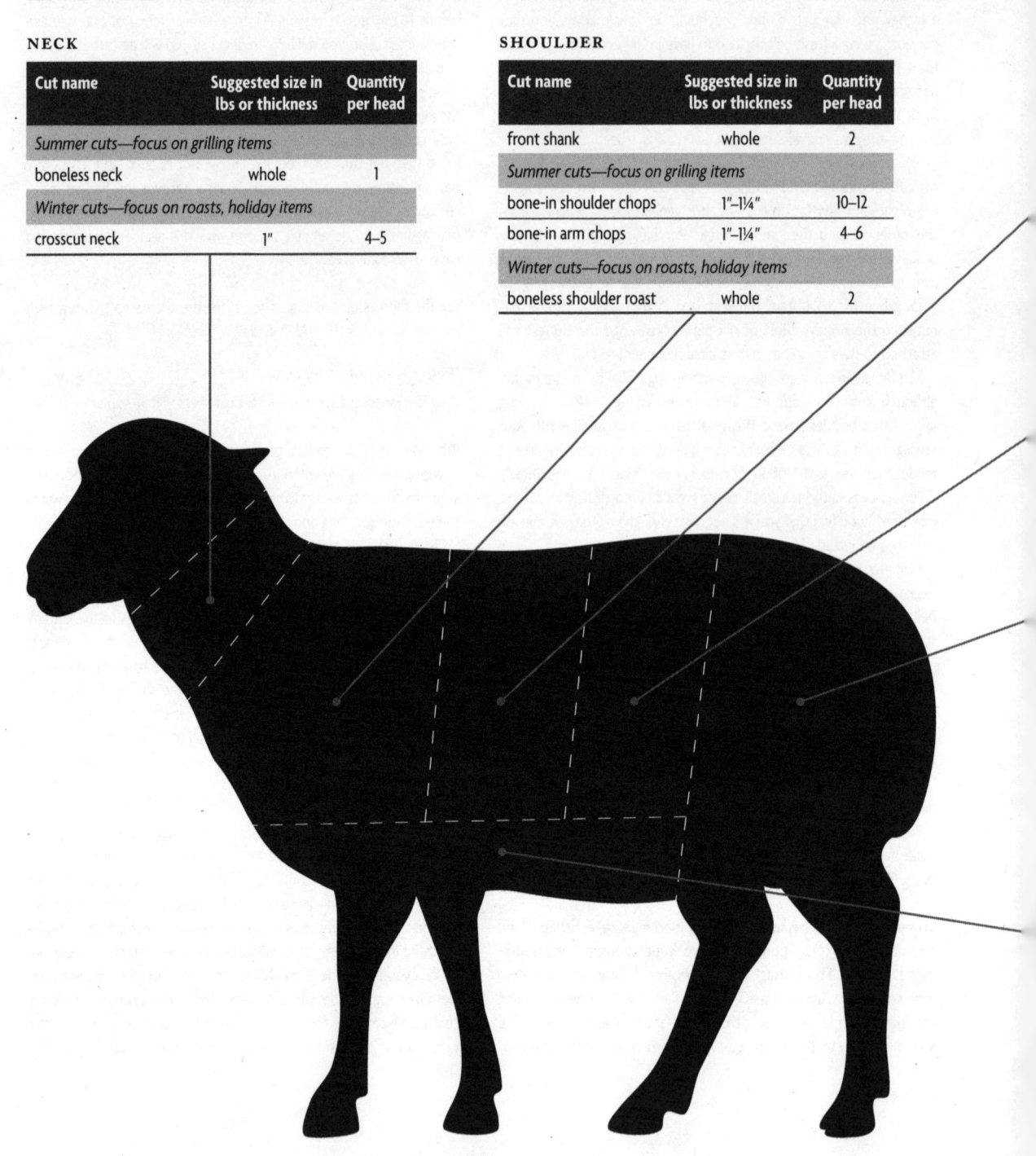

NECK

Cut name	Suggested size in lbs or thickness	Quantity per head
Summer cuts—focus on grilling items		
boneless neck	whole	1
Winter cuts—focus on roasts, holiday items		
crosscut neck	1"	4–5

SHOULDER

Cut name	Suggested size in lbs or thickness	Quantity per head
front shank	whole	2
Summer cuts—focus on grilling items		
bone-in shoulder chops	1"–1¼"	10–12
bone-in arm chops	1"–1¼"	4–6
Winter cuts—focus on roasts, holiday items		
boneless shoulder roast	whole	2

Sample Cutting Instructions

RIB

Cut name	Suggested size in lbs or thickness	Quantity per head
Summer cuts—focus on grilling items		
rib chops	1¼"–1½"	10–12
Winter cuts—focus on roasts, holiday items		
frenched rack	whole	2

LOIN

Cut name	Suggested size in lbs or thickness	Quantity per head
bone-in loin chops	1¼"–1½"	8–10
boneless sirloin chops	1¼"	4–6
English-style short ribs	2–3"	10–16

LEG

Cut name	Suggested size in lbs or thickness	Quantity per head
rear shank	whole	2
Summer cuts—focus on grilling items		
bone-in leg steaks	¾"–1"	6–8
boneless leg roast, butterflied	whole	1
Winter cuts—focus on roasts, holiday items		
bone-in leg roast	whole, or cut in half	1–2
boneless leg roast	whole	1

BREAST

Cut name	Suggested size in lbs or thickness	Quantity per head
Summer cuts—focus on grilling items		
riblets	individual whole rib	16
Winter cuts—focus on roasts, holiday items		
ribs, Denver cut	half	4

OFFAL

Cut name	Suggested size in lbs or thickness	Quantity per head
heart	trimmed, whole	1
tongue	whole	1
liver	½"	1–2 lbs
sweetbreads	whole	0.5 lbs
whole head	whole	1

TRIMMINGS

Cut name	Suggested size in lbs or thickness	Quantity per head
trim	sausage and ground	4–5 lbs

Although lamb and goat are different species, anatomically they are very similar. You can cut them exactly the same way. Below are some fairly complicated ways to break down your animals to provide a wide selection of cuts for your market. However, you may find that ground lamb and lamb sausage will be your most popular products. Each animal will yield very little grind, maybe just 4 to 5 pounds. One good way to increase the amount of grind and sausage you have available is to bone out and grind the entire carcass of older, more affordable animals, such as hogget and cull ewes and does. Or save the rib and loin of these animals and grind the rest.

Summer Lamb and Goat Cut Sheet Notes:

Neck: Young lamb and goat will not have much neck meat but an older yearling will. The tough muscles of the neck are full of flavor and lots of collagen. You can bone out the neck and make a nice little braising roast. This may not sell well in the summer, so you could freeze and save it for winter sales.

Shoulder: Bone-in shoulder chops offer a good item for grilling that is a lower-priced cut for budget-conscious consumers. The fat content in the shoulder makes for great flavor, the bone adds interest as well as saleable weight, and you get a fair amount of chops to sell.

Arm chops are small, but very tender and flavorful. Again, a good lower-priced chop option.

Foreshanks, aka front shanks, are not as meaty as the hind shanks, but are very flavorful and should make a good portion for someone with a small appetite, such as an older customer. Good for braising. Some consider this to be mainly a winter cut, but we find that customers ask for shanks year-round. We've found that it is one lamb or goat cut that people are very familiar with in the United States (thanks in part to the food service industry).

Rib: Rib chops are the classic lamb chop. They look like a little meat lollipop. When a restaurant serves lamb chops, nine times out of ten they are rib chops. These should sell themselves, and you can get a good price for them.

Loin: Bone-in loin chops are like mini T-bones—both adorable and delicious. Cut thick, they make a great cut for the grill, pan-seared, or roasted in the oven or broiler. Two or three per person is a good portion depending on the size of the lamb or goat.

The sirloin of a lamb or goat is very small. The bone-to-meat ratio is not very good. This is why Chris recommends getting the sirloin chops boneless. There is a lot of flavor, they're easy to cook, and they are a good middle-of-the-road cut price-wise.

Leg: Instead of going heavy on the leg roasts, in the summer you can cut leg steaks. These are excellent on the grill with most types of herbs and seasonings.

Butterflied leg is the best of both worlds—it can be sold as a boneless roast. However, if you unroll the roast as it's butterflied, it goes great on the grill for a cut that can be shared with a group.

Rear shank, aka hind shank, is the meatier of the two shank types. A larger portion with better meat than the foreshank.

Breast: This is really the belly cut, but is often called the breast. It is where you will get your "spare ribs" in pork terms. The butcher will cut straight between each rib to give you individual "riblets." These are great on the grill or broiled in the oven. Good with heavy seasoning or sauce. The belly meat can be cut off and turned into lamb bacon. (Yum.)

Offal: All of the offal is very tasty, but very small. These items are generally only used by high-end restaurants, if at all. Scour the markets for ways to unload these items, as you can get a good price for them and increase your margin on an already low-yielding carcass.

Trimmings: There won't be a lot of trim left from these carcasses, but what there is should sell relatively easily. If you want to get into value-added items from your lamb or goat, try making a sausage link, patty, or bulk with the trim. Lamb and goat make outstanding sausage in many different flavor types. Chris recommends a strong seasoning blend to complement the distinct flavor of the meat. Many people like making traditional sausages that were derived for lamb and goat, such as sausages from Africa and the Middle East where a lot of lamb and goat is consumed.

For the fabrication techniques for all of these cuts go to www.americanlamb.com or the NAMP guide.

Winter Lamb and Goat Cut Sheet Notes:

Only the variances from the summer cut sheet will be explained below.

Neck: A nice way to prepare the neck for winter sales is to crosscut it along each vertebra for easy portioning for soups or stews or to fortify stocks.

Shoulder: The boneless shoulder is one of the roasts available (there are not many to begin with) that will feed a large group. This cut should not be as pricey as leg roasts, and therefore will satisfy a customer looking for a good braising or roasting cut that doesn't break the bank. This cut can be made into stew meat, which is another nice option for winter stew season. Remember, lamb and goat are small creatures and you will only get four good-sized roasts from the whole carcass (two shoulder, two leg).

There is an "arm" cut, however it is very small. You'll find on the summer cut sheet that there are arm chops. These may not sell well in the winter, and therefore should be made into a cut that will sell well—for example, sausage. You will find more on sausage below.

Sample Cutting Instructions

Rib: Although you do not have to get the rib rack frenched, it makes for a much nicer presentation and you can charge a good bit more for it. If your customer base will not spend the money on this fancy cut, by all means go without frenching and get a standard rack. Either way, the rack is a very popular holiday cut, or even just a nice small roast for a few people. We had difficulty selling full racks due to the size and the price. You may want to cut them in half for a smaller presentation.

Leg: We have listed three ways of cutting the leg in the table above. The number of carcasses you're having processed may determine how many of each roast you'd like to have cut. Some folks want a smaller roast (half a bone-in leg), some want a big roast for the holidays, and some want boneless (easier to carve and cook). We recommend getting a few roasts of each style to satisfy a broad range of customer types.

Breast: You can have your processor trim these from all of the fat and gristle leaving a nice, square-cut rack of ribs. When you cut them in half (eight ribs cut into two pieces of four ribs each) you get Denver-cut ribs. These make a very nice dish for an appetizer or hors d'oeuvres.

Trimmings: As in the summer, turn your trim into ground or sausage. Some customers may question if they really want to try lamb for breakfast, but if your processor makes a good breakfast sausage, this is one of the best meats to eat with hash and eggs or on a breakfast sandwich. People eat heartier breakfasts in the winter, so have plenty of breakfast sausage on hand.

References

Alabama Cooperative Extension. 2007. *Reproductive Management of Sheep and Goats.* Publication: ANR-1316.

Alexandre, Joseph J. 2013. *Grassfed Dairy Steer Enterprise Analysis for Alexandre Family EcoDairy Farms.* Paper presented to satisfy Bachelor's of Science requirements, California Polytechnic State University.

Barton-Gade, P. 2008. "Effect of Rearing System and Mixing at Loading on Transport and Lairage Behaviour and Meat Quality: Comparison of Outdoor and Conventionally Raised Pigs." *Animal* 2 (6): 902–11.

Berlow, Ali. 2013. *The Mobile Poultry Slaughterhouse: Building a Humane Chicken Processing Unit to Strengthen Your Local Food System.* North Adams, MA: Storey Publishing.

Bhandari, B. D., J. Gillespie, G. Scaglia, and J. Wang. 2013. "Analysis of Pasture Systems to Maximize the Profitability and Sustainability of Grass-fed Beef Production." Selected poster prepared for Southern Agricultural Economics Association (SAEA) Annual Meeting, Orlando, Florida, February 3–5.

Brown-Brandl, T. M., Roger A. Eigenberg, and J. A. Nienaber. 2006. *Heat Stress Risk Factors of Feedlot Heifers.* Publication from USDA-ARS/UNL Faculty Paper 151.

Burden, Dan. 2012. Deer Ranching Profile webpage, Agricultural Marketing Resource Center, Iowa State University. Accessed December 28, 2014. www.agmrc.org/commodities__products/livestock/deer-venison-ranching-profile/.

Collins, Jim. 2001. *Good to Great. Why Some Companies Make the Leap . . . and Others Don't.* New York: HarperCollins.

Conner, M., M. Jaeger, T. Weller, and D. McCullough. 1998. "Effect of Coyote Removal on Sheep Depredation in Northern California." *Journal of Wildlife Management* 62 (2): 690–99.

Corah, Larry R. 2011. "Marketing Advantages for Calves with Performance and Source Information." Proceedings of Applied Reproductive Strategies in Beef Cattle, Boise, Idaho, September 30–October 1. Accessed March 30, 2014. beefrepro.unl.edu/proceedings/2011northwest/07_nw_marketing_corah.pdf.

Danforth, Adam. 2014a. *Butchering Beef: The Comprehensive Photographic Guide to Humane Slaughtering and Butchering.* North Adams, MA: Storey Publishing.

Danforth, Adam. 2014b. *Butchering Poultry, Rabbit, Lamb, Goat, and Pork: The Comprehensive Photographic Guide to Humane Slaughtering and Butchering.* North Adams, MA: Storey Publishing.

Davies W. L. 1939. "Fishiness as a flavour and a taint." *Flavours* 2: 18–21.

Ellis, M., A. J. Webb, P. J. Avery, and I. Brown. 1996. "The Influence of Terminal Sire Genotype, Sex, Slaughter Weight, Feeding Regime and Slaughter-House on Growth Performance and Carcass and Meat Quality in Pigs and on the Organoleptic Properties of Fresh Pork." *Animal Science* 62 (3): 521–30.

Fairlie, Simon. 2011. *Meat: A Benign Extravagance.* White River Junction, VT: Chelsea Green Publishing.

Geverink, N. A., A. Kappers, E. Van de Burgwal, E. Labmooij, J. H. Blokhuis, and V. M. Wiegant. 1998. "Effects of Regular Moving and Handling on the Behavioral and Physiological Responses of Pigs to Pre-Slaughter Treatment and Consequences for Meat Quality." *Journal of Animal Science* 76: 2080–85.

Goodman, Jack. 2008. *Grocery Shopping: Who, Where, & When.* The Time Use Institute white paper,

References

(October). Accessed May 15, 2014. timeuseinstitute.org/Grocery%20White%20Paper%202008.pdf.

Goodsell, Martha and Tatiana Stauton. 2010. *Resource Guide to Direct Marketing Livestock and Poultry*. Ithaca, NY: Cornell University Small Farms Program.

Gwin, L., and A. Thiboumery. 2013. *From Convenience to Commitment: Securing the Long-Term Viability of Local Meat and Poultry Processing*. NMPAN. Accessed April 1, 2014. http://hdl.handle.net/1957/38213.

Hahn Niman, Nicolette. 2014. *Defending Beef: The Case for Sustainable Meat Production*. White River Junction, VT: Chelsea Green Publishing.

Herring, Andy D. 2006. *Genetic Aspects of Marbling in Beef Carcasses*. Certified Angus Beef white paper.

Holderread, Dave. 2011. *Storey's Guide to Raising Ducks, 2nd Edition*. North Adams, MA: Storey Publishing.

Infovets. n.d. Management of Rams and Bucks. publication B538. Accessed April 3, 2014. www.infovets.com/demo/demo/smrm/B538.HTM

John F. Kennedy School of Government, Harvard University. 2006. *Social Capital Community Benchmark Survey*. Accessed April 15, 2014. www.ropercenter.uconn.edu/data_access/data/datasets/social_capital_community_survey.html

Johnson, R. J., D. L. Marti, and L. Gwin. 2012. *Slaughter and Processing Options and Issues for Locally Sourced Meat*. USDA, Economic Research Service publication.

Kim, D. H., P. N. Seong, S. H. Cho, J. H. Kim, J. M. Lee, C. Jo, and D. G. Lim. 2009. "Fatty Acid Composition and Meat Quality Traits of Organically Reared Korean Native Black Pigs." *Livestock Science* 120 (1–2): 96–102.

Koch, R. M., H. G. Jung, J. D. Crouse, V. H. Varel, and L. V. Cundiff. 1995. "Digestive Capability, Carcass, and Meat Characteristics of *Bison bison*, *Bos taurus*, and *Bos* × *Bison*." *Journal of Animal Science* 73: 1271–81.

Lebret, B. 2008. "Effects of Feeding and Rearing Systems on Growth, Carcass Composition and Meat Quality in Pigs." *Animal* 2 (10): 1548–58.

Lebret, B., H. Juin, J. Noblet, and M. Bonneau. 2001. "The Effects of Two Methods of Increasing Age at Slaughter on Carcass and Muscle Traits and Meat Sensory Quality in Pigs." *Animal Science* 72: 87–94.

Luce, William G., Joseph E. Williams, and Raymond L. Huhnke. 2007. *Farrowing Sows on Pasture*. Oklahoma Cooperative Extension Services, ANSI-3678.

MacDonald, James M, and William McBride. 2009. *The Transformation of US Livestock Agriculture: Scale, Efficiency, and Risks*. USDA Economic Research Service Bulletin 43.

Masamha B., C. T. Gadzirayi, & I. Mukutirwa. 2010. "Efficacy of *Allium Sativum* (garlic) in Controlling Nematode Parasites in Sheep." *International Journal of Applied Research in Veterinary Medicine* 8 (3): 161–69.

Mattocks, Jeff. 2014. "Grow and Grind Your Own Poultry Feed." Presentation given at Pennsylvania Association for Sustainable Agriculture Conference, State College, Pennsylvania, February 7.

Mintel Group Ltd. 2012. *Red Meat-US-Report*, (October). reports.mintel.com/display/590735/.

Moyer, Brian. 2006. "Direct Marketing Isn't about Selling, It's about Dreaming." In *Raising Poultry on Pasture: Ten Years of Success*, edited by Jody Padgham. Hughesville, PA: American Pasture Poultry Producers Association.

Muth, M. K., S. A. Karns, M. K. Wohlgenant, and D. W. Anderson. 2002. "Exit of Meat Slaughter Plants During Implementation of the PR/HACCP Regulations." *Journal of Agricultural & Resource Economics* 27 (1): 187–203.

Nation, Allen. 2002. *Farm Fresh: Direct Marketing Meats and Milk*. Ridgeland, MS: Green Park Press.

NMPAN (Niche Meat Processor Assistance Network). 2011. "Working Effectively with Your Processor." Webinar, accessed December 29, 2014. www.extension.org/pages/33477/nmpan-webinars#.VKBuk_8AA.

NMPAN. 2013. "Lorentz Meats: A Case Study." In *From Convenience to Commitment: Securing the Long-Term Viability of Local Meat and Poultry Processing* by Lauren Gwin and Arion Thiboumery. Accessed March 1, 2014. www.extension.org/sites/default/files/Lorentz%20Meats%20Case%20Study.pdf.

NMPAN. 2014. "Cost Analysis: Are You Making Money?" Webinar, accessed December 29, 2014. Accessed March 1, 2014. www.extension.org/pages/33477/nmpan-webinars#.VKBuk_8AA.

North American Elk Breeders Association. Rules and Regulations. Accessed April 11, 2014. www.naelk.org

Pollan, Michael. 2006. *The Omnivore's Dilemma: A Natural History of Four Meals*. New York: Penguin Press.

Posch J., G. Feierl, G. Wuest, W. Sixl, S. Schmidt, D. Haas, F. F. Reinthaler, and E. Marth. 2006. "Transmission of *Campylobacter* spp. in a Poultry Slaughterhouse and Genetic Characterisation of the Isolates by Pulsed-Field Gel Electrophoresis." *British Poultry Science* 47 (3): 286–93.

Practical Farmers of Iowa and Iowa State University. 2007. "Biosecurity, Pig Flow, and Introduction of Stock." In *Managing for Herd Health in Alternative Swine Systems: A Guide*, edited by Rick Exner. Accessed February 21, 2014. www.pfi.iastate.edu/OFR/Livestock/Herd_Health/Managing_for_Health.pdf.

Pugliese, C., F. Sirtori, S. D. Adorante, S. Parenti, A. Rey, C. Lopez-bote, and O. Franci. 2008. "Effect of Pasture in Oak and Chestnut Groves on Chemical and Sensorial Traits of Cured Lard of Cinta Senese Pigs." *Italian Journal of Animal Science* 8: 131–42.

Raes, K., S. De Smet, and D. Demeyer. 2004. "Effect of Dietary Fatty Acids on Incorporation of Long Chain Polyunsaturated Fatty Acids and Conjugated Linoleic Acid in Lamb, Beef and Pork Meat: A Review." *Animal Feed Science and Technology* 113 (1–4): 199–221.

Ruechel, Julius. 2006. *Grass-Fed Cattle: How to Produce and Market Natural Beef*. North Adams, MA: Storey Publishing.

Rust, S. R., and C. S. Abney. 2005. "Comparison of Dairy versus Beef Steers." In *Managing and Marketing Quality Holstein Steers*, edited by R. Tigner and J. Lehmkuhler. 161–74. Madison, WI: Wisconsin Agri-Service Association.

Salatin, Joel. 1995. *Salad Bar Beef*. Swoope, VI: Polyface Farm.

Savell, Jeffrey W. 2012. *Beef Carcass Chilling: Current Understanding, Future Challenges*. Beef Checkoff white paper.

Savory, Allen, and Jody Butterfield. 1998. *Holistic Management: A New Framework for Decision Making, 2nd Edition*. Washington D.C.: Island Press.

Schafer, David. 2006. "From Customers into Converts." In *Raising Poultry on Pasture: Ten Years of Success*, edited by Jody Padgham. Hughesville, PA: American Pasture Poultry Producers Association.

Schatzker, Mark. 2010. *Steak: One Man's Search for the World's Tastiest Piece of Beef*. New York: Penguin Group.

Shamel, Clarence. 1911. *Profitable Stock Raising*. New York: Orange Judd Company.

Stringer, W. C. 1970. "Influence of Nutrition on Pork Quality." In *Pork Carcass Quality: A Research Review*. Columbia, MO: Extension Division, University of Missouri.

Thomas, Heather S. 2009. *Cattle Health Handbook*. North Adams, MA: Storey Publishing.

Trimble, Lisa M., Walid Q. Alali, Kristen E. Gibson, Steven C. Ricke, Philip Crandall, Divya Jaroni, Mark Berrang, and Mussie Y. Habteselassie. 2013. "Prevalence and Concentration of Salmonella and Campylobacter in the Processing Environment of Small-Scale Pastured Broiler Farms." *Poultry Science* 92 (11): 3060–6.

Umberger, W. J., D. D. Thilmany McFadden, and A. R. Smith. 2009. "Does Altruism Play a Role in Determining U.S. Consumer Preferences and Willingness to Pay for Natural and Regionally Produced Beef?" *Agribusiness* 25: 268–85.

USDA Economic Research Service. 2013. Animal Products. Accessed June 1, 2014. www.ers.usda.gov/topics/animal-products.aspx

Walker, Randy. 2003. *Swine: Selection and Mating of Breeding Stock*. University of Florida IFAS Extension Publication RFAA083.

Wilson, Annie. 2001. "Romance versus Reality: Hard Lessons Learned in a Grass-fed Beef Marketing Cooperative." *Rural Papers Newsletter*, October. Accessed April 20, 2014. www.extension.iastate.edu/agdm/wholefarm/html/c5-220.html.

Wood, J. D., G. R. Nute, R. I. Richardson, F. M. Whittington, O. Southwood, G. Plastow, R. Mansbridge, N. da Costa, and K. C. Change. 2004. "Effects of Breed, Diet and Muscle on Fat Deposition and Eating Quality in Pigs." *Meat Science* 67 (4): 651–67.

Index

Note: page number followed by *ci* refer to the picture insert; page numbers followed by t refer to Tables.

100 Percent Grassfed, 13

abattoir, defined, 152. *See also* slaughtering and processing
accounts receivable, 277–278
acorns, 79–80
adult roundworms, 97
advertising. *See* marketing
AFO (animal feeding operations), 6
Afton Field Farm, 1*ci*–6*ci*, 9*ci*–11*ci*, 16*ci*
aggregated models, 202–208
aging loss, defined, 210
aging meat, 13*ci*, 127, 225–226
Agricultural Marketing Service, 13
all-in all-out systems, 96
amenable species, defined, 152
American Chinchilla rabbits, 130
American Consortium for Small Ruminant Parasite Control (ACSRPC), 68
American Grassfed Association (AGA), 13
American Guinea Hog, 15*ci*, 81t, 82
American Guinea Hogs, 2
American Humane Certified, 13
American Pastured Poultry Producers Association (APPPA), 19
American Yorkshire, 83
Ancona ducks, 46
Angus cattle, 106

animal feeding operations (AFOs), 6
animal health and welfare, 8
Animal Welfare Approved (AWA), 4*ci*, 12, 13, 14
ante-mortem inspections (AM), 163
anthelmintics, 51
antibiotic-resistant diseases, x
antibiotics, x, 6
Ark of Taste, 14–15
artificial insemination (AI), 51, 57, 85, 108, 113, 145–146
ascites, 38
aspergillosis, 38
auctions, 165
author photo, 16*ci*
automatic waterers, 34–35
Axis deer, 145, 146

Balancer cattle, 106
bankrupt worm, 67–68
barber pole worm, 64, 67
barrow, defined, 76
bears, 40
bedding material for poultry, 30
Beef Checkoff, 265
Beefmaster cattle, 9*ci*, 107
Bennett, Dan, 31
Berkshire, 81–82
Berlow, Ali, 160
biosecurity, 287–288
bison, 12*ci*, 14*ci*, 140–142
Black and White Baldies cattle, 106
Black Angus cattle, 106, 107t

Black Earth Meats, 180–181, 202–204
blackhead, 38
blast chilling, 225
bloat, 121
boars, 76, 85, 89–90, 92, 149–150. *See also* pigs
bobcats, 40
body condition score (BCS), 58, 109, 111
body-heat brooders, 31–32
Boer goats, 52t, 53
Bonsma, Jan, 108
Boondockers Farm, 46–47, 49
Bos indicus cattle breeds, 107
boxing, 243–244
Brahma cattle, 107
branding, 259–260, 262–263, 266–269
Brangus cattle, 106
breeding
 elk and deer, 144–146
 pigs, 84–87
 rabbits, 131–132
 sheep and goats, 56–59
 systems, 83–84
brewers' grains, 93
British cattle breeds, 106
Brix measurements, 119
Broad-Breasted Bronze turkeys, 14, 25t, 37
broiler, defined, 19
Broken Arrow Ranch, 238
brooding phase, 28–32
Brookshire Farm, 124

- 313 -

brown stomach worm, 67
brucellosis, 147
buck, defined, 51, 130
buckling, defined, 51
budgeting, 259–269
buffalo. *See* bison
bulls, 103, 112–113. *See also* cattle
butchering. *See* slaughtering and processing
Butchering (Danforth), 226
buying clubs, 180–181

cabrito, 71
Cadman, Jerica, 286
CAFO (confined animal feeding operations). *See* confined animal feeding operations
calf, defined, 103
California grazing climate, 104–105
Californian rabbits, 130
Callicrate Beef and Ranch Foods Direct, 189
Callicrate, Mike, 189
caprids, 51
caprine arthritis encephalitus (CAE), 69
carcass yield, 210, 222
Carman, Cory, 198–199, 200, 219–220
Carman Ranch, 195, 198–199, 219–220, 252, 258
castration, 70–71, 89–90
caterers, 192, 194–195, 197–199
cattle. *See also* grassfed beef
　beef quality, 123–125
　breeds, 9*ci*, 106–107
　bulls, 112–113
　CAFOs and, 6
　calving, 110
　concentrated feeds and, xi
　confinement *vs.* pasture and, xi–xii
　cow body types, 112
　cows and their calves, 109–110
　definitions, 103
　feeds and feeding, 10*ci*, 11*ci*, 113–118

fencing and pasture, 5*ci*, 122–123
finishers, 113
genetics for grass-based finishing, 107–109
handling and slaughtering, 123
health of, 119–122
heifers, 103, 112
Morris Grassfed Beef, 126–127
Northern grazing climates, 105
post-harvest handling, 125
regulations for, 153–158
shelter, 113
soil, nutrients and plants, 118–119
Southeast and year-round grazing climates, 104
Southwest and California, 104–105
watering, 118
weaning, 111–112
cattle panel hoop structures, 33
Center for Environmental Farming Systems, 206
certifications, 11–15
Certified Humane, 13
certified naturally grown (CNG), 12, 13
certified organic, 12
cervids. *See* elk and deer
cestodes, 68
Champagne rabbits, 130
Charolais cattle, 106
Chefs Collaborative, 192
Chester White, 83
chevon, 71
chick, defined, 19
chickens. *See also* poultry
　breeds, 25–28
　brooders, 1–2*ci*
　confinement *vs.* pasture, xi–xii
　ducks *vs.*, 47–49
　feed and feeding, 35, 37
　industrialized approaches, ix
　slaughtering, 14*ci*, 43–45
chicken tractors. *See* portable bottomless shelters

chlostridium, 69
chronic wasting disease (CWD), 143–144, 147
clabbered milk, 31
Clover Creek Farms, 55–56
coccidiosis, 30, 38–39, 68–69, 147
cockerel, defined, 19
cold storage, 185–186, 224–225, 247–253, 271
colostrum, 61, 76, 103
Columbia sheep, 53t
communication issues, 212–214
competition pricing, 279–280
composite cattle breeds, 106–107
concentrated feeds, x–xi
confined animal feeding operations (CAFOs)
　cattle and, 6
confinement systems, ix–xii, 5–6
The Conscious Carnivore, 202–203
Continental cattle breeds, 106
continuous service, 57
Coolbots, 250–251
Cooperative Interstate Shipment (CIS) program, 153, 155, 214
Core, Jennifer, 179
Cornell University Extension, 251
Cornish Cross chickens, 25–26, 30
Corriente cattle, 15
cost control, 273–274
cost of goods sold, 273, 275
cost-plus accounting, 278
cowpooling, 173, 189–190
cows, 103, 112. *See also* cattle
coyotes, 40, 42
Creme D'Argent rabbits, 130, 131
cross-breeds, 15, 84, 106–107
CSAs (community-supported agriculture), 160, 176–180
cured meats, 227
Curtis, Jennifer, 207
customer-centricity, 257–258, 270–272
customer demographics, 261, 263–265

Index

custom-exempt plants, 152
custom sales, 172–173
custom slaughter *vs.* federally inspected slaughter, 153–158
cut and wrap, butchering or fabrication, 152
cutting instructions, 217–218, 296–309
cutting yield, 210, 221–223

Danforth, Adam, 72, 226
Davis, Walt, 113
day-old chicks, 28–29
day range shelters, 22–24, 33–34
Deck Family Farm, 4*ci*
deep bedding or hoophouse production, 76–77
deer. *See* elk and deer
Defending Beef (Niman), 2
defense chicks, ix–x
demand-side pricing, 278–279
demographics, 261, 263–265
Devon cattle, 106
dewormers, 51
Dexter cattle, 106
Dick's Kitchen, 199
direct-marketing, 15–17. *See also* marketing
direct-market meat production, 6–7
dirt lots, 80
disbudding, 71
displaying meat, 185–186
disposition of cattle, 108–109
distributors, 200–202
doe, defined, 51, 130
doeling, defined, 51
Doering family, 246–247
dog guards, 3*ci*, 40–41
dog predators, 40
dogs for herding, 99
donkeys, 40, 41
Dorpers, 52*t*, 72
drafts, 29
drake, defined, 19
drift lambing, 60
drought, 126

ducks
  Ancona breed, 3*ci*, 46–47
  baths, 4*ci*
  breed characteristics, 25*t*
  chickens *vs.*, 47–49
  feed and feeding, 37–38
  food trends and, 19–20
Durand, Bartlett, 202–203
Duroc hogs, 15, 83

Eatwild.com, 6
Ecological Farming Conference, 1
economically relevant traits (ERTs), 108
egg production, xi–xii. *See also* chickens; poultry
electric fencing, 8*ci*, 39, 63–67, 98–99
elk and deer
  breeds and breeding, 144–146
  feeding and watering, 146–147
  fencing, 148
  handling and slaughter, 148–149
  health, 147–148
  Nicky USA, 149–150
  overview, 142–143
  post-harvest handling, 149
  shelters, 146
  special regulations, 143–144
EMT conduit houses, 33
EQIP (Environmental Qualities Incentive Program), 103
ethical meats, 7–11
ewe, defined, 51
ewe lamb, defined, 51
exotic cattle breeds, 106
exotic droppings, 31
expected progeny differences (EPDs), 108

fabrication, 216
Fainting goats, 52*t*
Fairlie, Simon, 2
Fallow deer, 145
FAMACHA score, 58, 67–68, 73
farmers markets, 181–188

farmstands, 174–176
Farms with a Future, 2, 277
farm tours, 16*ci*
farrowing, 87–89
fatty acids, 224
FDA (US Food and Drug Administration), 152. *See also* Food Safety Inspection Services (FSIS)
federally inspected slaughter *vs.* custom slaughter, 153–158
feed additives, x
feed and feeding
  for broiler chickens, 35
  of cattle, 10*ci*, 11*ci*, 113–118
  of elk and deer, 146–147
  ethical meats and, 9–10
  grit, 34
  for pigs, 93–95
  for poultry, 30–31, 35–37
  of rabbits, 12*ci*, 133–135
  for sheep and goats, 63
fencing
  for cattle, 5*ci*, 122–123
  for elk and deer, 148
  for pigs, 7*ci*, 8*ci*, 97–99
  for poultry, 39, 42–43
  for sheep and goats, 63–67
financial management
  cash flow and, 277–278
  cost control and, 273–274
  financing and, 277
  growing larger, 285–286
  incorporation and, 284
  insurance and, 283–284
  labor management and, 284–285
  pricing strategies, 278–283
  profitability and, 275–277
finishers, 113
Firsthand Foods, 206–207
fladry, 41–42
Flagollet, Olivier, 179
flash freezing, 219–220
Flemish Giant rabbits, 130–131, 133
fleshing ability, 108

flock, defined, 51
Florida White rabbits, 130
Flynn, Daave, 198–199
flystrike, 69–70
fodder systems, 19
food co-ops, 191–194
Food Safety Inspection Services (FSIS), 152, 153, 155, 163, 216, 244–246, 270
food trucks, 175
foot rot, 70
forage, 63, 69, 94, 104–105, 113–114
The Foragers, 181
forest-fed pork, 79
Franzten, Tom, 16, 77, 85, 86, 99
Freedom Ranger, 25t, 27, 30
free range stationary shelters, 24–25
free ranging, xi–xii
free samples, 258
freezer space, 13*ci*, 166, 169–174, 247–253, 271
fresh *vs.* frozen meat, 247–253
friction points, 270–272
frozen *vs.* fresh meat, 247–253
fryer, defined, 130
FSIS (Food Safety Inspection Services), 152, 153, 155, 163
Fuller, Chris, 190, 195, 273

Galloway cattle, 106
game birds, 161
geese, 3*ci*, 25t, 37–38
Gelbvieh cattle, 106, 107t
gestating sows, 85–87
Giant Chinchilla rabbits, 130, 133
giblets, 44–45
gilts, 76, 84–85, 91–92, 96
Global Animal Partnership (GAP), 14
Gloucestershire Old Spot, 81t, 82
goats. *See* sheep and goats
Good to Great (Franzten), 16
Graese, Lee and Mary, 236–237
grain production, xi
grain subsidies, xi

Grandin, Temple, 123, 169
grants, 277
grassfed beef, xi–xii, 113, 123–125, 193–195, 198–199, 270. *See also* cattle
The Grassfed Gourmet, 1, 184
grazing management, 115–117
Gregoire, Evan, 46–47
grit, 34
grocery stores, 191–192
grow-out phase, 32–34
guard animals, 40–41, 55
guinea hens, 40
Gunthorp Farms, 89–90, 94–95, 101, 230, 235, 252
Gunthorp, Greg, 37, 85–86, 88, 89, 90, 92, 97, 101
Gwin, Lauren, 200, 286

hair breeds, 53–54
Halverson, Marlene, xi
Hampshire, 83
hand-mating service, 57
hanging weight, 210, 221–223
hardening off, 31–32
Harris, Graeme, 134
Harris, Kathleen, 170, 204–205, 220
Harris, Will, 24, 27, 34, 113, 188, 234–235
Hazard Analysis and Critical Control Points (HACCP) plans, 157, 170, 225, 232
health
 of cattle, 119–122
 of elk and deer, 147–148
 ethics and, 8
 of pigs, 95–97
 of rabbits, 134–136
heating systems, 1*ci*, 29
heat stress, 136
hedgehog concept, 16
heifers, 103, 112. *See also* cattle
heirloom varieties, 14
Hereford cattle, 106, 107t
Heritage Breed, 14

heritage breed chickens, 25t, 27–28, 37
heritage pigs, 81–82
heritage turkeys, 25t
Hettie Belle Farm, 179–180
Higgins, Greg, 219–220
historical perspectives, ix–x
hogs. *See* pigs
Holderread, David, 46
Holmes, Cody, 284
Holstein cattle, 106
HomeGrownCow.com, 6
honor system, 174–175
hoof trimming, 70
hoophouses, 6*ci*, 34, 76–77
housing. *See* shelters
HRI sales, 160
Hudson Mohawk Resource Conservation and Development Council, 204
Humane Farm Animal Care program, 13
humane handling and slaughter, 163
Humane Livestock Handling (Grandin), 123

ice, 252–253
immigrants, 166, 183
Improvest, 90
inbreeding, 84
incorporation, 284
incubators, ix
industrialized approaches, ix
inspections, 152, 162–163
institutional sales, 200
insurance, 283–284
intermittent service, 57
internet sales, 188–190
inventory management, 248
irrigation, 10
Island Grown Farmers Cooperative (IGFC), 205–206, 238

Jacob sheep, 53, 71
Jamison Farm, 72, 197–199, 225, 235, 237, 252

Index

Jamison, John and Sukey, 235
Jeffries, Walter, 172
jerkies, 227
Jersey cattle, 106, 107t
Joel-Salatin-style shelters, 32–33
The Joy of Cooking, 1
jug, defined, 51

Katahdin hair sheep, 52t, 55–56
ketosis, 70
kid, defined, 51
kidding and lambing, 59–61
Kiko goats, 52t, 53
killing. *See* slaughtering and processing
kindling, defined, 130
kit, defined, 130
Knippel, Doug, 138–140
Kornstein, Rachel, 46–47

labeling, 244–247
Label Submission and Approval System, 245
labor management, 284–285
lactation, 89, 109
lamb, defined, 51
La Montanita Co-op, 193–194
land management, 9
Large Black, 81t, 82
Lasater Ranch, 9ci, 110
Latham, Geoff, 149–150
Leachman Cattle Co., 127
leader-follower grazing system, 10ci, 78
leadership strategies, 256–257, 267
liability insurance, 283–284
lice, 39
Limousin cattle, 106
line breeding, 84
Little Farms LLC, 14ci
liver flukes, 68
Livestock Conservancy, 14
live weight, 210, 221–223
llamas, 40, 41
LocalHarvest.com, 6
locker plants, 166, 169–174, 248–252

logos, 268
Long Island Meatshare, 173
Lopez Island Community Land Trust, 205
Lorentz Meats, 229, 231
Lorentz, Mike, 173
lung worm, 68
lysine, 93

mail order sales, 188–190
Maine-Anjou cattle, 106
Malan, Faffa, 68
malignant catarrhal fever (MCF), 142
malocclusion, 135
mange, 97
Manitoban elk, 144, 145t, 146
manure handling, 137
The Maple Grille, 13ci, 196–197
marbling cattle breeds, 106, 108
market hog, 76
marketing. *See also* value-added products
 aggregated models, 202–208
 of beef, 265–266
 buying clubs, 180–181
 CSAs, 176–180
 customer-centricity, 257–258
 customer demographics, 261, 263–265
 differentiation and, 269–270
 ease of buying and, 270–272
 extending the season, 166–168
 farmers markets, 181–188
 farmstands, 174–176
 finding new customers, 258–259
 freezer meat, 166, 169–174
 institutional sales, 200
 internet and mail order, 188–190
 meat market trends, 260–261
 networking and, 272
 nutrition and, 270
 overview, 165–166
 principles of, 255–257
 Red Meat Market, 262–263

 restaurants and caterers, 192, 194–195, 197–199
 retail butcher shops, 190–191
 ritual slaughter, 168–169
 small retailers and food co-ops, 191–192
 strategic messaging and branding, 266–269
 team building, 259–260
 wholesalers and distributors, 200–202
 willingness to pay and, 266
Massa Organics, 183–184
mast, 79
mastitis, 70
Mattocks, Jeff, 34, 35–37
McNally, Janet, 53, 60–62, 69
Meat: A Benign Extravagance (Fairlie), 2
The Meat Locker, 251
meatpacking industry, 5
meat processors, 212–218, 220–221
meat science, 223–224
Meat Suite, 251
methicillin-resistant *Staphylococcus aureus* (MRSA). *See* MRSA
Metzer Farms, 48–49
milk, 31, 93
Miller Livestock Co., 112, 190
Misty Brook Farm, 174–175, 188
mites, 39
mob grazing, 78
mobile poultry-processing units (MPUs), 15ci, 160–161, 237–239
Mobile Poultry Slaughterhouse (Berlow), 160
mobile slaughter units (MSUs), 206, 237–239
Morris Grassfed Beef, 126–127
Morris, Joe, 105, 111, 123, 126–127, 177
Morris, Julie, 126
mountain lions, 40, 41
moveable mini-barns, 33–34
MRSA (methicillin-resistant *Staphylococcus aureus*), x

Murray Grey cattle, 106
mycoplasma-induced arthritis, 142
Myotonic goats, 52t, 53, 72–73

name brand, 267
Navajo-Churro sheep, 15, 53
networking, 259, 272
New Zealand rabbits, 130
Niche Meat Producer Assistance Network (NMPAN), 152–153, 212–214, 230
Nicky USA, 130, 143, 149–150, 201
night penning, 41
Niman, Nicolette Hahn, 2
Niman Ranch, ix
Northeast Livestock Processing Service Company, 204–205
Northern grazing climates, 105
Northstar Bison, 14ci, 188, 189, 236–237
Northwest Red Worms, 138–140
nutrition, xii, 270
nutrition labeling, 246–247, 270

Ofte, Rod, 117, 123
one-litter farrowing, 85
on-farm processing
    abattoirs, 234–237
    cleanup, 4ci
    logistics, 229–234
    marketing and, 175–176
open cow, defined, 103
Ossabaw Island hogs, 15
outcrossing, 84
overhead expenses, 275

packaging, 15ci, 16ci, 217, 241–244, 268
Packers and Stockyards Act, 232
paper wrapping, 243
parasites
    in cattle, 120
    in pigs, 96–97
    poultry and, 39
    sheep and goats and, 64, 67–69
pasture-based rotational system, 77–79

Pasturella, 136
pasture raising, xi–xii, 3ci, 10ci, 11ci, 19–20, 78
payment methods, 271
Pekin ducks, 46
Perry, Bob, 192, 200, 202, 226
Pharo Cattle Company, 104, 108
Pharo, Kit, 104, 108, 127
Philly Meatshare, 173
pigs
    boars, 92
    breeding, 83–87
    breeds, 80–83
    castration, 89–90
    concentrated feeds and, xi
    deep bedding or hoophouse production, 76–77
    definitions, 76
    dirt lots and, 80
    farrowing, 87–89
    feeds and feeding, 10, 36, 93–95
    fencing, 7ci, 8ci, 97–99
    gilts, 91–92
    growers and finishers, 91
    Gunthorp Farm, 101
    handing and sorting, 99–100
    health, 95–97
    lactation, 89
    meat quality, 100–101
    newborns, 89
    pasture, 94
    pasture-based rotational system, 77–79
    regulations for, 153–158
    shelters, 6ci, 7ci, 92–93
    watering, 95
    weaning, 90–91
    woodlots, 79–80
plucking, 44
pneumonia, 96–97
pododdematitis, 135–136
Polyface Farm, 131
Pordomingo, Anibal, 113, 124
Pork and Beef Checkoff, 261
Pork Intake Development Study, 263

portable bottomless shelters, 2ci, 20–22, 32–33
portable electric net fencing, 39
post-harvest handling, 45
post-mortem inspections, 163
poult, defined, 19
poultry. *See also* chickens
    Boondockers Farm, 46–47, 49
    breeds, 25–28
    brooding phase, 28–32
    concentrated feeds and, xi
    definitions, 19
    diseases, 38–39
    ducks *vs.* chickens, 47–49
    feed and feeding, 30–31, 35–37
    fencing, 39, 42–43
    free range stationary shelters, 24–25
    grit, 34
    grow-out phase, 32–34
    industrialized approaches, ix
    pasture-based production overview, 19–20
    portable bottomless shelters and, 20–22
    post-harvest handling, 45
    predators, 31, 39–43
    processing, 45
    regulations for, 158–162
    semi-portable day range shelters, 22–24
    slaughtering, 43–45
    watering systems, 29, 34–35
poultry inspection programs (PIPs), 161
Poultry Products Inspection Act (PPIA), 158, 160
Practical Farmers of Iowa (PFI), 96
Predator Friendly certification, 14
predators, 10–11, 31, 39–43
Premier 1 Supplies, 64
Prevatte, Tina, 206–207
price sheets, 217
pricing, 275–283
Privett Hatchery, 26
PRIZM, 263–264

Index

processing. *See* slaughtering and processing
profit margins, 166, 275–277
PSE (pink, soft, exudative) pork, 101
public health veterinarians (PHV), 163
Public Law 90-492, 151–152, 158–160, 161–162
pure breeding, 83–84

rabbitry, defined, 130
rabbits
   breeding, 131–132
   breeds, 130–131
   definitions and overview, 129–130
   feeding, 12*ci*, 133–135
   handling and slaughtering, 137
   health, 135–136
   manure handling, 137
   Northwest Red Worms, 138–140
   post-harvest handling, 140
   regulations and, 161
   shelters, 132–133
   watering, 135
rabbit tractors, 12*ci*, 132, 133
Rainbow Broiler, 25t, 27
Rambouillet sheep, 53t
ram, defined, 51
ram lamb, defined, 51
Ranching Full-Time on 3 Hours a Day (Holmes), 284
Rancho Veal Corp, 216
range riding, 41
rats, 31
Real Food Challenge, 200
Red Angus cattle, 106, 107t
Red deer, 145
Red Meat Market, 262–263
Red Poll cattle, 106
Red Wattle, 81t, 82
Reese, Duane, 86
regulations
   for cattle, pigs, sheep, goats, 153–158

definitions, 152
inspections and, 162–163
overview, 151–153
for poultry, 158–162
state processing rules, 289–295
respiratory infections, 136
restaurants, 192, 194–195, 197–199
retail butcher shops, 190–191
ritual slaughter, 168–169
Rocky Mountain elk, 145t, 146
ROI pricing, 280
Roosevelt elk, 144–145
rotational service, 57

Salad Bar Beef (Salatin), 277, 285
Salatin, Joel, 7, 170, 277, 285
Salers cattle, 106
sales options. *See* marketing
Salmonella, xi–xii
Sarver Heritage Farm, 246–247
sausages, 226–227
Savanna goats, 52t
Savell, Jeffrey, 225
Scaglia, Guillermo, 124
scalding, 43
scales, 170–171
Schaeding, Josh, 196–197
Schafer, David, 29, 30, 43, 45
Schatzker, Mark, 124
Scottish Highlander cattle, 106
scours, 121
scrapie, 70
season extension, 166–168, 271–272
self-handling instructions, 246
semi-portable day range shelters, 22–24
Sericea lespedeza, 69, 72
set stock lambing, 60
sheep and goats
   body modifications, 70–71
   breed characteristics, 52t
   breeding, 56–59
   breeds, 53–54
   Clover Creek Farms and, 55–56
   definitions, 51

differences between, 54, 56
feeding and watering, 63
fencing, 63–67
handling equipment, 62
illnesses and diseases, 69–70
kidding and lambing, 59–61
parasites, 64, 67–69
regulations for, 153–158
shelters for, 5*ci*, 62
slaughtering, 13*ci*, 71–73
Vanguard Ranch, 72–73
weaning, 61–62
sheep, defined, 51
Shell, Timothy, 31
shelters
   for cattle, 113
   cattle panel hoop structure, 33
   day range, 22–24, 33–34
   for elk and deer, 146
   free range stationary, 24–25
   hoophouses, 6*ci*, 34, 76–77
   Joel-Salatin-style, 32–33
   moveable mini-barns, 33–34
   for pigs, 6*ci*, 7*ci*, 92–93
   portable bottomless, 20–22, 32–33
   for rabbits, 132–133
   for sheep and goats, 5*ci*, 62
Shields, Sara, xi
shipping meats, 188–189
Shorthorn cattle, 106, 107t
shrink wrapping, 242–243
Sika deer, 145
Simmental cattle, 106
slaughtering and processing
   aging, 127, 225–226
   of cattle, 123
   cooling carcasses, 224–225
   custom *vs.* federally inspected, 153–158
   cutting instructions, 217–218, 296–309
   cutting yield, 210, 221–223
   day of, 211–212
   definitions, 210
   of elk and deer, 148–149
   ethical issues, 4*ci*, 11, 163

finding a good meat processor, 214–218
meat science, 223–224
mobile poultry-processing units, 15*ci*, 160–161, 237–239
mobile slaughter units (MSUs), 206, 237–239
on-farm processing, 4*ci*, 175–176, 229–237, 234–237
of poultry, 14*ci*, 43–45
preparing for, 209–211
processor relations, 212–214
of rabbits, 137
ritual, 168–169
of sheep and goats, 13*ci*, 71–74
state rules, 289–295
value-added products, 226–227
working with meat processors, 220–221
slogans and jingles, 268
Slow Cornish chickens, 26
Slow Food, 14, 192
smoked meats, 227
sore hocks, 135–136
Southeast grazing climates, 104
Southern Sustainable Agriculture Working Group (SSAWG), 55
Southwest grazing climates, 104–105
sow, defined, 76
Spanish goats, 52t, 53
Spanish Iberian pigs, 80
species selection, 11
squab, 158
state inspection, 153
state processing rules, 289–295
Steak (Schatzker), 124
steer, defined, 103
stewer, defined, 130
straight run chicks, 28–29
strategic messaging, 266–269
stress, 121–122, 224, 238
Suffolk sheep, 53t
Sugar Mountain Farm, 172

Sweetgrass Beef Cooperative, 193–194

Tails and Trotters, 150
Tallgrass Prairie Producers Cooperative, 207–208
Talmadge-Aiken (TA) plants, 153, 155, 214
Tamworth, 81t, 82
tapeworms, 68
Tarantaise cattle, 106
target return-on-investment (ROI) pricing, 280
Targhee sheep, 53t
team building, 259–260
teeth, 58
tenderness of meat, 224
terminal breeding, 83
tetany, 120–121
Thiboumery, Arion, 229
TLC Ranch, 1
tourism, 175
transporting animals, 209–211, 215
trends, 260–261
tuberculosis, 142–144, 147
turkeys, 3*ci*, 25t, 37
Turner, Renard, 72–73

UPC codes, 245–246
USDA grade, 210
USDA-inspected plants, 153
USDA National Nutrient Database, 246
US Food and Drug Administration (FDA), 152. *See also* Food Safety Inspection Services (FSIS)

vacuum sealing, 242
value-added products, 187, 217, 226–227, 233
value chains, 202
Vanguard Ranch, 72–73, 175
ventilation, 29

Vermont Salumi, 175
vultures, 40

warren, defined, 130, 133
watering systems
for cattle, 118
for elk and deer, 146–147
for pigs, 95
for poultry, 29, 34–35
for rabbits, 135
for sheep and goats, 63
weaners, 76, 90–91
weight live *vs.* final weight, 221–223
wet feather, 39
wether, defined, 51
whipworms, 97
white muscle disease, 70
White Oak Pastures, 27, 175–176, 188, 233, 234–235, 252
Whitetail deer, 145
Whole Foods Market, 10, 14, 90
wholesalers, 200–202
Wildlife Friendly certification, 14
Wilhelms, Mark, 262–263
Williams, Allen, 112, 118, 119, 124, 125
willingness to pay, 266
Wilson, Chris, 55–56
Wilson, Dan, 77, 78, 88–89, 92, 95, 99
wolves, 40, 41–42
woodlots, 79–80
wool breeds, 53–54
word of mouth advertising, 169, 173–174, 258–259
World War II, ix
worm raising, 132–133, 138–140

yearling, defined, 51, 103
year-round grazing climates, 104
yield grade, 210

Zebu cattle, 107
zoning, 229–230

About the Authors

Rebecca Thistlethwaite, the coauthor of *The New Livestock Farmer,* is also the author of *Farms with a Future* (Chelsea Green, 2012). She runs Sustain Consulting, which specializes in food and farm issues, working with both nongovernmental organizations and for-profit businesses. Her website is http://rebeccathistlethwaite.com/.

Rebecca also operates a small farm and a community farmstand in Oregon with her husband and coauthor, **Jim Dunlop,** and their daughter, Fiona. They previously owned TLC Ranch in Watsonville, California, where they raised organic, pastured livestock and poultry, selling to direct markets across Northern California.

the politics and practice of sustainable living
CHELSEA GREEN PUBLISHING

Chelsea Green Publishing sees books as tools for effecting cultural change and seeks to empower citizens to participate in reclaiming our global commons and become its impassioned stewards. If you enjoyed *The New Livestock Farmer*, please consider these other great books related to agriculture and food.

DEFENDING BEEF
The Case for Sustainable Meat Production
NICOLETTE HAHN NIMAN
9781603585361
Paperback • $19.95

THE GOURMET BUTCHER'S GUIDE TO MEAT
How to Source It Ethically, Cut It Professionally, and Prepare It Properly
COLE WARD with KAREN COSHOF
9781603584685
Hardcover with CD • $49.95

FARMS WITH A FUTURE
Creating and Growing a Sustainable Farm Business
REBECCA THISTLETHWAITE
9781603584388
Paperback • $29.95

THE SMALL-SCALE POULTRY FLOCK
An All-Natural Approach to Raising Chickens and Other Fowl for Home and Market Growers
HARVEY USSERY
9781603582902
Paperback • $39.95

For more information or to request a catalog, visit **www.chelseagreen.com** or call **(800) 639-4099**.